Smaller World, Bigger Issues

Hale Kırer Silva Lecuna (ed.)

Smaller World, Bigger Issues

Growth, Unemployment, Inequality and Poverty

Bibliographic Information published by the
Deutsche Nationalbibliothek
The Deutsche Nationalbibliothek lists this publication in the Deutsche
Nationalbibliografie; detailed bibliographic data is available online at
http://dnb.d-nb.de.

Library of Congress Cataloging-in-Publication Data
A CIP catalog record for this book has been applied for at the
Library of Congress.

Printed by CPI books GmbH, Leck

PONTIFICIA **UNIVERSIDAD CATÓLICA** DEL PERÚ

ISBN 978-3-631-80205-2 (Print)
E-ISBN 978-3-631-80225-0 (E-PDF)
E-ISBN 978-3-631-80226-7 (EPUB)
E-ISBN 978-3-631-80227-4 (MOBI)
DOI 10.3726/b16147

© Peter Lang GmbH
Internationaler Verlag der Wissenschaften
Berlin 2019
All rights reserved.

Peter Lang – Berlin · Bern · Bruxelles · New York · Oxford · Warszawa · Wien

All parts of this publication are protected by copyright. Any
utilisation outside the strict limits of the copyright law, without
the permission of the publisher, is forbidden and liable to
prosecution. This applies in particular to reproductions,
translations, microfilming, and storage and processing in
electronic retrieval systems.

This publication has been double-blind peer reviewed.

www.peterlang.com

Preface and Acknowledgements

Our book, entitled *Smaller World, Bigger Issues: Growth, Unemployment, Inequality, and Poverty*, contains the topics that came up on the day mankind existed and that have become more and more deep issues in the globalized world.

Growth is one of the most debated topics in economics. According to the International Monetary Fund's World Economic Outlook Report (October 2018), the growth rate in developed countries is projected to expand by 2.4 % in 2018, while this value is 4.7 % for developing countries. One of the most important questions here is whether growth is sustainable or not and how much it contributes to development. Within this framework, this book includes the studies about economic growth and development in the first section.

In the "Regional Risks for Doing Business Report '18" published by World Economic Forum, unemployment or underemployment are ranked first in the global top 10 risks list. The expectations of the labor market have changed particularly with the great technological advances in our era and the need for manpower has almost disappeared in many sectors. Along with the experienced and expected economic crises, the unemployment problem comes up as a bomb that is ready to explode. In this context, in the second section, unemployment, types of it, its effects and labor markets are evaluated from different perspectives.

Inequality in income, gender, health, education is one of the subjects that the society struggles with since the day it exists. In fact, it is not possible to separate completely different inequalities from each other. Increasing inequality in income and wealth distribution manifests itself in education and health. Gender inequality is more common in the countries where income and wealth inequality is high, but it is seen in different levels all over the world. In the World Inequality Report '18, while the Middle East is considered to be the most unequal region in the world, inequality in Europe is the lowest. In this respect, it is seen that inequality is widespread and an inevitable problem in many ways all over the world. However, it is important to analyze deeply and take precautions before it causes major troubles.

One of the biggest problems of the modern world is the issue of poverty. According to the United Nations' Sustainable Development Goals Report '18, the worst forms of poverty still remain today, although there has been a serious decline in the rate of extreme poverty since the 1990s. Today, while billions of people struggle with poverty in the world, approximately 600 million people, 8 % of the world's population, live in extreme poverty. In this context, in the last section, the researches on the inequality and poverty from various perspectives are included.

As a result, the purpose of this book is to bring together the studies of academicians and researchers who are competent in their fields to enlighten the readers and make recommendations for policymakers to reduce these issues.

In this context, I would like to thank all the authors, contributors and researchers for their devotion and diligence. I would also like to express my deep gratitude to Professor Dr. Burak Darıcı for his enthusiastic encouragement and useful critiques of this editorial book. I would also like to extend my thanks to Asst. Prof. Dr. Ufuk Bingöl for his help in providing software during the review period. My grateful thanks are also extended to all the families and children who are the source of motivation for our authors with their patience and support during the conduct of these valuable works.

Finally, I wish to thank my spouse Elias Esteban Silva Lecuna for all his support and constructive recommendations throughout the study.

Sincerely,
Editor; Dr. Hale KIRER SILVA LECUNA.

Table of Contents

List of Authors 11

SECTION 1: Growth and Development

Serçin Şahin
1. Product and Process Innovations, Market Structure and Economic Growth: A Literature Review 15

Semanur Soyyiğit
2. China's Neo-Mercantilist Policies and Growth Process 31

Begüm Erdil Şahin and Deniz Dilara Dereli
3. An Evaluation on Middle-Income Trap in Turkey 45

Sertaç Hopoğlu
4. Are Youth Labor, Trade Openness and Foreign Direct Investment Effective in Economic Growth of CIVETS Countries? A Panel Causality Analysis 63

Tuncer Gövdeli
5. Tourism, Oil Prices and Economic Growth in the Mediterranean Countries: Bootstrap Panel Granger Causality Analysis 81

Habibe Günsel Doğrul, Mediha Mine Çelikkol, and İlhan Korkmaz
6. The Analysis of the Relationship Between Creative Class, Financial Development and Regional Innovativeness in Turkey 97

SECTION 2: Labor Markets and Unemployment

Umut Akduğan and Seyhun Doğan
7. Youth Unemployment: An Empirical Analysis 117

Mehmet Güçlü
8. Industrial Revolution and Labor Market 135

Kemal Eker, Görkem Bahtiyar, and Hasan Bakır
9. Systemic Policies Towards Attracting Skilled Labor: An
Investigation on Turkey 153

H. Işıl Alkan
10. The Gender Impact of Last Global Crisis on Labour Markets 165

Mehmet Kenan Terzioğlu
11. Poverty, Income Inequality, Unemployment and Human Capital in
the Democratization Process 181

Ufuk Bingöl
12. Employment Policy Goals in Turkish Government Programs in
Terms of Social Policies: Justice and Development Party Governments 199

SECTION 3: Inequality and Poverty

Ayşe Aylin Bayar, Bengi Yanık-İlhan, and Nebile Korucu-Gümüşoğlu
13. The Effects of Elders' Earnings on Turkish Income Inequality 219

Mehmet Akif Destek
14. Liberalization, Globalization and Income Inequality in Emerging
Economies 235

Kıymet Yavuzaslan
15. The Attitude of Income Inequality of Individuals: An Experimental
Economics Approach 251

Mustafa Şit and Erdal Alancıoğlu
16. Macroeconomic Factors Determining Income Distribution: An
Analysis on Mist Countries 263

Selçuk Çağrı Esener
17. The Forgotten Unit of Income, Expenditure and Wealth
Chain: Remembering the Taxation of Wealth and Our Fight Against
Inequality 279

Nilüfer Yörük Karakılıç
18. Social Perspectives of Gender-Based Discrimination
Perception: An Empirical Study on University Students 295

Hale Kirer Silva L. and Rüya Eser
19. Energy Consumption, Carbon Emissions and Income Inequality in Turkey 315

List of Figures 327

List of Graphs 329

List of Tables 331

List of Authors

Ayşe Aylin Bayar,
Assoc. Prof. Dr., Istanbul Technical University

Begüm Erdil Şahin,
Asst. Prof. Dr., Istanbul Kultur University

Bengi Yanık-İlhan,
Assoc. Prof. Dr., Altınbaş University

Deniz Dilara Dereli,
Asst. Prof. Dr., Istanbul Kultur University

Erdal Alancıoğlu,
Dr., Harran University

Görkem Bahtiyar,
Asst. Prof. Dr., Uludağ University

H. Işıl Alkan,
Asst. Prof. Dr., Ondokuz Mayıs University

Habibe Günsel Doğrul,
Asst. Prof. Dr., Dumlupınar University

Hale Kırer Silva Lecuna,
Asst. Prof. Dr., Bandırma Onyedi Eylül University

Hasan Bakır,
Asst. Prof. Dr., Istanbul Kultur University

İlhan Korkmaz,
Dr., Dumlupınar University

Kemal Eker,
Dr. National Defence University

Kıymet Yavuzaslan,
Asst. Prof. Dr., Aydın Adnan Menderes University

Mediha Mine Çelikkol,
Asst. Prof. Dr., Dumlupınar University

Mehmet Akif Destek,
Asst. Prof. Dr., Gaziabtep University

Mehmet Güçlü,
Assoc. Prof. Dr., Ege University

Mehmet Kenan Terzioğlu,
Assoc. Prof. Dr., Trakya University

Mustafa Şit,
Asst. Prof. Dr., Harran University

Nebile Korucu-Gümüşoğlu,
Asst. Prof. Dr., Istanbul Kultur University

Nilüfer Yörük Karakılıç,
Asst. Prof. Dr., Afyon Kocatepe University

Rüya Eser,
Dr., Mimar Sinan University

Selçuk Çağrı Esener,
Asst. Prof. Dr., Bandırma Onyedi Eylül University

Semanur Soyyiğit,
Assist. Prof. Dr., Erzincan Binali Yıldırım University

Serçin Şahin,
Dr., Yildiz Technical University

Sertaç Hopoğlu,
Asst. Prof. Dr., İskenderun Technical University

Seyhun Doğan,
Prof. Dr., İstanbul University

Tuncer Gövdeli,
Asst. Prof. Dr. Atatürk University

Ufuk Bingöl,
Assis. Prof. Dr., Bandırma Onyedi Eylül University

Umut Akduğan,
Assist. Prof. Dr., Trakya University

SECTION 1: Growth and Development

Serçin Şahin

1. Product and Process Innovations, Market Structure and Economic Growth: A Literature Review[1]

Abstract: The relationship between market structure, innovative activities, and economic growth is among the most intensively researched topics in economics. Studies on the problem reached a diverse set of conclusions on the nature of this relationship. The main reason for this diversity is the differences in the depiction of competition and innovative activities in different studies. A group of studies approached competition as an effort to create niche product varieties to obtain market power. However, another group of studies described it as seizing the market power of an incumbent firm by producing a better version of the existing product variety. Mostly, innovative activities are regarded as either product or process innovation. However, some investigated the issue with the assumption that firms invest in process and product innovations simultaneously. In this chapter, the literature on the relationship between market structure and innovation is reviewed in its historical flow.

Keywords: Innovation, Market structure, Economic growth

1 Introduction

The static and dynamic allocation efficiency of market structure—and competition—is one of the central questions of economics. Basically, two main paradigms dominate the literature. Formerly, the microeconomic theory's paradigm prevailed, which considers the differences among firms as a short-term phenomenon resulting from temporary "imperfections" in the market mechanism. These imperfections, such as imperfect information, short-term technological rents, and market power, cause allocative inefficiency. However, as profit-maximizing firms update their expectations, differences among the firms would fade away, and the economy will converge to a perfect competition equilibrium in the long run. According to this view, perfect competition ensures allocative efficiency both in the short and the long term (Mazzucato, 2000).

---

1 This chapter is prepared by updating the "Literature Review" chapter of the author's Ph.D. dissertation, titled "The Relationship Between Optimal Innovation Combinations of Firms, Market Structure, and Economic Growth" (2010).

On the other hand, in his seminal book in 1942, Schumpeter challenged this paradigm and defined competition as a disequilibrium process in which firms face a constant pressure of differentiating themselves from the others in order to survive (Mazzucato, 2000). The market power that firms acquire as a result of being different from the other firms, or the hope of acquiring it, causes them to engage in innovative activities or to adopt new technologies, and hence, constitute the engine of technological progress. Therefore, there is a positive relationship between market power and innovative activities (Van Cayseele, 1998). Schumpeter asserts that, apart from being unrealistic, perfect competition is also a suboptimal phenomenon because highly concentrated market structures are more favorable to technological progress and economic growth. The long-term welfare growth achieved as a result of high innovation rate in imperfectly competitive markets more than compensates the welfare losses resulting from the short-term allocation inefficiencies (Kamien and Schwartz, 1975).

In essence, studies on the relationship between market structure, innovative activities, and economic growth are the reflections of the debate between these two paradigms on the dynamic efficiency of market structures. This chapter reviews the literature and presents the ideas on the issue in a historical flow, without any intention of being exhaustive. The second section presents the models that try to explain economic growth with perfect competition, and the third section with imperfect competition. Finally, the fourth section concludes.

2 Early Models of Economic Growth with Perfect Competition

Arguably, the model developed by Solow and Swan separately in 1956 has been the pioneering and the most influential work in the economic growth literature. Apart from drawing the route of the literature, it has also been used as a benchmark to evaluate the models that followed. The Solow-Swan model focuses mainly on how the growth rate of output per capita is related to the accumulation of production inputs, such as capital, labor, and "knowledge" (technology).

In the model, the total amount of output that can be produced in an economy using the available levels of inputs is represented with an aggregate production function. Each period, economic agents save an exogenously given, fixed fraction of this output to make capital investments and consume the rest. The amount of capital investment and depreciation determines the next period level of capital. Thus, the dynamics of capital accumulation is endogenized in the model. However, the growth rates of other factors, namely, labor and knowledge, are exogenously given.

The Solow-Swan model predicts that the economy will eventually converge to a "balanced-growth path" on which all the variables grow at a constant rate. The growth rate of output and capital are determined by the growth rates of labor, and technology, which are exogenous to the model. However, the landscape completely changes when the analysis is conducted with per capita variables. In this case, the model converges to a steady state in which output per capita and capital–labor ratio remain constant regardless of the saving rate, namely, per capita growth eventually ceases. This result means that capital accumulation cannot be the engine of long-term sustainable per capita output growth.

With the growth accounting approach Solow developed in his 1957 study, he decomposed growth according to its sources. In this approach, the part of the growth that cannot be explained by the accumulation of production factors is attributed to the increase in total factor productivity and referred to as the "Solow residual". This residual is regarded as representing the contribution of technological progress on economic growth. The US Bureau of Labor Statistics used the growth accounting approach to decompose the growth rate of USA from 1948 to 2010 and found that Solow residual, namely technological progress, accounts for more than half of the growth in the referred period (Jones and Vollrath, 2013). This result demonstrates the vital role played by technological progress in economic growth.

Cass (1965) and Koopmans (1965) built infinite horizon, general stochastic equilibrium models based on Ramsey (1928), in which the preferences of households endogenously determine the saving rate. However, the endogenization of the saving rate did not change the main result of the Solow-Swan model regarding long-run growth. These so-called "neoclassical growth models" also concluded that capital accumulation alone could not generate sustainable economic growth; the long-run growth rate is determined by the rates of population growth and technological progress.

After these striking results, the economic growth literature principally oriented towards finding an answer to the question of how long-term, sustainable per capita output growth can be achieved. The neoclassical growth theory relies on Walrasian perfectly competitive equilibrium analysis in which production factors are compensated according to their marginal products. Since the marginal product of capital diminishes as the level of capital stock increases, growth eventually halts (Aghion and Howitt, 1998). The simplest way of sustaining growth within the perfect competition framework is to introduce a mechanism that ensures the marginal product of capital to remain fixed (Grossman and Helpman, 1991b). In the models that are developed with this approach, the production function takes the form of $Y=AK$ in reduced form. Therefore, regardless of the level of the capital stock, the marginal product of capital remains equal to "A". Accordingly, transition dynamics to steady state never ends, and per capita output growth is sustained in the long run (Jones and Vollrath, 2013). These models are referred to as "AK models" after the form of their production function in reduced form (Aghion and Howitt, 1998).[2]

Harrod (1939) and Domar (1946) studies are the archetypes of AK models. These studies argue that capital stock will always go hand in hand with an increase in employment, because of the constant existence of unemployment in an economy. Therefore, a rise in the level of capital stock will always increase output (Aghion and Howitt, 1998).

According to Frankel (1962), Cobb-Douglass type production function is more appropriate for modeling the production processes of individual firms, in which production factors are compensated based on their marginal products. On the other hand, Y=AK type production function can explain some empirical findings for economies such as the stability of the share of aggregate output devoted to investment and the stability of economic growth rate. Frankel argues that these two functions can be reconciled with each other. Namely, if the production functions of individual firms are extended with a "level of development" parameter, which is assumed to be a function of the total capital stock in the economy, then the aggregate production function would take the form of $Y=AK$. In this case, as individual firms increase their capital stock, they increase the aggregate output of the economy both directly by increasing their own production and indirectly by increasing the productivity of the whole economy with their contribution to the total capital stock. Notably, in the presence of such indirect effects, capital accumulation can generate sustainable, long-run per capita output growth. Frankel exemplified these indirect effects of capital accumulation with technological progress, increase in labor quality, and organizational improvements.

Arrow (1962) introduced "learning-by-doing" as another indirect effect of capital accumulation. The capital goods–producing firms enhance their productivity as they encounter problems and discover better production methods in the production process. As it was for Frankel (1962), this productivity increase that occurs as a by-product of capital goods production does not remain limited to

---

2 In the beginning, AK models are called "Endogenous Growth Models" due to the fact that they can create long-term sustainable per capita output growth, different from neoclassical growth models (Jones and Vollrath, 2013). However, this qualifier has mostly come to be used for the studies that endogenize technological progress later on. In this study, the latter, more common convention is adopted.

the firm that makes the production but spreads to the whole economy. Therefore, the aggregate knowledge stock grows, and the productivity of the economy increases.

Romer (1986) emphasized that Arrow's (1962) inference is a consequence of the public good property of knowledge. Firstly, knowledge is a non-rival good. Once produced, different firms can use the knowledge simultaneously for different ends, without reducing its usability by other firms (Acemoglu, 2009). Secondly, knowledge is a non-excludable good; namely, the firm that produces the knowledge cannot prevent other firms from using it by patenting or keeping it secret, at least partially. Having these properties, the knowledge produced by a firm would spill over to the whole economy and create a positive externality that increases the productivity of all firms. Different from the previous studies, Romer (1986) treated knowledge as a kind of capital good and incorporated it into the production function as a separate production factor. Knowledge is produced by an R&D production function with diminishing returns property, similar to the capital goods production in other studies. The produced knowledge enhances not only the production possibilities of the firm by increasing its private knowledge stock but also those of all other firms by contributing to the aggregate knowledge stock of the economy. The economy-wide positive externalities more than compensate the decrease in the marginal product of knowledge for individual firms with the increase in their private knowledge stock. Therefore, knowledge exhibits increasing marginal product property, and the output production function exhibits increasing returns to scale in all factors of production. By this means, the continuous growth of knowledge stock enables long-term per capita output growth in the economy.

Romer (1986) is a pioneering study in the economic growth literature because it incorporated intentional knowledge production by economic agents in a perfect competition framework and presented that the continuous growth of knowledge stock along with knowledge spillovers can generate sustainable per capita output growth (Acemoglu, 2009).[3]

---

3 Another group of study incorporates human capital, instead of technological progress, and its externalities into the model in order to prevent the marginal product of capital from diminishing. In this type of models, there is a separate production function for human capital, and positive externalities emerge as a result of the interaction of highly skilled workers (Barro and Sala-i-Martin, 2004). Since this chapter focuses on technological progress, models that explain economic growth with human capital are not covered here. See Uzawa (1965) and Lucas (1988) as examples.

3 Endogenous Technological Change and Imperfect Competition

In the models reviewed so far, markets were assumed to be perfectly competitive, and firms' incentives for obtaining market power were not explicitly taken into consideration. However, the last 200 years of the world economy shows that the extraordinary economic growth that occurred in the period is primarily due to innovations that arose as a result of intentional R&D activities of economic agents (Akcigit, 2017).

Firms' R&D expenditures constitute a fixed cost and cause the emergence of increasing returns to scale property in production. Because of this property, firms' average costs are always higher than their marginal costs. If firms charged a price equal to their marginal costs as in perfectly competitive markets, they could never cover the fixed costs of production and would suffer a loss. In this case, the firms would not have an incentive to make any innovation and to enter into the market in the first place. Therefore, increasing returns to scale requires imperfect competition, in which firms can charge prices that are higher than their marginal costs (Jones and Vollrath, 2013).

The AK models mentioned above maintained the perfect competition assumption and incorporated knowledge with public good properties in order to avoid the technical difficulties of modeling imperfect competition in the general equilibrium framework. After Spence (1976), Dixit and Stiglitz (1977) and Ethier (1982) developed a formal framework to deal with imperfect competition, "endogenous growth models" have emerged, which explain economic growth with the intentional R&D activities undertaken by firms in the hope of obtaining market power (Jones and Vollrath, 2013).

3.1 Product (horizontal) innovation and economic growth

Product innovations can be defined as the introduction of horizontally differentiated product varieties that provide new services, and that are not perfect substitutes to the existing varieties.[4] New product varieties can be either final goods that provide utility directly to consumers or intermediate goods that

---

4   In this chapter, it is assumed that product and process innovations correspond to the introduction of horizontally and vertically differentiated goods, respectively. However, there is no consensus in the literature on this issue. It is also common to define product innovations as improving the quality of the product varieties (vertical innovation) or process innovations as the introduction of differentiated intermediate goods (horizontal innovation). These studies make the distinction according to the effects of innovation on the production process. However, in the convention adopted here, the

increase the efficiency of the production process. These new varieties open niche markets to firms and provide them with the market power with which they can charge a price above their marginal costs.

The most important study that explains endogenous economic growth with product innovations is Romer (1990). There are four inputs to production in his model: capital, labor, human capital, and technology. Among these, human capital represents the rival, and technology represents the non-rival types of knowledge. Both human capital and technology can be accumulated indefinitely. When a new unit of technology, defined as a new product variety, is produced, the productivity of the whole economy increases. There are three sectors in the economy: final goods sector, intermediate goods sector, and R&D sector. Final goods sector produces a homogeneous final good using labor, capital, and human capital with a Dixit-Stiglitz (1977) type technology. Final goods can either be consumed by households or can be used by the intermediate goods sector in the production of capital goods. However, the intermediate goods sector also needs a new product design to make production. These designs are produced in the R&D sector using human capital and the aggregate knowledge stock of the economy and sold to the firms in the intermediate goods sector with an indefinite patent. The payments made for the patents constitute a fixed cost for intermediate goods production, and, hence, causes increasing returns to scale in this sector. Intermediate goods firms cover this fixed cost by charging a price above the marginal cost of production thanks to the market power they obtained due to the fact that intermediate goods varieties are not perfect substitutes. However, since entry is free, firms gain zero economic profit in the long run. Thus, the intermediate goods sector has a monopolistic competition structure, and technological progress arises as a result of the market incentives provided by monopoly rents.

On the other hand, since technology is assumed to be a non-rival and non-excludable type of knowledge, it spills over to all agents in the economy and increases the aggregate knowledge stock of the economy. The aggregate knowledge stock enters into the R&D production function linearly, meaning that every new product variety also enhances the productivity of human capital in the R&D sector. Therefore, the economy exhibits increasing returns to scale as a whole, which generates sustainable per capita output growth in the long run (Jones and Vollrath, 2013; Barro and Sala-i-Martin, 2004).

---

distinction is made according to the location of new product varieties in the product spectrum comparing to the previous varieties.

Grossman and Helpman (1991b) set forth a different interpretation of Romer (1990) study. Instead of a separate final goods sector producing a homogenous final good with a Dixit-Stiglitz type production function, Grossman and Helpman assumed that households have a Dixit-Stiglitz type utility function; namely, they gain utility directly from the variety of final goods. In this case, Romer's (1990) intermediate good firms turn into monopolistically competitive final goods firms, which produce final goods varieties with the blueprints they bought from the R&D sector. Grossman and Helpman showed that their model reached the same results as Romer (1990). Rivera-Batiz and Romer (1991) developed another variety of Romer's (1990) model, in which R&D production uses final products rather than human capital. Since R&D production requires just physical capital good investment, this model is referred to as the lab-equipment model (Acemoglu, 2009).

In all of the models mentioned above, there is a negative relationship between competition and economic growth, because of the need for market power to cover the fixed cost of innovations (Aghion, Akcigit, and Howitt, 2014). However, Bucci (2005) showed that if labor, the total amount of which is fixed, is used in all the three sectors of the economy as an input, an inverse-U type relationship emerges between competition and economic growth.

3.2 Process (vertical) innovation and economic growth

Another group of models explains the economic growth with process innovations. Process innovations can be defined as the introduction of vertically differentiated product varieties that provide more product services per unit cost, and that are perfect substitutes to the existing product varieties (Grossman and Helpman, 1991b). Process innovations can take the form of either decreasing production costs or increasing the quality of existing product varieties (Gopalakrishnan, Bierly, and Kessler, 1999; Acemoglu, 2009).

Since consumers have a clear preference for the new product varieties that are created by process innovations, new varieties render the old product varieties obsolete; therefore, firms that make the product innovations seize the market power of the producers of old varieties and attain monopoly rents. However, this market power is not for an indefinite period as for horizontally differentiated product varieties, and it will only last until another firm replaces them by creating a better variety with another process innovation. In this sense, models that explain economic growth with process innovations formalize the "creative destruction" idea of Schumpeter (1942), and, therefore, they are referred to as "Schumpeterian Growth Models".

Aghion and Howitt (1992) introduce the seminal paper in this vein and present a "quality-ladder" model in which firms create higher-quality intermediate goods varieties with process innovations. There are three sectors in the economy. Final goods sector produces a homogeneous final goods using unskilled labor and intermediate goods with a constant-returns-to-scale technology. Intermediate goods varieties are used in the intermediate goods sector using skilled labor and product variety designs. Finally, the R&D sector produces higher quality intermediate goods designs randomly, which is represented with a Poisson process. The arrival rate of innovations is a positive function of the amount of skilled and specialized labor invested in the R&D sector. The knowledge created with innovations spills over to the economy and enhances productivity in the final goods sector. Entrepreneurs that make a successful innovation acquires an indefinite patent right on the product design and start production of the variety in the intermediate sector as a monopoly. Since new product varieties represent more efficient methods in production, they render the old varieties obsolete; therefore, the innovating firm replaces the old firms and acquires the market power until the next innovation. The higher the degree of market power that firms hope to acquire, the higher would be the average growth rate. In other words, competition and economic growth are negatively related. Grossman and Helpman (1991a) and Segerstrom, Anana, and Dinopoulos (1990) studies also developed similar models. Grossman and Helpman (1991b) introduced a different interpretation of Aghion and Howitt's (1992) model, in which process innovations reduce the cost of production rather than increase the quality of the product, and showed that the same results would arise.

Later studies on the subject tried to reconcile the Schumpeterian paradigm with the empirical findings on the existence of a positive or an inverse-U relationship between competition and economic growth. Aghion and Howitt (1996) decomposed the R&D process into two separate pieces as "research" and "development". Research opens new horizons with the discovery of fundamental paradigms or invention of new production methods. On the other hand, development realizes these opportunities by creating concrete plans for the production of new product varieties or developing new product lines. Aghion and Howitt (1996) argue that research efforts increase as the adaptability of workers in the development process rise. Because, as the degree of substitution increases between the two, developers leave the old production lines and join the research process more quickly. The resulting increase in research would cause a rise in the growth rate. Therefore, it is more likely to have a positive relationship between competition and growth in industries in which developers are more mobile between product lines, comparing to the ones in which mobility is restricted due to the existence of fixed costs specific to product lines.

Aghion, Dewatripont, and Rey (1999) model the case in which the incentives of agents in the decision processes are different from those of the firm. According to them, intense competition reduces the slack in the market by making managers feel the pressure of failure and liquidation of the firm. Hence, managers become more eager to adopt new technologies and the growth rate increases.

Aghion, Harris, Howitt, and Vickers (2001) emphasize the tacit character of knowledge. If firms are required to discover the knowledge in innovations of their rivals by their own R&D efforts, then the standard leapfrogging assumption in the Schumpeterian approach must be replaced with a step-by-step technological progress assumption. In this case, for a firm to outrun the industry's technology leader, it must first close the gap step-by-step by making the previous innovations of the leader with its own R&D efforts. However, once a firm catches the leader, not any patent can protect the leader from Bertrand competition. Since neck-to-neck competition renders life more difficult for firms in such industries, firms have a constant incentive to engage in R&D to surpass their rivals. Therefore, it is more likely to have a strong positive relationship between competition and growth in the industries where tacit knowledge is a restrictive barrier to imitation rather than the ones in which patent protection is the only barrier.

Aghion, Bloom, Blundell, Griffith, and Howitt (2002) investigate the effects of "leveled" and "unleveled" competition in the product markets. Two firms engage in Bertrand competition with differentiated products in industries where entry and exit are blocked. Innovations are assumed to be step-by-step as in Aghion et al. (2001) and enable firms to climb up the quality ladder. In this case, two opposite effects are in operation. First one is the "escape competition effect", which describes the firms' incentive to increase the distance with its rivals. The second one is the "Schumpeterian effect", which refers to the firms' incentives to appropriate the monopoly rents. In general, the escape competition effect dominates the Schumpeterian effect in low levels of competition, and vice versa in high levels of competition so that an inverse-U type relationship emerges between competition and economic growth.

3.3 Process and product innovations and economic growth

Empirical observations indicate that firms invest in both process and product innovations simultaneously. For instance, Scherer (1965) found that the US firms invest three dollars in product innovations for each dollar they invest in process innovations. On the other hand, Rosenkranz (2003) suggested that Japanese firms allocate 60 % of their R&D investments to process innovations and 40 % to process innovations.

Process and product innovations affect economic growth in different ways. Essentially, process and product innovations have different knowledge spillover rates. Ornagi (2006) analyzed the Spanish manufacturing firms for the period of 1990–1999 and found that product innovations diffuse more than process innovations. On the other hand, process and product innovations also interact with market structure and market characteristics in different ways. Lunn (1986) analyzed the US firms for the years 1976–1977 and found that there is a positive correlation between market concentration and process innovations, but no correlation between market concentration and product innovation. These observations suggest that it is crucial to differentiate between process and product innovations and to consider that firms invest in both types of innovations simultaneously, while investigating the relationship between market structure and economic growth.

Lately, some studies in the literature made such an analysis. Firstly, Benedetti-Fasil (2009) developed a general equilibrium model to analyze the effects of process and product innovations on the firm size distribution and economic growth. Firms produce differentiated goods in a monopolistically competitive market. Each firm is different from the others in the sense of product quality and cost structure. Firms increase the quality of their products with product innovations and decrease production costs with process innovations. Each period, firms that cannot make profit leave the market, and new firms that imitate the incumbent firms enter into the market. Imitation acts as a knowledge spillover mechanism in the economy. The assumption that new entrants are more productive than the incumbents on average enables sustainable growth in the long run. Benedetti-Fasil (2009) also calibrated her model for Spanish industrial firms and found that the resulting firm size distribution offers a reasonable explanation for the data. According to her model, process innovations account for 70 %, and product innovations account for 30 % of the economic growth. The only downside of this model is that both product and process innovations correspond to vertical product differentiation.

Akcigit and Kerr (2018) also constructed a general equilibrium model in which multiproduct firms engage in both process and product innovations. Their process and product innovation definitions correspond to vertical and horizontal product differentiations, respectively. Firms develop new product varieties with product innovations and replace the firms that are already producing that variety as in Schumpeterian growth models. Besides, firms also make process innovations to improve the quality of the product varieties that are already in their portfolio. Akcigit and Kerr calibrated their model with the firm and innovation data from US Census Bureau and US Patent Office and found that the model's equilibrium firm size distribution closely matches with the US firm size distribution. Results suggest that the contribution of product innovation on US economic growth exceeds that of process innovation.

4 Conclusion

Endogenous growth literature exhibits that firms' intentional R&D investments are the engine of economic growth. Firms undertake costly R&D activities in the hope of acquiring monopoly rents in monopolistically competitive markets. The non-rival and non-excludable nature of knowledge plays a vital role in the sustainability of long-run growth.

There are different, even conflicting, findings in the literature regarding the nature of the relationship between market structure and economic growth. The different conceptions of innovative activities and competition can be considered as the main reason behind the diversity of the results. In the models based on product (horizontal) innovation, the competitive process is depicted as acquiring monopoly rents from the introduction of niche product varieties, without affecting the market power of incumbent firms. On the other hand, in the models based on process (vertical) innovation, the competitive process is depicted as appropriating the market power of incumbent firms by introducing a better product variety that is a perfect substitute to the existing one. The latter construction crystallizes the "creative destruction" idea of Schumpeter.

In all the studies mentioned above, innovative activities are treated as either product innovation or process innovation. However, empirical findings suggest that firms engage in both types of innovations simultaneously. Process and product innovations interact with economic growth and market structure in different ways. Therefore, models that incorporate these two types of innovative activities are more likely to achieve a better understanding of the relationship between market structure, innovative activities, and economic growth.

References

Acemoglu, D. (2009). *Introduction to Modern Economic Growth*. Princeton: Princeton University Press.

Aghion, P. & Howitt, P. (1992). A Model of Growth Through Creative Destruction. *Econometrica*, 60(2), 323–351.

Aghion, P. & Howitt, P. (1996). Research and Development in the Growth Process. *Journal of Economic Growth*, 1, 49–73.

Aghion, P., Akcigit, U., & Howitt, P. (2014). *What Do We Learn from Schumpeterian Growth Theory? Handbook of Economic Growth*, Volume 2B. Amsterdam, North-Holland: Elsevier.

Aghion, P. & Howitt, P. (1998). *Endogenous Growth Theory*. Cambridge, MA: MIT Press.

Aghion, P., Bloom, N., Blundell, R., Griffith, R., & Howitt, P. (2002). *Competition and Innovation: An Inverted-U Relationship*. IFS Working Papers, No.02/04, Institute for Fiscal Studies (IFS), London.

Aghion, P., Dewatripont, M., & Rey, P. (1999). Competition, Financial Discipline and Growth. *Review of Economic Studies*, 66(4), 825–852.

Aghion, P., Harris, C., Howitt, P., & Vickers, J. (2001). Competition, Imitation and Growth with Step-by-Step Innovation. *Review of Economic Studies*, 68(3), 467–492.

Akcigit, U. (2017). Economic Growth: The Past, the Present, and the Future. *Journal of Political Economy*, 125(6), 1736–1747.

Akcigit, U. & Kerr, W. R. (2018). Growth Through Heterogeneous Innovations. *Journal of Political Economy*, 126(4), 1374–1443.

Arrow, K. (1962). The Economic Implications of Learning by Doing. *Review of Economic Studies*, 29, 155–173.

Barro, R. J. & Sala-i-Martin, X. (2004). *Economic Growth*. Cambridge, MA: MIT Press.

Benedetti-Fasil, C. (2009). *Product and Process Innovation in a Growth Model of Firm Selection*. European University Institute Working Papers, ECO 2009/30.

Bucci, A. (2005). An Inverted-U Relationship Between Product Competition and Growth in an Extended Romerian Model. *Rivista di Politica Economia*, 95(5), 177–206.

Cass, D. (1965). Optimum Growth in an Aggregate Model of Capital Accumulation. *Review of Economic Studies*, 32, 233–240.

Dixit, A. K. & Stiglitz, J. E. (1977). Monopolistic Competition and Optimum Product Diversity. *American Economic Review*, 67, 297–308.

Domar, E. D. (1946). Capital Expansion, Rate of Growth and Employment. *Econometrica*, 14, 137–147.

Ethier, W. J. (1982). National and International Returns to Scale in the Modern Theory of International Trade. *American Economic Review*, 72, 389–405.

Frankel, M. (1962). The Production Function in Allocation and Growth: A Synthesis. *American Economic Review*, 52, 995–1022.

Gopalakrishnan, S., Bierly, P., & Kessler, E. H. (1999). A Reexamination of Product and Process Innovations Using a Knowledge-Based View. *The Journal of High Technology Management Research*, 10(1), 147–166.

Grossman, G. M. & Helpman, E. (1991a). Quality Ladders in the Theory of Growth. *Review of Economic Studies*, 68, 43–61.

Grossman, G. M. & Helpman, E. (1991b). *Innovation and Growth in the Global Economy*. Cambridge, MA.: MIT Press.

Harrod, R. (1939). An Essay in Dynamic Theory. *Economic Journal*, 49, 14–33.

Jones, C. I. & Vollrath, D. (2013). Introduction to Economic Growth, third edition. New York: W.W. Norton and Company, Inc.

Kamien, M. & Schwartz, N. I. (1975). Market Structure and Innovation: A Survey. *Journal of Economic Literature*, 13(1), 1–37.

Koopmans, T. C. (1965). *On the Concept of Optimal Economic Growth*. In The Econometric Approach to Development and Planning. Amsterdam: North-Holland, 225–295.

Lucas, R. E. (1988). On the Mechanics of Economic Development. *Journal of Monetary Economics*, 22, 3–42.

Lunn, J. (1986). An Empirical Analysis of Process and Product Patenting: A Simultaneous Equation Framework. *The Journal of Industrial Economics*, 34(3), 319–330.

Mazzucato, M. (2000). *Firm Size, Innovation and Market Structure: The Evolution of Industry Concentration and Instability*. Cheltenham, UK: Edward Elgar Publishing.

Ornagi, C. (2006). Spillovers in Product and Process Innovation: Evidence from Manufacturing Firms. *International Journal of Industrial Organization*, 24, 349–380.

Ramsey, F. (1928). A Mathematical Theory of Saving. *Economic Journal*, 38, 5443–559.

Rivera-Batiz, L. A. & Romer, P. M. (1991). Economic Integration and Endogenous Growth. *Quarterly Journal of Economics*, 106, 531–555.

Romer, P. M. (1986). Increasing Returns and Long-Run Growth. *Journal of Political Economy*, 94, 1002–1037.

Romer, P. M. (1990). Endogenous Technological Change. *Journal of Political Economics*, 98(5), 71–102.

Rosenkranz, S. (2003). Simultaneous Choice of Process and Product Innovation When Consumers Have a Preference for Product Variety. *Journal of Economic Behavior & Organization*, 50, 183–201.

Scherer, F. M. (1965). Firm Size, Market Structure, Opportunity, and the Output of Patented Inventions. *The American Economic Review*, 55(5), 1097–1125.

Schumpeter, J. A. (1942). *Capitalism, Socialism and Democracy*. London: Harper and Brothers.

Segerstrom, P. S., Anana, T. C. A., & Dinopoulos, E. (1990). A Schumpeterian Model of the Product Life Cycle. *American Economic Review*, 80, 1077–1091.

Solow, R. M. (1956). A Contribution to the Theory of Economic Growth. *Quarterly Journal of Economics*, 70, 65–94.

Solow, R. M. (1957). Technical Change and the Aggregate Production Function. *Review of Economics and Statistics*, 39, 312–320.

Spence, M. (1976). Product Selection, Fixed Costs, and Monopolistic Competition. *Review of Economic Studies*, 43, 217–235.

Swan, T. W. (1956). Economic Growth and Capital Accumulation. *Economic Record*, 32, 334–361.

Uzawa, H. (1965). Optimum Technical Change in an Aggregative Model of Economic Growth. *International Economic Review*, 6, 18–31.

Van Cayseele, P. J. G. (1998). Market Structure and Innovation: A Survey of the Last Twenty Years. *De Economist*, 146(3), 391–417.

Semanur Soyyiğit

2. China's Neo-Mercantilist Policies and Growth Process

Abstract Although the 21st century started as a century in which global trends took an increase, and expectations for 'liberalization' were on the rise, practices were developed in the opposite direction. The protectionist policies that countries embraced in the times of crisis came to the agenda again with the 2008 global crisis. This protectionist policy has been implemented by the US administration towards many countries (mainly towards China) as a 'mercantilist policy' since 2017. The tension that was 'verbal' in 2017 started to be felt as of the beginning of 2018. While China responds to these practices of the United States by saying that "there will not be any winners in a trade war", it is seen that its own global policies are neo-mercantilist as well. Because, while China follows policies towards energy mercantilism in Central Asia, the Middle East, Latin America and Africa, it follows innovation mercantilism policies towards developed countries such as the USA. In this context, the aim of the chapter is to draw attention to the contradiction between China's liberalization advocates and its policy practices in the growth process and to 'unequal' competition conditions in international economic relations; in addition, to evaluate the Belt-Road Initiative, covering the regions in which it has implemented mercantilist policies.

Keywords: Neo-mercantilism, Economic growth, Trade policies, Export-based growth, Chinese mercantilism, Belt-Road Initiative

1 Introduction

From the emergence of the classical school to the emergence of the liberal and conservative paradigms in the economic environment, especially from the end of the 20th century to the beginning of the 21st century(the period beginning from the 1980s to the 2008 crisis), globalization of the liberalization policies has been maintained (or even increased). Free circulation of goods and services, financial assets and production factors across the countries reached the maximum rates, however, global contraction and increased risk perception led the countries to mercantilist policies through the 2008 crisis. As of 2017, it has been observed that the USA under Trump administration has started to implement mercantilist policies against many countries (mainly China) and counterparties have responded with similar policies in order to protect their economies. On the other hand, mercantilist policies have another aspect. This is seen in China's policies on the countries of Central Asia, the Middle East, Latin America and Africa.

Well then, does the development of China promised on the basis of the principle of mutual interest in these regions come true? Or is it a policy that only one side carries interest? Will the "Belt and Road Initiative", which was announced by China in 2013 and actively covering the mentioned regions, really contribute to the growth and development of these countries? Or is it just a neo-mercantilist policy tool that will serve China's energy security and market search? In this section, firstly mercantilist and neo-mercantilist thoughts are briefly reviewed, then an evaluation is made on China's growth process and globally applied mercantilist policies.

2 From Classical Mercantilism to Neo-Mercantilism

The protectionist approach in foreign trade began when the European colonies settled in the American continent (mainly Spanish and Portuguese) after the discovery of the New World. This period of classical mercantilism, which continued until the industrial revolution, led economic policies for nearly 300 years. Within the framework of these practices, economic policies were seen as a part of national goals. In this period of diplomatic and military superiority, wealth was of great importance in achieving these goals. Thus, the rich of foreign trade will create a rich society power; and this power would protect the commercial power (Frieden, 2012). In this environment, mercantilism had established expansionist colonial policies and trade wars without questioning (Wolfe, 1981).

With the emergence of the classical school, the period of liberal economy (where the government interference in the economy was minimal) had begun. However, as Rodrik (2013) points out, the history of economics is in fact a struggle of the liberal and mercantilist schools of thought, which represent two opposing ideas. In this context, although liberalism is the dominant paradigm that directs today's economic structure, mercantilism still exists (Rodrik, 2013).

The developments in the sense of globalization, especially from the last quarter of the 20th century until today, have revealed a structure in which economies are very integrated. Within the framework of this structure, the mercantilist policies implemented by countries in the period also differ from the classical mercantilism. After the dominance of neoliberal policies, the structure of mercantilism has changed; these policies serve as a trade strategy at the national level, but are stretched to create a global impact. The "neo" prefix of mercantilism corresponds to the transformation from the development of classical mercantilism to economic development and also to the adoption of prices in the market (Okeke, 2016).

In this new paradigm called neo-mercantilism, the market mechanism is not opposed; however, in order to protect the interests of the country, the effort to shape the national and international functioning of the markets comes to the fore. For this purpose, the countries that follow neo-mercantilist policies are trying to control by identifying the biggest and strategic sectors in the economy through the public economic institutions or institutions acting as government agencies (Ziegler and Menon, 2014).

Neo-mercantilism, which is based on the idea that the closed national economic structure cannot be realized within the current global economic system, supports a policy regime that includes export incentives, reduction of imports, control of capital movements and exchange rate decisions by the central government. The aim is to implement policies to create external surpluses. In this sense, neo-mercantilism aims to maximize the benefit to be achieved through the high and stable price of goods sold by the country, stability of foreign supply and increase of exports (accompanied by decreasing imports) (Okeke, 2016).

3 China's Growth Process

Since its establishment in 1949, the People's Republic of China (PRC) has followed policies within the framework of the planned economy system. These policies have been effective in the planned and stable development of the Chinese economy, but have begun to create obstacles to growth and wealth increase. For this reason, the 1970s was a period of reforms in China to change the planned economic system. In the process following this period, China's economic growth is divided into periods; the transition period to the socialist market economy in the period of 1978–1996, the period of recession in 1997–2002 period, the period after 2003 can be expressed as the period of growth strategy (Saray and Gökdemir, 2007).

The year 1978 is a turning point in the Chinese economy, because China's market orientation started on this date. Following the Third General Assembly of the Chinese Communist Party (CCP), the Chinese government has started the reforms. First, it has been stated that one of the most missing aspects of economic management is the excessive concentration of power, with a focus on state-owned enterprises, which are considered to be one of the weakest sides of the economy. From this it was decided to give more authority to local administrators and economic organizations (Coase and Wang, 2015). One of the biggest problems of the Chinese Socialist Economy, which was about the performance of state-owned large enterprises, was an accepted situation before the Third General Assembly and some attempts had already started even before this meeting. However, these reforms gained momentum after the board meeting. According to these applications, enterprises would have the authority to assign their own mid-level managers, to retain some of their profits and to produce above the amounts specified in the state plan. In this context, 73 % of rural farms were collectivized in 1981 and 80 % of state companies were allowed to keep profits in the company. However, the fact that businesses keep their profits in themselves has had an impact on the government's tax revenues. The fiscal deficit was accompanied by high rates of inflation. Because price controls were removed in order to encourage production, price increases were realized and this increase was higher than the wages. Following these developments, the Chinese government decided to stop the reforms for a while. At the time of the reforms pausing, a transformation has taken place in the surrounding regions where state control was weak (Coase and Wang, 2015; Savaş, 2016).

The most important of these transformations was in the *agricultural sector*. During the period of Mao, the agricultural sector was severely damaged and the people in this sector were poor. In the 1978 statement, this was accepted and a number of measures were taken in 1979. In the same period of transformation in agriculture, another transformation process was experienced. This was the *rural industrialization* that emerged with the increase of town and village initiatives. Town and village initiatives are of great importance for the development of competitive environment in China. These initiatives were always seen by the government as a bad competitor of 'state-owned enterprises'; they did not have the same advantages as state enterprises in access to raw materials, energy, consumer markets and credit; they have been subject to discriminatory policies. Compared to state-owned enterprises, these enterprises, which operate with lower technology and labor-intensive nature, have become competitors of state-owned enterprises after a while. These initiatives, which gather talented labor with high wages, have also entered into intense investments with these profits (because they can also keep profit) and have led to the emergence of a competitive environment. Therefore, thanks to the town and village initiatives, the industry in China has ceased to be an area operating only under the sovereignty of the state, and the Chinese economy has met with competition (Coase and Wang, 2015).

China's traditional development model can be expressed as investment-driven and export-based economy. In China, where national savings and investments were initially met, foreign investments began to be addressed after the 1978 revolution. In 2003, China became the world's largest foreign investment destination. Most of this foreign investment went to foreign multinational corporations' establishments in China. This was largely due to the desire to

Tab. 2.1: Foreign Direct Stock in China (2014–2017). Source: United Nations Conference on Trade and Development, https://unctad.org/sections/dite_dir/docs/wir2018/wir18_fs_cn_en.pdf (3.03.2019).

	2014	2015	2016	2017
Share in FDI to developing countries	12,8	14,1	14,7	14,4
Share in FDI worldwide	4,3	4,8	4,9	4,7

benefit from cheap labor in China and to export low-cost production to world markets (Savaş, 2016).

Approximately 42 % of foreign investments in China in 1985–2005 period were from Hong Kong. Meanwhile, investments from the US, Canada, Japan and the EU to China accounted for 25 % of total foreign investments in the same period (Savaş, 2016). As seen from Tab. 2.1, the share of China's foreign direct investment stock in the foreign direct investment stock in developing countries is 14.4 % as of 2017; while its share in the total foreign direct investment in the world has been realized as 4.7 %.

Hong Kong has become a country that paves the way for investments in China. When China first pursued open policies to foreign investments in 1979, there were considerable uncertainties regarding the rules and policies to be followed. In the early days, the majority of foreign investments in China were from Hong Kong. In these investments, Hong Kong's investors' desire to invest in a country with past ties was big. At the same time, the fact that these investors had links to reduce the risk and uncertainty of the investment was also influential in their early investments in Hong Kong. These investments have contributed to the integration of China and in particular South China into the supply chain and production system. At the same time, infrastructure investments that enabled China to connect to the global economy were also coming from Hong Kong. To sum up, entrepreneurs who were willing to invest in this country, whose uncertainty about the regime and the practice of the treaties were yet uncertain, were Hong Kong entrepreneurs (Enright, 2017). Although the importance of investments from Hong Kong to China is still high, the purpose of investments has changed. Statistics about Hong Kong, also known as a tax heaven, do not provide complete information on how much of the foreign investment in the country has been redesigned in China; but in the period 1998–2013, it is stated that investments in China account for about 34 % of foreign direct investments in Hong Kong. Undoubtedly, for some purposes such as the utilization of low taxes by foreign firms, the objectives of investing in China through Hong Kong are effective (Enright, 2017).

The Chinese government has always placed industrialization on the top of the development agenda. In order to stimulate investments, domestic commodity prices including energy prices and environmental costs were decreased. Moreover, the government has provided an artificial low interest rate environment by interest rate regulations. Under the influence of low commodity prices and low interest rates, an implicit welfare transfer from the household and commodity supplier to the commodity user was realized. When the low demand – high investment duo is formed in the economy, the country has tended to export in order to absorb this excess capacity. In parallel with this process, the Chinese government implemented a low exchange rate policy to support export growth (Zhang, 2011). On the other hand, the development model implemented by China has brought some negative results as well as high growth rate (Zhang, 2011). The first is that the growth of private consumption is quite low compared to the growth in investment and exports. This situation will cause excess capacity in case insufficient foreign demand due to a global economic turmoil and this excess capacity will affect negatively many other economic processes in case of not to be able to absorb it. Another negative reflection of China's development model is expressed as the unbalanced development of sectors. This indicates that some branches of the manufacturing industry in China have developed very much, while the development in the service sector is very limited.

In addition to this development model in China, the fact that the country became a member of the World Trade Organization (WTO) in 2001 was also an important issue. As a result of these developments, China has become a significant producer and exporter of the global economy and an important energy importer on the other hand. Due to these two changes, the sustainability of both the energy security and the commercial channels has become a priority for the Chinese economy (Balcı, 2018). The fact that energy became a decisive influence on Chinese foreign policy took place after the 1990s. In the country that exhibited a self-sufficient appearance during the Cold War period, energy security became an important concept in meeting the increasing needs in parallel with the increasing global performance after 1990. Within the framework of energy security, Middle East, Africa and Latin America are the regions where China is paying particular attention to (Atagenç, 2012).

Following a significant growth momentum with these developments, following the global economic crisis the contraction in demand in the US, the EU countries and Japan, which are China's largest trading partners, caused a slowdown in growth in the Chinese economy (Karagöl, 2017; Bocutoğlu, 2017).

As seen in Graph 2.1, the growth rate of China has increased continuously in the post-2000 period; with the 2008 crisis, the decline in growth rate has begun.

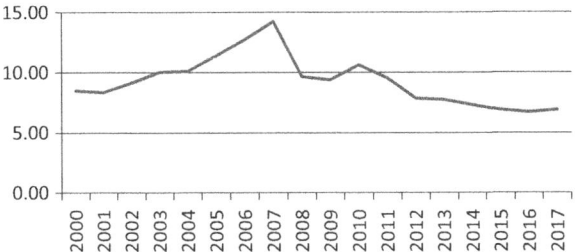

Graph 2.1: China's Growth Rate Between 2000 and 2017 (Annual %).
Source: The World Bank, https://data.worldbank.org/indicator/NY.GDP.MKTP.
KD.ZG?locations=CN (18.02.2019)

In order to eliminate this slowdown in the growth rate, banks lowered their loan rates, and State-Controlled Entities (SCEs) started to invest in infrastructure by turning to off-shore practices to create new jobs (Bocutoglu, 2017). China has also seen that the implementation of the Transatlantic Trade and Investment Partnership Agreement between the US and the EU is a factor in the decline in export rates (Karagöl, 2017). The Belt and Road Initiative is also stated to be a project developed in order to sustain capital accumulation in China and to increase the demand for the goods of the country and thus increase the growth momentum (Bocutoğlu, 2017; Balcı, 2018). On the other hand, this initiative of China is also part of the mercantilist practices of the country. As of the end of 2018, when 123 countries on the project route and the income level of these countries are examined, it can be seen that these countries (except for a group of countries which can be considered as minorities) are composed of significantly low- and middle-income countries; it is seen that the initiative includes geographically resource-rich African, Central Asian and Middle Eastern countries.

4 Chinese Mercantilism and Belt and Road Initiative

The 21st century features a period of interesting developments. While China defends globalization by means of statements and actions towards liberalization, the USA stands out with its protectionist policies. The Belt and Road Initiative, which was announced by the President of China in 2013, is a project that extends to Asia, Africa and Latin America today. This project, which is expressed as a combination of China's development strategy and its projects, creates a synergy and the principle of mutual benefit is considered as one of China's advocacy for globalization (Li, 2017). In addition, China's President Xi Jinping expressed at the

Davos 2017 Summit that the approach to globalization as a "Pandora's box" was wrong; not the globalization, but the search for "excessive profit" should be held accountable for the global crisis; and that no sides in a trade war can be a winner. In spite of these explanations of Jinping, in the recent period, the statements that brought the protectionist policies to the forefront started to rise. In response to these protectionist policy discourses of the most liberal economy of the world, China's advocacy of globalization, which is governed by the one-party system by the CCP, is an indication of an unusual period. In this sense, it is important that the Belt and Road Initiative, which is claimed by China to be based on the win-win concept, is considered as a whole by China's global policies.

One of the countries in which China has implemented neo-mercantilist policies is the United States. The trade war between China and the United States, which was verbally signaled in 2017, began to manifest itself in the concrete practices since the beginning of 2018. In the National Security Strategy of the United States published in December 2017, China was referred to as a strategic competitor. It has been declared that protectionism is important in terms of the Chinese imitating the innovation-based structure in the USA, while maintaining the long-term competitive advantage of the country (Mayer Brown Consulting, 2018). While China supports globalization, it can be stated that the policies it implements are neo-mercantilist. The US equivalent of these policies is also pointed as *innovation mercantilism*.

It is a fact that innovation is effective in ensuring long-term growth acceleration of countries. However, the innovation policies implemented by the countries in the direction of economic growth may bring with them threats as well as opportunities. Ezell (2011) mentions four possible situations that can occur in this process. The first one of these is "good" policies, which simultaneously benefit the country and other countries that implement innovation policy; the second one is "ugly" policies that benefit the country implementing the innovation policy at the expense of other countries; the third one is "bad" policies that seem to be good for the country implementing the policy, but which do not benefit both that country and other countries; and finally, "self-destructive" policies that do not benefit the country that enforces the policy but benefit other countries instead. The ideal application is: it is the implementation of "good" policies that provide a climate based on win-win concept for everyone by developing the innovation structure of the country simultaneously with other countries by providing a spreading effect and enabling them to turn to similar policies. However, this innovation-based growth by some countries is implemented with the beggar-thy-neighbor approach. The fundamental of this approach is the mercantilist policies. These policies, called 'innovation mercantilism', are defined as policies based on the development of high-wage industries and business lines by reducing the welfare of the global economy (Ezell, 2011). Within the framework of this policy, which China is one of the main practitioners, exchange rate, market and intellectual property rights manipulation are applied to increase the country's innovation-based exports while also lead to unfair competition. These policies raise the countries, which are implementing countries, to the high added-value production stage of the global production chain by means of trade-disrupting practices. Although this type of modern protectionism is based on the regulations behind the country, not the tariffs, in order to protect the local firms, the aim and impact of the policies implemented do not change: to replace the foreign goods and services with the domestic ones and to increase unfairly the export of these goods and services (Cory, 2018). Gerwin (2018) stated that, in the 10 years since China's accession to the WTO, the country has shown significant deterioration through practices such as dumping and exchange rate manipulation in sectors with low technology, such as furniture and textiles, and consequently caused a loss of 2.4 million jobs. In the current situation, Gerwin (2018) pointed out that there is a similar situation in high technology industries through innovation mercantilism, but this time the situation is more serious; the mentioned industries provide jobs to 45 million people and equal to the 39 % of the US GDP.

Accordingly, it is mentioned that the policies pursued by China in South America, Central Asia, Middle East and Africa also serve mercantilist applications; and these are called "energy mercantilism". Although China advocates the market economy and the implementation of the liberal regime in international trade, when it comes to energy policy, it moves towards mercantilist policy practices in every geography. China National Petroleum Corporation (CNPC), China Petroleum and Chemical Corporation (SINOPEC) and China National Offshore Oil Corporation (CNOOC) were guiding China's energy policy. In the early 1990s, a change was made in China's energy policy. With this change, China supported the national oil companies to cooperate with non-Chinese companies by acquiring the assets of foreign firms. Based on this, from 1997 onwards, these three large Chinese state oil companies began to move outside the border (Ziegler and Menon, 2014).

Lind and Press (2018) associated the practices of China in the form of energy mercantilism to the policies of the United Kingdom at the beginning of the 20th century and explain it by citing that period. According to this, at the beginning of the 20th century, the British government was greatly disturbed by the importation of oil by the United Kingdom from the Netherlands and particularly the dependence of the Royal Army on overseas oil. The Netherlands was under the influence of Germany, and in the event of a dispute, it could also cut exports to Britain. At that time, Winston Churchill insisted on the idea that the British government should be the controller of oil sources by eluding the idea of acting market based on imports of critical inputs such as oil. Following this period, the UK acquired 51 % of the British-Iranian Oil Company. This policy was followed by other countries. According to Lind and Press (2018), China's energy mercantilist policies are based on the energy-dependent nature of the Chinese economy, and the extreme fragility of the Chinese economy in response to any disruption in energy supply. In this way, China alleviates the effects of a coercive policy of the US on China through the South China Sea or Taiwan. It is also worth mentioning here that the increasing demand for energy and raw materials with the economic development of China has given the country monopsony power in these markets. A well-known example is that China, where Australia dispatches approximately 80 % of its raw materials, has the monopsony power and bargaining power that it holds as the world's largest importer of iron ore (Chaulia, 2014).

China's presence in South America remained quite weak until the 2000s, as the region was regarded as the "backyard" of America. On the other hand, in the 2000s, it is seen that China has more active commercial activities in this region. Both export and import relations in the region have developed in volumetric terms. On the other hand, when we look at the nature of trade, China imports raw materials from Latin American countries. On the other hand, goods exported from China to Latin America and the Caribbean are final goods. These competitively priced final goods are considered as a threat to the manufacturing industry in these countries (Farnsworth, 2011).

The Central Asian region is a region in which China, Russia and the US are competing for political influence and natural resource control. This region, which is rich in oil and natural gas, has quite a less amount of reserves compared to the Middle East, Africa and Latin America, but it is important in terms of providing an alternative to these sources of unstable supply. The trade of oil, as well as natural gas, determined by geography as well as technology, determines the geopolitical dimension of the strategies applied by the countries in question in the region. According to this, the US wants to reduce Russia's influence; Russia wants to preserve its monopoly power on the export route of these two commodities, while China follows a strategy to diversify its supply network (Ziegler and Menon, 2014). This region has a characteristic that China, as well as Russia, is implementing mercantilist policies. The activities of CNPC, SINOPEC and CNOOC, the three largest Chinese state-owned companies, in Central Asia have been shaped around Kazakhstan, Turkmenistan and Uzbekistan. Thus,

China purchased shares in state-owned oil companies in Kazakhstan, completed the construction of the petroleum pipelines, which was 1348-mile-long oil pipeline from Kazakhstan to China in 2008, and the 1240-mile-long natural gas pipeline (through Uzbekistan and Kazakhstan) from Turkmenistan to China in 2008. Meanwhile, Chinese oil companies have also been advocates of overseas investments in recent years. Thus, in Central Asia, Africa and Latin America, China also obtains equity shares (Ziegler and Menon, 2014). When the relations between the Middle East countries and China are examined, it is seen that relations developed to meet the energy needs of China. The Middle East is an important energy source for China, especially in terms of oil. In this sense, China's presence in the region is also economic and commercial. After Iraq opened its oil fields to foreign companies in the late 2000s, China became a foreign partner in five oil fields. It is also noted that the infrastructure and communication investments that China carries out in the region has increased the exportation of consumer goods and machinery in the region. In addition to these, China's interest in the region is to strengthen its presence and influence in this geostrategically important region, while protecting China's sovereignty and territorial integrity and internal stability and (although it appears not likely) protecting a region that is neighboring to China. Middle East countries are among the countries in which China is included in the Belt and Road project (Scobell et al., 2018).

Finally, when China's policies on the African continent are considered, the fact that the country's economic growth process is directed towards mercantilist policies is encountered. After the establishment of the PRC in 1949, its relations with Africa supported the anti-colonial, African independence movements at first. This relationship brought to mind the "solidarity of the third world countries". After the Bandung Conference, which was held by Indonesia's efforts in 1955, China's relations with the new independent African countries were realized within the framework of equality, mutual interest and non-interference in other states. However, with China's transition process to market economy in 1978, its approach to Africa has been based on economic foundations. At the end of the 20th century, the nature of China's economic and commercial activities with Africa became unilateral, neo-colonial and mercantilist. In the current situation, the basis of China's African strategy has four fundamental interests (Scobell et al., 2018): (i) access to basic resources, in particular petroleum and natural gas; (ii) to establish an export market for China's final goods; (iii) to provide legitimacy as a global power in international politics; and (iv) adequate political stability and security for the protection of citizens, economic and commercial interests.

The Belt and Road Initiative, which was put forward by China in 2013 and is intended to revive the old Silk Road, also covers the aforementioned regions, which are said to have implemented policies towards China's energy mercantilism. In this sense, serious criticism is directed at the initiative. It is stated by Xi Jinping that the initiative, based on mutual trust, equality, mutual interest, inclusiveness, mutual learning and win-win principles, is a soft power policy that serves China's world leadership (Phillips, 2018). For this reason, this initiative, which consists of developing countries with a high degree of middle income, needs to be evaluated seriously by every country involved in or taking part in the initiative. However, it would be very difficult to conduct a detailed evaluation of the initiative. World trade is an inherently complex system where mutual commercial partnerships are quite complex by any means. The main reason why there are not too many empirical studies in the literature is this complex relations network.

5 Conclusion

While China engages in rhetoric of liberalization on one hand, it follows neo-mercantilist policies on the other. These policies of China, which are called "innovation mercantilism" in developed countries and "energy mercantilism" in developing countries, are being criticized. Therefore, the Belt and Road Initiative introduced by Xi Jinping in 2013 is also being approached with suspicion by some.

Although the project has been described by China as a win-win principle, it is stated that it is a policy tool for "aiming to provide energy security" and "finding new markets for itself" in some areas based on China's geographical scope. In this sense, the evaluation of the initiative should be done very carefully by the countries that are involved in it. In fact, if this initiative is by large extent an attempt to serve China's market search and energy security objectives, there will be a possibility in the long term for this project to not contribute to the development processes of other countries.

On the other hand, such a study is difficult because of the comprehensive nature of the initiative and the complexity of mutual business activities between countries. This is the reason why there are few empirical studies on the subject in the literature.

References

Atagenç, Ö. (2012). Çin ve Hindistan'ın Deniz Stratejisi ve Hint Okyanusu'nda Güç Mücadelesi. *Bilge Strateji*, 4(6), 135–166.

Balcı, Z. (2018). Çin'in İpek Yolu Projesi. *İNSAMER*, No. 52. Retrieved from https://insamer.com/rsm/files/CININIPEKYOLUPROJESI.pdf (11.08.2019)

BBC, Retrieved from https://www.bbc.com/news/world-asia-china-38650487 (20.02.2019)

Belt and Road Portal, Retrieved from https://eng.yidaiyilu.gov.cn/info/iList.jsp?cat_id=10076 (14.12.2018).

Bocutoğlu, E. (2017). Çin'in 'Bir Kuşak – Bir Yol' Projesinin Ekonomik ve Jeopolitik Sonuçları Üzerine Düşünceler. *International Conference on Eurasian Economies 2017*, 265–270. Retrieved from https://www.avekon.org/papers/1995.pdf (11.08.2019)

Chaulia, S. (2014). *Politics of the Global Economic Crisis – Regulation, Responsibility and Radicalism*. India: Routledge, 1st published.

Coase, R. &Wang, N. (2015). *Çin Nasıl Kapitalist Oldu?*. Ankara: BigBang Yayınları, 1. Baskı.

Cory, N. (2018). The Worst Innovation Mercantilist Policies of 2017. *Information Technology&Innovation Foundation*, Retrieved from http://www2.itif.org/2018-worst-mercantilist-policies-2017.pdf (24.02.2019).

East Asia Forum, Retrieved from http://www.eastasiaforum.org/2016/09/02/whats-driving-chinas-one-belt-one-road-initiative/ (16.02.2019).

Enright, M. J. (2017). *Developing China: The Remarkable Impact of Foreign Direct Investment*. New York: Routledge, 1st published.

Ezell, S. (2011). Fighting Innovation Mercantilism. *Issues in Science and Technology*, 27(2), 83–90.

Farnsworth, E. (2011). The New Mercantilism: China's Emerging Role in the Americas. *Current History*, 110(733), 56–61.

Frieden, J. (2012). The Modern Capitalist World Economy: A Historical Overview. In *Oxford Handbook of Capitalism*, edited by Dennis Mueller. New York: Oxford University Press, 1–15.

Gerwin, E. (2018). Confronting China's Threat to Open Trade. Washington, DC: Progressive Policy Institute.

Karagöl, E. T. (2017). Modern İpek Yolu Projesi. *SETA Perspektif*, 174. Retrieved from http://www.evreninsirlari.net/dosyalar/136_s14_02.pdf (11.08.2019).

Li, Y. (2017). Belt and Road: A Logic Behind the Myth. In *China's Belt and Road: A Game Changer?*, edited by Alessia Amighini, Milano: ISI, 13–33.

Lind, J. & Press, D. G. (2018). Markets or Mercantilism?. *International Security*, 42(4), 170–204.

Mayer Brown Consulting. (2018). The Stakes in the US-China Trade War, Retrieved from https://www.mayerbrown.com/files/

Publication/2e0bfbbe-ee56-4bac-afdb-5141d4b9fe7b/Presentation / PublicationAttachment/f9cb82cd-9f60-4737-9cc6-5392138dc095/20180726-MBC-The-Stakes-in-the-US-China-Trade-War_EN.pdf(27.10.2018).

Okeke, D. (2016). *Integrated Productivity in Urban Africa: Introducing the Neo-Mercantile Planning Theory*. Switzerland: Springer International Publishing.

Phillips, R. (2018). Mercantilism with Chinese Characteristics: Creating Markets and Cultivating Influence. *China Economic and Security Review Commission*, Retrieved from https://www.uscc.gov/sites/default/files/Phillips_USCC%20Testimony_17Jan2018.pdf (3.03.2019).

Rodrik, D. (2013). The New Mercantilist Challenge, Retrieved from http://www.fabricedefever.com/pdf/Rodrik%20-%20The%20new%20mercantilist%20challenge.pdf (16.02.2019).

Saray, M. O. & Gökdemir, L. (2007). Çin Ekonomisinin Büyüme Aşamaları (1978–2005). *Journal of Yaşar University*, 2(7), 661–686.

Savaş, V. F. (2016). *Bilmediğimiz Çin*. Ankara: Efil Yayınevi, 1. Baskı.

Scobel, A., Lin, B., Shatz, H. J., Johnson, M., Hanauer, L., Chase, M. S., Cevallos, A. S., Rasmussen, I. W., Chan, A., Strong, A., Warner, E. & Ma, L. (2018). *At the Dawn of Belt and Road – China in the Developing World*. Santa Monica, CA: RAND Corporation.

United Nations Conference on Trade and Development. (2018). World Investment Report, Retrieved from https://unctad.org/sections/dite_dir/docs/wir2018/wir18_fs_cn_en.pdf (3.03.2019).

Wolfe, D. A. (1981). Mercantilism, Liberalism and Keynesianism: Changing Forms of State Intervention in Capitalist Economies. *Canadian Journal of Political and Social Theory*, 5(1), 69–96.

World Bank, Retrieved from https://data.worldbank.org/indicator/NY.GDP.MKTP.KD.ZG?locations=CN (18.02.2019).

World Bank, Retrieved from https://datahelpdesk.worldbank.org/knowledgebase/articles/906519-world-bank-country-and-lending-groups (14.12.2018).

Zhang, M. (2011). The Transition of China's Development Model. In *G20 Perceptions and Perspectives for Global Governance*, edited by Wilhelm Hofmeister. Singapore: Konrad-Adenauer-Stiftung, 51–56.

Ziegler, C. E. & Menon, R. (2014). Neomercantilism and Great-Power Energy Competition in Central Asia and the Caspian. *Strategic Studies Quarterly*, 38(4), 17–41.

Begüm Erdil Şahin and Deniz Dilara Dereli

3. An Evaluation on Middle-Income Trap in Turkey

Abstract The middle-income trap has come to the forefront as a topic frequently discussed in recent years. How long the countries in the middle-income group are staying at this level of income is important, and the failure to pass to the upper-income group is defined as the middle-income trap. Countries included in income groups are classified with different approaches. The middle-income trap has also been a hot topic particularly in Turkey, which was included in the upper-middle income group in 2005. In this chapter, first middle-income trap process is reviewed and then current situation in Turkey is discussed. In addition, the factors affecting the process of middle-income trap are discussed and policy proposals are presented to get out of the middle-income trap. In particular, structural reforms that Turkey must achieve in order to get out of this trap are commented.

Keywords: Middle-income trap, Economic growth, Development, Turkey

1 Introduction

The middle-income trap is frequently high on the agenda in the economic literature. The middle-income trap, which is essentially a growth problem, is described as the inability of a middle-income country to reach the upper-income level. Growth is occurred through three main sources: capital accumulation, population increase and technology. Although capital accumulation and population increase are not sustainable, their occurrence is relatively easier. Long-term growth is achieved through a human capital structure based on education and technological progress (Taşçı and Özsan, 2014: 1). Low-income countries producing labor-intensive goods with the help of simple technologies transferred from abroad have cost advantages. The income level increases in the short term in parallel with the changing productivity, and after the middle-income level, the lack of employment in agriculture decreases and the wage increase and cost advantage is lost. These countries reach the position that is not competitive with low-income countries with cheap labor advantages as well as high-income countries producing high technology (Çobanoğulları & Eroğlu, 2017: 258–259).

The middle-income trap, which was introduced into the literature for the first time by Intermit Gill and Homi Kharas in their report called "An East Asian Renaissance: Ideas for Economic Growth", explains the slower growth of middle-income countries than low-income and high-income countries. According to the report, it is stated that the countries may get out of the middle-income trap with a three-stage transformation. In the first stage, specialization in production and employment takes place; in the second stage, innovative policies and accelerated Research and Development (R&D) activities come into question; and in the third stage, reforms in the education system are carried out to present a technology-oriented and well-supported human capital (Gill and Kharas, 2007: 17–18; Yavuz, 2017: 79).

Turkey was stuck in low middle-income level for many years, and since 2005, has risen to the upper middle-income level. However, it has not yet succeeded in moving to an upper-income level and maintains its position at the level of middle-income countries. In this respect, Turkey has gained importance especially in debates on the middle-income trap. Therefore, it is necessary to evaluate the current situation in terms of the middle-income trap and to determine the policies to be implemented in order to move to a higher income group.

2 Middle-Income Trap

Middle-income trap is defined as the level of income per capita in which the countries in the middle-income group cannot exceed this level and cannot pass to the high-income group. According to the mainstream view of growth and growth sources, countries move from agricultural production to consumer's goods in the first phase of growth and accelerate growth. The labor force in the agricultural sector is moving towards new industries; with the increasing profits, capital concentration and an increase in growth rates are observed. However, as the countries approach the middle-income group, declines in increasing profits begin and the growth rates slow down with the aging technology. At this stage, the source of growth shifts from production to productivity, and factors such as R&D activities and human capital gain importance (Alçın and Güner, 2015: 30).

Low-income countries, which are rich in natural resources and labor, perform labor-intensive production through the imported technologies they provide, and grow rapidly with the cost advantage of cheap labor. As a result of the increasing urbanization, the labor market shifts from agriculture to industry and services and the demand for capital increases. Rising profits, cheap labor and abundant natural resources and the exports of produced goods by these means bring the growth to the level of middle-income countries. Rising incomes and shifting production from agriculture to manufacturing industry where productivity is higher increase productivity. At this point, the continuation of production with the old technologies, the decrease of the return of capital and the inability to switch to productivity-based production structures cause the countries to be

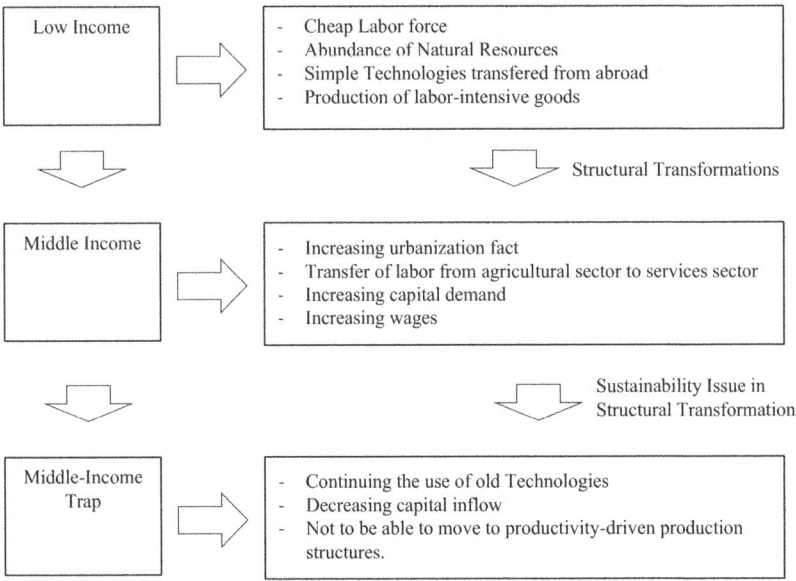

Fig. 3.1: Middle-Income Trap Process. Source: Ünlü and Yıldız, 2018a: 4.

trapped. When countries fall into the middle-income trap, they fall behind in international competition, lose their advantages in the industrial goods they are superior to in the competition with low-income countries because of cheap labor force, and cannot compete with innovation-oriented growth of high-income countries that export high value-added products. Therefore, these countries need to develop strategies based on technology rather than a resource-based approach and establish an economic, political and culture-based change. In particular, the change in the production structure through the innovation and human capital axis plays a crucial role (Ünlü and Yildiz, 2018b: 49). All this process can be seen in Fig. 3.1.

According to Ohno, the stages of industrialization and as a result of this, middle-income trap should be evaluated in five stages, which are illustrated in Fig. 3.2. As a matter of fact, not all countries are able to make progress on the same course, while some countries are waiting in the zero phase, some other countries are in the middle-income trap by staying in the second stage because they cannot renew the human capital. In the first phase, zero-income countries with problems such as war, political issues and economic problems are located. This position, which is based on agriculture-based economy, uniform export and

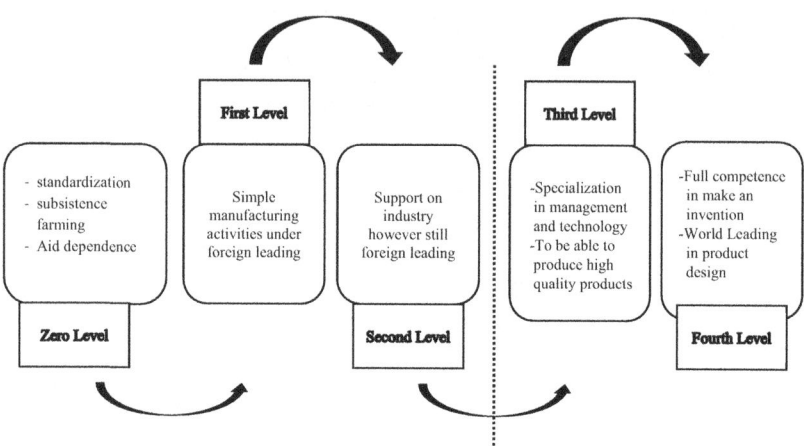

Fig. 3.2: Stages of Industrialization and Middle-Income Trap. Source: Ohno, 2009: 28; Çaşkurlu and Arslan, 2014: 74.

foreign aid, represents a stage in which there is no manufacturing activity. In the first stage, the tools and equipment needed for production are imported, foreigners carry out the design, production and marketing activities and supply the technology and the related country constitutes the unskilled labor force and industrial area. In the second stage, foreign direct investments increase and the production of parts and subsidiary product in the related country become possible with the increasing production. While the competitiveness of these assembly firms is increasing, the industry has been growing quantitatively with the increase in domestic input supply; however, the production proceeds in the management of foreigners and there is no increase in wages and income. In the third stage, while the industrial human capital increases, knowledge and quality are internalized and production passes to local units. Intrinsic value creation increases and the country becomes an exporter of industrial products with high competitiveness. The last stage countries have the ability to create new products and manage the global market (Ohno, 2009: 26–28; Çaşkurlu and Arslan, 2014: 73).

A number of indicators are useful in determining the middle-income trap issue in the economy. In the case of middle-income trap, some or all of the negativities such as the low level of savings and investment in the country, the slowdown in the expansion of the manufacturing industry, the decrease in diversification in the industry, and the weakening of the working conditions in the labor market can be observed. In order to determine the middle-income trap, firstly, it has been examined how long the country has been in the middle-income level. The average growth performances of countries should be investigated during this period and the middle-income trap classification is made according to the performance of the countries concerned. Countries are evaluated in the context of lower middle-income trap and upper middle-income trap according to their income levels. If a country cannot leave the lower middle-income group for 28 years or more, it is assumed that the country has been caught by the lower middle-income trap. Being caught by the upper middle-income trap for a country is the situation that this country cannot exceed the upper middle-income level for 14 years or more (Çetinkaya, 2016: 3).

The reasons for falling into the middle-income trap of countries and what should be done to get out of this trap are examined in the literature within the framework of different approaches. In the supply-side approach, it is stated that the middle-income trap has been caught since the productivity increase do not contribute to the growth processes and low growth rate is seen as a result of growth policies of economies where R&D and technology are not taken sufficiently into account. The demand-side approach is based on the supply-side approach, but also advocates the need to analyze the middle-income trap over demand and emphasizes the importance of consumption and investment expenditures in avoiding middle-income trap. In the inequality approach, which tries to explain the relationship between middle-income trap and inequality, the Kuznets curve approach examines whether income inequality results in a middle-income trap or not and claims that inequality will worsen even more if the mentioned issues are not solved in this middle-income countries and growth will be slower steadily. Another approach, the phases of economic development approach, which is developed by Ohno (2009), Aoki (2011), Zhuang et al. (2012), Tho (2013) and Dewitte (2014) emphasized that the developmental stages of the countries are different from each other and the need to increase the use of technology and education and R&D policies in order to enable economies to come out of the middle-income trap and to be included in the high-income group. In the political economy approach, which states that institutional infrastructure should be improved in order to get rid of the trap, political instability in the country, injustice in the distribution of power and income, non-dissemination of democratic culture, weight in public economy negatively affect the growth process and cause middle-income trap. The slowdown in growth approach, which is established by Eichengreen et al. (2012) and Aiyar et al. (2013) within the framework of a number of predetermined criteria, is based on determining the slowdown growth and the analysis of the factors that lead to the trap. The threshold value approach, in which the threshold year numbers are determined to find out the state of the middle-income trap of the country, is proposed by Felipe et al. (2012). Another approach introduced by Woo (2012) is the Catch-Up Index (CUI) approach, where the countries are categorized by the index derived from the ratio of national income per capita of countries to national income per capita in the US (Ünlü and Yıldız, 2017: 111).

According to Egawa (2013), the middle-income trap should not be confused with the case of slowdown in growth, which is likely to appear in every income level. Factors such as the lack of adequate input increase, the rise in wages and the weight of public investments, the current problems of middle-income countries, the dependence on manufacturing industry exports, the regional income inequalities, the income inequalities in the households, and the aging have led to the middle-income trap, while inadequacies in input-driven growth and natural resources, global warming, increase in the number of migrants and excessive devaluation practices slow the growth rate. Factors causing middle-income trap do not cause any problems in low-income groups and cause trap problems in middle-income countries. The factors that prevent the escaping from the trap and controlling the slowdown in growth are similar. Since the middle-income trap is a structural problem, it can only be solved in the long run, and thus the factors that prevent the escape from the trap are more comprehensive. Measures that may be taken for the deceleration of growth can be created with appropriate short- and medium-term policy instruments, and many measures, incentives and reforms are needed to move a country with a certain level of income to the advanced-income level (Ünlü and Yıldız, 2018b: 47–48).

3 Middle-Income Trap in the Turkish Economy

Turkey, in 1955, made the transition to low middle-income group and remained in this level until 2005. In the post-2005 period, Turkey moved to high middle-income group, however, has remained in this group until today. When evaluated in general terms, Turkey has been in the middle-income countries group for the last 64 years, with 50 years of this period in a low middle-income group.

Economic growth has three main sources in terms of production factors. These are capital accumulation, which means the rise in production capacity, increase in employment and growth in productivity. In the early stages of development, capital accumulation and employment increase play an important role. However, after the per capita income level exceeds 10,000 dollars, sustainability of the high-level income increase depends on productivity growth. Therefore, the contribution of the increases in capital stock and employment to

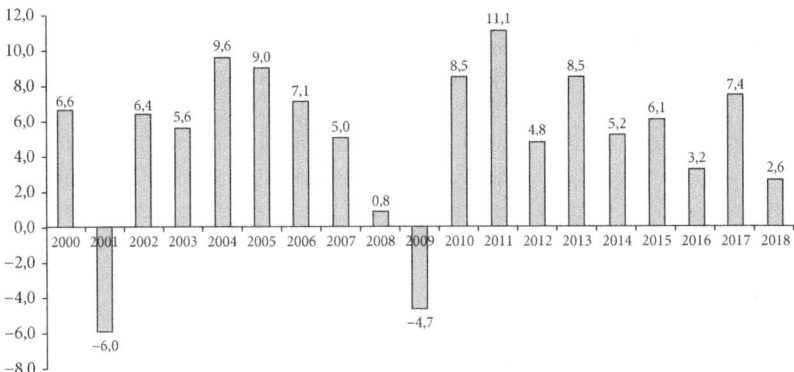

Graph 3.1: Turkey GDP Growth Rates (2003–2018). Source: General Directorate of Budget and Financial Control, * Data for 2018 were taken from TurkStat.

growth in the middle-income level is slowing down. When the value-added per employee increase is low, the per capita income growth rate decreases (Gürsel and Soybilgen, 2013: 1–2).

Turkey's economy abandoned import substitution growth model in the post-1980 period and has started export-led growth model. In the 1980–1990 periods, legal regulations were made for the implementation of free market economy and the share of the state in the economy was reduced. With the foreign exchange legislation adopted in 1989, perfect convertibility was achieved and the conditions of free market economy were provided pretty well (Kaya et al., 2015: 830). The 1990s were the years of economic crises and fluctuations in real growth rates. The economy in these years could not be stabilized due to high inflation, grey economy, 1997 Asian Crisis, sudden fluctuations in exchange rates and increased terrorist incidents. The 1990s are considered to be lost years in this respect (Yaşar and Gezer, 2014: 141).

These structural problems during 1990s in Turkey's economy caused two more major crises in November 2000 and February 2001. After these crises, many structural and legal arrangements have been made within the scope of the Transition to Strong Economy Program and thus a stabilization environment has been created in the economy (Kaya et al., 2015: 830). With the new policy measures that started to be implemented after the 2001 crisis, economic growth (Graph 3.1) and thus the national income per capita level have increased significantly (Şahin et al., 2015: 225). The per capita income of around 4,000 dollars in 2000 increased to 10,000 dollars by 2008.

After the high growth rates in the economy following the 2001 crisis, the success achieved by the transition to the upper middle-income group in 2005 was questionable—whether it could be sustainable and switch to the high-income group or not (Nişancı et al., 2015: 232). Although Turkey passed to the upper middle-income class, there had been important facts which took attention, the growth rate remained below the average in 2007 and decreased 0.8 % by the effect of the global crisis in 2008 and to a negative value of −4.7 % in 2009. During the recovery period after the recession in the global economy, high-growth rates of 8.5 % and 11 % were achieved in 2010 and 2011, respectively. However, below-average growth rate in the coming years has brought Turkey's middle-income trap to fall into a discussion agenda.

Various approaches have been developed in the literature to determine the presence of a middle-income trap. In the capture index approach, the concept of middle-income trap is examined according to be in the 20 % of the per capita income of the USA (Alkan and Ümit, 2018: 105). Gross domestic product (GDP) per capita in Turkey and the United States between the years 1990 and 2018 are included in Tab. 3.1.

According to the data in Tab. 3.1, it is seen that from 1990 to 2007, Turkey could not reach 20 % of the income per capita in USA. Since 2008, it can be said that it has reached middle-income level. In the five-year period between 2008 and 2014, the average per capita income in Turkey over the one in the USA remained around a 20.20 % level. In recent years, this ratio decreased to 18.9 % in 2016 and to 17.7 % in 2017. In order to decide whether an economy has fallen into the middle-income trap or not, it must be in the middle-income group for a long time.

The increase in per capita incomes of countries is considered as an indicator of the increase in welfare levels. However, if these increases do not provide continuity, per capita income levels may remain at the same level for many years or may go back (Yeldan et al., 2012: 151). In Turkey, after reaching the middle-income level, middle-income trap can be mentioned since the per capita income.

Besides the decline in the economic growth rate, other indicators of the economies facing middle-income trap are the low level of savings and investments in these countries. In addition, the development of the manufacturing industry, and finally the conditions of labor market in the weakening of the situation is in question (Alkan and Umit, 2018: 107).

Savings are the source of investments. However, savings are required to be in a certain level to be converted to investment. Otherwise, the country needs to finance inadequate savings through foreign loans. In this case, it creates a burden on the budget and increases the interest payments. In fact, increasing savings

An Evaluation on Middle-Income Trap in Turkey 53

Tab. 3.1: Comparison of Turkey and the US Per Capita GDP (1990–2017). Source: World Bank, GDP per capita (current US $).

Years	Turkey GDP per capita ($)	USA GDP per capita ($)	TR/USA (%)
1990	2.794	23.954	11,70
1991	2.736	24.405	11,20
1992	2.842	25.493	11,10
1993	3.180	26.465	12,00
1994	2.270	27.777	8,20
1995	2.898	28.782	10,10
1996	3.054	30.068	10,20
1997	3.144	31.573	10,00
1998	4.497	32.949	13,60
1999	4.108	34.621	11,90
2000	4.317	36.450	11,80
2001	3.120	37.274	8,40
2002	3.660	38.166	9,60
2003	4.718	39.677	11,90
2004	6.041	41.922	14,40
2005	7.384	44.308	16,70
2006	8.035	46.437	17,30
2007	9.710	48.062	20,20
2008	10.851	48.401	22,40
2009	9.036	47.002	19,20
2010	10.672	48.375	22,10
2011	11.341	49.794	22,80
2012	11.720	51.451	22,80
2013	12.543	52.782	23,80
2014	12.127	54.697	22,20
2015	10.985	56.444	19,50
2016	10.863	57.589	18,90
2017	10.546	59.532	17,70

converted to investment in the economy of the country contribute to economic growth. In order not to enter the middle-income trap or to get out of this trap, the savings in the national economy need to increase (Yıldız, 2015: 159). The ratio of domestic savings to GDP in Turkey between the years 1990–2017 is shown in Graph 3.2.

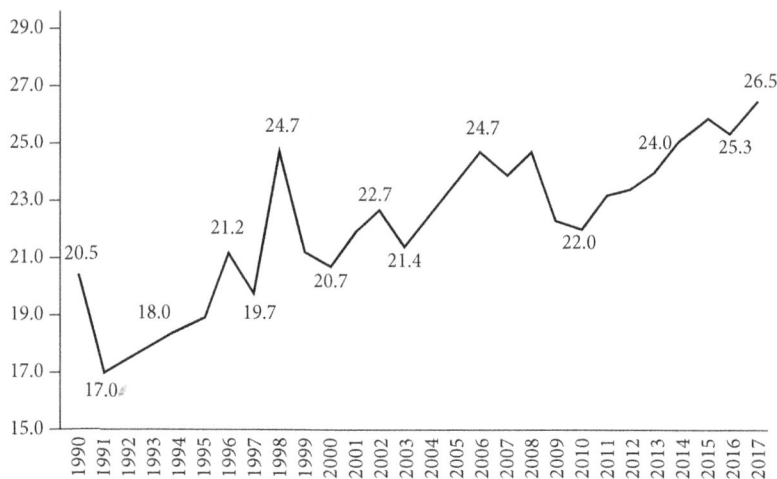

Graph 3.2: The Ratio of Domestic Savings to GDP in Turkey (1990–2017).
Source: World Bank.

The graph shows that Turkey's total domestic savings rate, the average between the years 1990–2017, was realized as 22.2 %. It is observed that the said ratio has been affected by the economic crisis, has increased after the global crisis and reached the highest value in 2017. However, although the rise in recent years, Turkey's savings rate is still far below the medium high-income countries savings rate (Özsan et al., 2017: 7).

Increasing savings depends on some factors: variables in the economy (interest, inflation, terms of trade and public saving rates), financial variables (financial degree of freedom); income and growth variables (GDP per capita and economic growth); demographic variables (population structure, dependency ratio and women's labor force participation ratio); external variables and uncertainties (external shocks and crises). For this reason, developing countries and those who want to get out of the middle-income trap should try to increase their savings by taking these factors into consideration (Yıldız, 2015: 159).

Another condition of competition with other developed economies in the globalized world is based on having a qualified and well-trained labor force, while increasing the quality of the labor force depends on education. As the education system develops, human capital also develops and thus qualified labor force can have more flexibility, mobility and entrepreneurship. In this case, unemployment in the country's economy decreases and employment opportunities expand thanks to new jobs (Çakmak and Gümüş, 2005: 61; Yıldız, 2015: 161).

Tab. 3.2: The Comparison of Turkey and Developed Countries in Terms of Selected Indicators. Source: World Bank 2016, *United Nations Human Development Reports, ** TURKSTAT 2017.

Indicators	Turkey	High-Income Economies
R & D/GDP (%)	0,96**	2,51
Number of Patents	6230	841,458
Knowledge Intensive Employment Share in Total (%)	19,73	39
Share of High-Tech Products in Exportation (%)	2,5	16,5

When the unemployment rate in Turkey is evaluated, a high level of unemployment rate over 10 % is encountered. Moreover, when we look at the distribution of employment, a service-intensive structure is noteworthy. The labor force participation rate of women remains low compared to developed countries.

Turkey, who is ranked among upper middle-income countries group, remains at a low level in high-tech production due to following factors: R&D expenditures share in national income, number of patents, knowledge-intensive employment share in total, and the share of high-tech products in exportation. According to this, the comparison of Turkey and developed countries in terms of selected indicators are displayed in Tab. 3.2.

The share of R&D expenditures in national income is below 1 % in Turkey, while at the level of 2.5 % in high-income economy. This ratio is quite low compared to high-income countries. The number of patents that we can define as an indicator of innovation activities in our country is quite low compared to high-income countries. Another factor that distinguish Turkey from high-income countries is the share of high-tech products in exportation, and this stated ratio is 2.5 % in Turkey, while 16.5 % in developed economies. A further factor is the low share of knowledge intensive employment in total employment. While the share of knowledge intensive employment in total employment is 19.73 % in Turkey, this rate rises to about 40 % in high-income countries. These indicators reveal that Turkey has significant problems in producing high-tech products.

It is important to grant patents to individuals and companies to promote innovation. However, in developing countries like Turkey, it is observed that the number of patents is inadequate. The deficiencies in property rights and patent management increase the likelihood of countries falling into the middle-income trap. On the contrary, the protection and promotion of these rights contribute to the development of innovative activities (Çobanoğulları and Eroğlu, 2017: 265).

4 Recommendations for Turkey to Get Out of the Middle-Income Trap

Countries in the middle-income trap are unable to compete with high-income countries because they cannot produce and export high value-added products, but also lose comparative advantage in the industrial sector, as wages are relatively low in low-income countries. Therefore, these countries need to make a transition to technology-driven growth process in order to give importance to technological production and to make growth sustainable. To be successful in this process, economic, political and cultural factors should support the change as well as economic targets. In the structural reform process, there are two important factors affecting the production structure of countries. These are the dynamics of economic growth: innovation and human capital (Ünlü and Yıldız, 2018a: 49).

Another policy proposed in the fight against the middle-income trap is the restructuring of public policies implemented by the states. In this context, reforms to improve infrastructure investments in the country, to increase the protection of property rights, to develop R&D activities and information technologies are proposed (Agenor et al., 2012: 1–2). In order to get out of the trap, an export-driven growth policy is needed together with technological development. This process should be supported with foreign investments and sustainability should be ensured with the increase in domestic demand. In addition, productivity increase should be ensured by considering capital and labor productivity (Şahin et al., 2015: 229).

Total factor productivity is an important indicator of economic growth. It is used to explain the growth differences occurring in the development processes of countries and shows which factors are effective in ensuring the growth (Vergil and Abasız, 2008: 160). According to Easterly and Levine (2001), the main determinant factor in the development of countries is the total productivity increase. They stated that the effect of capital and labor force is less on the increasing income gap between countries, while it results from total factor productivity. In the long term, it is argued that the rate of savings in the national economy, human capital and investments in R&D expenditures, taxes, technology level and innovation capacity, infrastructure investments, financial development and competitiveness are important for the sustainability of economic growth (Yildiz, 2015: 165).

With the increase of productivity in the country's economy, the transition to the production process of high value-added products and services should be made. Especially by increasing the support and incentives for R&D activities, it should be tried to change the foreign-dependent structure in imports. The creation of new targets in the development phase and the provision of product diversity in foreign trade is another important element. Therefore, there is a need for qualified labor force at this stage, not for cheap labor. The measures to be taken towards the development of human capital contribute to the development of high value-added and innovation-based strategies (Kaya et al., 2015: 833).

Technology is the most important issue to be considered in developing countries. Technology and related R&D activities are the most effective means of transition to the knowledge-based society stage. The development of technological production level minimizes external dependence. Through technology, production, and qualified workforce, investments can be utilized in the most appropriate way. In addition, high technology will contribute to the increase in revenues from exports. However, unfortunately, the share of R&D expenditure in national income in Turkey is not even 1 %. High-income countries allocate more shares to these expenditures (Erkoç, 2015: 191). In recent years, the importance of developing a Regional Innovation Strategy has started to be emphasized. According to this, R&D activities are carried out through the incubation centers to be established in the regions and the resulting production information can create activities supporting the development and competitiveness of local capital in regional priority areas. Within this scope, it provides a competitive advantage especially for developing regions (Alçın and Güner, 2015: 43).

Another important strategy that should be applied for Turkey is to increase the domestic savings. Due to insufficient savings in our country, investments remain limited and economic growth rates are affected by this situation. In this process, it is necessary either to give up investments or to apply for foreign resources (Yıldız, 2015: 159). Moreover, since to reach high-income level and to increase production level in Turkey would be only possible by qualified labor, it is needed to give more attention to education (Erkoç, 2015: 190). When the development of education system in a qualitative manner in terms of total factor productivity is considered, the education system in Turkey should be restructured in parallel with the developments in the world based on information technology. Thanks to the innovations that will be carried out in a structure that supports science and industry policy implementations, improvement can be provided for the qualified labor-force needed (Alçın and Güner, 2015: 33).

South Korea, which had a similar process to Turkey in terms of development in the 1960s, was able to rise to developed countries level by showing a rapid development in the upcoming years. When South Korea's development process is evaluated, there are some important lessons that can be taken into consideration. In particular, the policies of the country in terms of education and R&D

activities and the role of the state in achieving the target are important in this respect. In order to increase the capacity of technology in the industrial sector in the country, the emphasis was placed on the development of higher education and R&D activities. In this way, the number of higher education graduates in South Korea, which was 5 % in 1960, increased to over 60 % in recent years, while the share of R&D expenditures in the national income was 4 per 1,000 in 1960 and increased up to 3 % in the last period (Alçin, 2010: 134).

In 1960, South Korea included information technology in education during the restructuring process of the education system and in this context also adopted a Law on Science and Support. Following the adoption of the law, practices for the development of the industrial sector were encouraged. In this way, a qualified labor force in the country has started to occur. South Korea is now able to produce demand-oriented technology in the world and even has many global brands that can be described as technology giants. Behind this success, the restructuring of the education system in terms of priority sectors and the support of the state by the laws and incentives introduced has been influential (Uyanık, 2015: 182).

When the overall Turkish economy is considered, the factors that led to the middle-income trap are seen as more likely because of the structural problems. Therefore, the way to get out of the trap is based on structural reforms. In addition to labor and capital, a large number of factors such as education, infrastructure, goods and services markets, the status of the labor market, technological capacity, innovation capacity and the increase in these capacities, affect the economic growth rates of countries and thus the per capita income level. In this respect, middle-income trap countries must realize structural reforms in production and produce high-tech and innovative products. These structural reforms are a necessity especially in this process where globalization increases international competition (Paus, 2014).

5 Conclusion

The middle-income trap, in which a middle-income country cannot pass from this level of income to an upper-income level, is a growth problem. The income level of countries is important for the emergence of middle-income trap. Parallel to the stages of industrialization, the transfer of labor from agriculture to industry and the growth achieved in the front phase have been replaced by a decline in the production as a result of the old technologies. Countries that have reached middle-income level cannot carry themselves to an upper-income level unless they restructure their production with innovative approaches and realize high value-added production with the use of advanced technology. On the one hand, these countries are unable to compete with the low-income countries with the advantage of cheap labor, while on the other hand they are unable to compete with high-income countries.

The importance of knowledge is increasing and it becomes important in terms of innovation and competition. In parallel with the developments in science and technology, knowledge-based production is the determinant of economic growth. In this way, developed countries have managed to produce high value-added products and provided economic growth due to increasing export rates. In the developing countries like Turkey, the reason of unstable growth is the very low share of these high value-added products in exportation.

When the economies such as Japan, Korea and Singapore, which got out from the middle-income trap in a short time and reached the upper-income level, are investigated, it is seen that these nations are countries that can produce high value-added products and compete with developed countries (Yıldız, 2015: 166). In this context, the developing countries that want to get out of the middle-income trap, including Turkey, must carry out structural reforms in many areas to exit from it. First of all, efforts should be made to close the savingsinvestment gap by increasing savings rates in the country, and then to develop human capital for qualified labor by giving priority to R&D and technological investments. Because these countries can only increase productivity by this means and achieve sustainable economic growth and raise national income per capita level by producing innovative products and exporting them to foreign markets.

Turkey was placed in the low middle-income level for nearly 50 years and since 2005 it elevated to high medium-income level. However, it continues to maintain its position in middle-income countries, which have not yet succeeded in reaching an upper-income level. In fact, in 1960, the per capita national income of South Korea was at the same level with Turkey, and today it has nearly doubled the one in our country and has managed to rise to high-income countries category. The impact of South Korea's structural reforms, especially on information and communication technologies, has been influential in this success. It is necessary for Turkey to make the necessary reforms by producing structural solutions as immediate as possible in order to reach high-income countries' levels.

In order to escape from the middle-income trap, Turkey needs a sustainable growth policy. In this context, first of all, high value-added products should be produced and especially foreign-source dependency for intermediate goods should be reduced. Priority sectors should be identified in exports and innovation activities should be encouraged in these sectors. Government incentives for companies investing in technology should be increased. For the development of R&D activities, private sector and public cooperation should be focused on. Education policies in Turkey are also very important to ensure the development. Turkey's education system is in the transition to a knowledge-based society, so it needs to make radical and permanent reforms. The development of advanced and innovative products is only possible with a well-trained and qualified labor force. The lack of qualified labor force trained in this field prevents the establishment of high-tech industries. For this reason, education policies should be re-evaluated in order to get out of the middle-income trap, and a structure based on producing information and technology should be made.

References

Agenor, P. R. & Canuto O. (2012). Middle-Income Growth Traps. World Bank Policy Research Working Paper, 6210, 1–33.

Aiyar, S., Duval, R., Puy, D., Wu, Y., & Zhang, L. (2013). Growth Slowdowns and the Middle Income Trap. IMF Working Paper, No. WP/13/71.

Alçın, S. (2010). *Teknolojik Determinist Kalkınma Aracı Olarak Teknoekonomi Politikaları*. İstanbul: Tarem Yayınları.

Alçın, S. & Güner, B. (2015). Orta Gelir Tuzağı: Türkiye Üzerine Bir Değerlendirme. *Marmara Üniversitesi İ.İ.B. Dergisi*, 37(1), 27–45.

Alkan, H. I. & Ümit, A. Ö. (2018). Orta Gelir Tuzağının Türkiye Açısından İncelenmesi ve Tuzaktan Çıkış Stratejileri. *Manas Sosyal Araştırmalar Dergisi*,7(4), 97–112.

Aoki, M. (2011). The Five-Phases of Economic Development and Institutional Evolution in China and Japan. ADBI Working Paper Series, No: 340.

Çakmak, E. &Gümüş, S. (2005). Türkiye'de Beşeri Sermaye ve Ekonomik Büyüme: Ekonometrik Bir Analiz (19602002). *Ankara Üniversitesi Siyasal Bilgiler Fakültesi Dergisi*, 60(1), 59–72.

Çaşkurlu, E. & Arslan C. B. (2014). Orta Gelir Tuzağından Çıkışa Odaklanma: Ürün Tuzağı (Ürün Boşluğu) ve Demiryolu Taşımacılık Sektörü. *Maliye Dergisi*, 167, 71–92.

Çetinkaya, Ş. (2016). Orta Gelir Tuzağı Tehlikesi ve Türkiye Üzerine Bir Analiz. *Akademik Sosyal Araştırmalar Dergisi*, 4(25), 213–222.

Çobanoğulları, G. & Eroğlu, E. (2017). Orta Gelir Tuzağından Çıkış: Türkiye Örneği. Maliye Araştırmaları Dergisi, 3(3), 257–268.

Dewitte, R. (2014). Middle Income Trap and Export Sophistication: Assessment and Economic Policy Implications. Unpublished Master Thesis. Ghent: Ghent University Faculty of Economic and Business Administration.

Easterly, W. & Levine, R. (2001). It Is Not Factor Accumulation: Stylized Facts and Growth Models. *The World Bank Economic Review*, 15, 177–219.

Egawa, A. (2013). Will Income Inequality Cause a Middle Income Trap in Asia? Bruegel Working Paper, 2013/06. Retrieved from https://bruegel.org/wp-content/uploads/imported/publications/WP_2013_06.pdf (11.08.2019).

Eichengreen, B. Park, D., & Shin, K. (2012). When Fast-Growing Economies Slow Down: International Evidence and Implications for China. *Asian Economic Papers*, 11(1), 42–87.

Erkoç, Ç. (2015). Orta Gelir Tuzağında Türkiye: 2023. *Sosyoekonomi*, 23(26), 187–194.

Felipe, J., Abdon, A., & Kumar, U. (2012). Tracking the Middle Income Trap: What Is It, Who Is in It, and Why? *Levy Economics Institute of Bard College*. Working Paper, No: 715, 1–59.

Gill, I. & Kharas, H., (2007). An East Asian Renaissance: Ideas for Economic Growth. World Bank Publications. Washington, DC: The World Bank.

Gürsel, S. & Soybilgen, B. (2013). Türkiye Orta Gelir Tuzağının Eşiğinde. *Betam Araştırma Notu*, 13 (154), 1–7.

Kaya, Z., Tokucu, E., Aykırı, M. & Durmuş, C. (2015). Türkiye Ekonomisinde Orta Gelir Tuzağı ve Ödemeler Bilançosu Kısıtı. *International Conference on Eurasian Economies*, 830–840. Retrieved from https://www.avekon.org/papers/1251.pdf (11.08.2019)

Nişancı, M., Gerni, M., Türkmen, A. & Emsen, Ö. S. (2015). Türkiye Ekonomisinin Orta Gelir Tuzağına Düşüp Düşmediğine Dair Tartışma: Kur Değerlenmesi Çerçevesinde Bir Bakış. *International Conference on Eurasian Economies*, 232–242. Retrieved from http://www.avekon.org/papers/1260.pdf (11.08.2019)

Ohno, K. (2009). Avoiding the Middle-Income Trap: Renovating Industrial Policy Formulation in Vietnam. *ASEAN Economic Bulletin*, 26(1), 25–43.

Özsan, A. G., Erdem, B. P. & Ata, S. (2017). *Türkiye'de Yurt İçi Tasarrufların Ve Tüketimin Gelişimi*. Ankara: Kalkınma Bakanlığı.

Paus, E. (2014). Latin America and Middle Income Trap. *Financing For Development Series*, 250, 1–57.

Şahin, İ., Başer, K. & Karanfil, M. (2015). Orta Gelir Tuzağı Üzerine Ampirik Bir Çalışma: Türkiye Örneği (19802013). *Uluslararası Alanya İşletme Fakültesi Dergisi*, 7(2), 225–235.

Taşçı, K. & Özsan, M. E. (2014). Türkiye'de Bölgeler itibariyle Orta Gelir Tuzağı. *İşveren*, 52(2), 1–11.

Tho, T. V. (2013). The Middle-Income Trap: Issues for Members of the Association of Southeast Asian Nations. *VNU Journal of Economics and Business*, 29(2), 107–128.

Uyanık, C. C. (2015). Orta Gelir Tuzağı ve Türkiye'nin Konumu Açısından Bir Değerlendirme. *Sosyoekonomi*, 23(26), 175–186.

Ünlü, F. & Yıldız, R. (2017). Orta Gelir Tuzağını Açıklayan Teorik ve Ampirik Yaklaşımlar, Erciyes Üniversitesi İktisadi ve İdari Bilimler Dergisi, 49, 87–115.

Ünlü, F. & Yıldız, R. (2018a). Orta Gelir Tuzağının Belirlenmesi: Ekonometrik Analiz. *Uluslararası Yönetim İktisat ve İşletme Dergisi*, Cilt 14(1), 1–20.

Ünlü, F. & Yıldız, R. (2018b). Orta Gelir Tuzağının Belirleyicileri: Diskriminant Analizi. *Atatürk Üniversitesi İktisadi ve İdari Bilimler Dergisi*, 32(1), 45–64.

Vergil, H. & Abasız, T. (2008). Toplam Faktör Verimliliği, Hesaplanması ve Büyüme İlişkisi: Collins Bosworth Varyans Ayrıştırması. *Kocaeli Üniversitesi Sosyal Bilimler Enstitüsü Dergisi*, 16(2), 160–188.

Woo, W. T. (2012). China Meets the Middle-Income Trap: The Large Potholes in the Road to Catching-up. *Journal of Chinese Economic and Business Studies*, 10(4), 313–336.

World Bank. Retrieved from https://blogs.worldbank.org/opendata/new-country-classifications-income-level-2018-2019, (01.03.2019).

Yaşar, E. & Gezer, M. A. (2014). Türkiye'nin Orta Gelir Tuzağına Yakalanma Riski ve Bu Riskten Kurtulma Önerileri. *Maliye Dergisi*, 167, 126–148.

Yavuz, E. (2017). Maliye Politikası Bağlamında Türkiye'de Orta Gelir Tuzağı Sorunsalının Analizi. *Social Sciences (NWSASOS)*, 12(2), 78–101.

Yeldan, E., Taşçı, K., Voyvoda, E., & Özsan, M. E. (2012). *Orta Gelir Tuzağından Çıkış: Hangi Türkiye?* İstanbul: Türkonfed.

Yıldız, A. (2015). Orta Gelir Tuzağı ve Orta Gelir Tuzağından Çıkış Stratejileri. *Fırat Üniversitesi Sosyal Bilimler Dergisi*, 25(2), 155–170.

Zhuang, J., Vandenberg, P., & Huang, Y. (2012). Growing Beyond the Low-Cost Advantage: How the People's Republic of China Can Avoid the Middle Income Trap. Philippines: Asia Development Bank.

Sertaç Hopoğlu

4. Are Youth Labor, Trade Openness and Foreign Direct Investment Effective in Economic Growth of CIVETS Countries? A Panel Causality Analysis

Abstract The group of emerging economies consisting of Colombia, Indonesia, Vietnam, Egypt, Turkey and South Africa is known by the acronym CIVETS. Fast growth, high-youth employment and open economies are some of the characteristics used to define CIVETS. This chapter aims to investigate the causality relationships between youth employment, trade openness and foreign direct investment, and economic growth in the panel of CIVETS countries. Emirmahmutoglu-Kose (2011) panel causality test was used for this purpose. Findings indicate bidirectional causality between youth employment, trade openness and foreign direct investment, and growth. The analyses of individual relations produced mixed results. While there is unidirectional causality from youth employment to growth in Egypt and Turkey, there is unidirectional causality from growth to youth employment in South Africa. There are no causal relationships from foreign direct investment to growth in any CIVETS country. However, there is unidirectional causality from growth to foreign direct investment in Indonesia, South Africa and Turkey.

Keywords: CIVETS, Economic growth, Trade openness, Youth employment, Foreign direct investment

1 Introduction

Since the time Brazil, Russia, India and China have been defined as "BRIC" by the economists of Goldman-Sachs to guide investors at the beginning of the 2000s (Anonymous, 2018a), other international financial institutions that want to direct their investors have started to group and define the rising economies with various names and abbreviations such as N11, MINT, MIST, VISTA, Fragile Five and so on. Using fashionable abbreviations to define emerging economies has been criticized in terms of their misleading of investors by increasing the expectation of guaranteed profits without reflecting the domestic dynamics of the grouped countries (Cohn, 2014) and of that the other groups do not carry the potential of the diplomatically and economically blocked BRICS countries (Khanna, 2013).

The Economist Intelligence Unit (The EIU) of the Economist magazine in 2008 defined the emerging economies Colombia, Indonesia, Vietnam, Egypt, Turkey and South Africa as CIVETS, by referring to the initials of their names in English and the term was made popular by HSBC economists at the beginning of 2010. Since the word CIVETS in English is also the name of a species of a bearcat, the abbreviation also implies that these countries in the future have the potential to become the new tigers of the world's economy (Anonymous, 2018b). In addition to the high growth rates, relatively free markets, openness of markets, export-led policies, rapid industrialization and financial fragility that can be seen in other emerging economies, CIVETS countries have been defined by diversified economic bases, relatively high capital accumulation, high young populations, improving political stability compared to the previous periods, and the importance they give to higher education. However, despite the export-led policies, the dynamism of CIVETS economies is not dependent on foreign demand and exports of goods, as in many other emerging economies, and public and private sector debts are relatively low (Anonymous, 2018c).

In the literature, the number of studies on CIVETS is limited. While a significant portion of these studies focus on financial markets in CIVETS countries (e.g. Saleem, Al-Hares and Ahmed, 2016), there are also studies on exchange rates (Almudhaf, 2014), innovation performance (Yi, Qi and Wu, 2013), foreign economic policy (Guerra-Baron and Mendez, 2015) and the relationship between investments and poverty (Sökmen, Açcı and Taşar, 2017).

In the definition of CIVETS, the young population and not being overly dependent on foreign demand are shown as common characteristics. On the other hand, it is thought that the popularization of these countries with fashionable abbreviations as well as the advantages they provide increases the foreign direct investments to these countries. Therefore, the aim of this chapter is to contribute to the literature by investigating the causality relations between youth employment, openness to foreign trade and foreign direct investment (FDI) with economic growth in CIVETS countries.

In the next section of the chapter, a review of selected studies from the literature on emerging economies and country groups is given. In the third section, the data is first defined and then the methodology is specified, and then the econometric analysis and findings are explained. The analysis of the results is given in the fourth and last section of the chapter.

2 Literature Review

There is an increasing number of studies in recent years, on various economic dynamics of emerging economies grouped by different names and abbreviations. Particularly, a considerable literature has been forming on FDI in BRICS countries and emerging economies. Some studies selected from the literature are summarized in Tab. 4.1.

Tab. 4.1: Literature Summary. Source: Author's own construction.

Authors	Countries	Period	Subject	Method	Result
Jadhav (2012)	BRICS	2000–2009	Determinants of FDI in BRICS countries	Multiple Regression	Real GDP and trade openness have a positive impact on FDI flows.
Zeren and Ari (2013)	G7	1970–2011	The relationship between trade openness and growth in G7 countries	Dumitrescu and Hurlin (DH-2012) Panel Casuality Test	There is a bidirectional relationship between trade openness and growth.
Basu, Barik and Arokiasamy (2013)	BRICS	1991–2009	Demographic determinants of economic growth	Fixed Effects Model	The ratio of working-age population to population is effective in growth.
Agarwal (2015)	BRICS	1989–2012	Relationship between FDI and Growth	Granger Causality Test	There is unidirectional causality from FDI to GDP in the long-term.
Topallı (2016)	BRICS-T	1982–2013	The relationship among FDI, trade openness and growth	Emirmahmutoglu and Kose (EK-2011) Causality Test	There is unidirectional causality from growth to FDI, and bidirectional causality between growth and trade openness and between FDI and trade openness.
Dritsakis and Stamatiou (2016)	13 new member countries that accessed the EU after 2004	1995–2013	Trade openness and growth relationship	Panel Granger Causality Test	There is unidirectional causality from trade openness to growth in both short and long term.
Apergis and Cooray (2016)	22 rising economy	1960–2013	Convergence of openness and FDI in rising economies	Phillips and Sul (2007) Convergence Test	Emerging economies do not constitute a homogeneous group in terms of openness and FDI and different groups may be mentioned according to different levels of development.

(continued on next page)

Tab. 4.1: (continued)

Authors	Countries	Period	Subject	Method	Result
Doğanay and Değer (2017)	21 Rising Economy	1996–2014	Relationship between FDI and Export	Westerlund (2007) Panel Cointegration Test	When cross-sectional dependence is considered, there is only a cointegration relationship between FDI and manufacturing industry exports.
Şaşmaz ve Yayla (2018)	34 OECD Countries	2000–2016	The relationship between FDI and development	DH (2012) Panel Causality Test	There is unidirectional causality from FDI to per capita income.
Öncü and Çelik (2018)	BRIC-T Countries	1998–2016	The relationship between FDI and growth	DH (2012) Panel Causality Test	There is a one-way causality from economic growth to FDI.
Özcan, Özmen and Özcan (2018)	18 Rising Economics	1992–2015	The relationship between openness and growth	Konya (2006), EK (2011), and DH (2012) Panel Causality Tests	The DH (2012) test points to causality from growth to openness across the panel. EK (2011) and Konya (2006) tests indicate different causality relations for the countries studied.
Sahin (2018)	BRICS-T	1993–2013	FDI-international trade-financial development relationships	Konya (2006) Panel Causality Test	Except China, there is bidirectional relationship between openness and financial development in the countries in the panel.
Altıner, Bozkurt and Toktaş (2018)	10 Rising economies	1990–2015	The relationship between globalization and economic growth	EK (2011) Causality Test	While there is no causality from economic globalization to economic growth, there is unidirectional causality from political globalization to economic growth and from economic growth to social globalization.

Tab. 4.1: (continued)

Authors	Countries	Period	Subject	Method	Result
Sezer (2018)	BRICS-T	1992–2017	Foreign Trade-FDI relationship	DH (2012) Panel Causality Test	While there is bidirectional causality between FDI and export, there is unidirectional causality from imports to FDI.
Swamy and Narayamuthy (2018)	BRICS	2001–2015	Factors affecting the FDI flow in BRICS countries.	Panel Granger Causality Test	There is bidirectional causality between FDI and Gross Domestic Product (GDP) growth rates, however there is no relationship between openness and FDI.

3 Data and Methodology

3.1 Data and methodology

The ratio of working population aged 1524 to the population (LBF) calculated by the International Labor Organization (ILO) (as a proxy of the youth employment), the ratio of net FDI inflow to GDP (as a proxy of the foreign direct investments (NETFDI)), the share of trade in GDP (as a proxy of the trade openness (TRADE)) and GDP growth rate (GDP) are included as variables in the study.

All data are gathered from the World Bank's "World Development Indicators—WDI" database and are the annual data for the period 2000–2017. Since all data were expressed in ratios, no modifications on data were made and the data were used in their original form in the analysis. The causality relationships shown in Tab. 4.2 are tested in the analysis:

Whether the variables have cross-sectional dependency was tested in the first step of the analysis. Then, the homogeneity of the panel is then tested by the procedure developed by Pesaran and Yamagata (PY-2008 hereafter). The stationarity of the variables is analyzed by Pesaran (2007) CIPS (cross-sectionally augmented IPS) test. For causality analysis, the panel causality test developed by Emirmahmutoglu and Kose (EK-2011 hereafter) is used.

Tab. 4.2: Causality Relationships Tested in the Analysis.
Source: Author's own construction.

LBF → GDP
NETFDI → GDP
TRADE → GDP
GDP → LBF
GDP → NETFDI
GDP → TRADE

3.2 Test of cross-sectional dependence

In econometric analysis, whether there is cross-sectional dependence in the panel is tested first, followed by a test of whether the slope of the variables in the panel is homogeneous. These two steps are important because they affect the choice of methods to be followed in the next steps of analysis and the interpretation of the findings from these methods.

In the first step of econometric analysis whether there is cross-sectional dependence in the series forming the panel is investigated. The analysis continues with the first-generation tests if there is no cross-sectional dependence in the series, and with the second-generation tests if there is cross-sectional dependence in the series. Breusch-Pagan (BP-1980 hereafter) CD_{LM1} LM test is based on the correlation of the sum of residual squares of cross-sectional residues and can be calculated as follows:

$$CD_{LM1} = T \sum_{i=1}^{N-1} \sum_{j=i+1}^{N} \hat{\rho}_{ij}^2. \tag{1}$$

$\hat{\rho}_{ij}$ is the ordinary correlation coefficient between the residues obtained from the Ordinary Least Squares (OLS) method of each equation. $\hat{\rho}_{ij}$ can be calculated as follows:

$$\hat{\rho}_{ij} = \frac{\left(\sum_{t=1}^{T} \hat{v}_{it} \hat{v}_{jt}\right)}{\sqrt{\left(\sum_{t=1}^{T} \hat{v}_{it}^2\right)} \sqrt{\left(\sum_{t=1}^{T} \hat{v}_{jt}^2\right)}} \tag{2}$$

In this equation, u_{it} indicates the residuals of OLS. This test, which is proposed by BP (1980) can be used in cases where N is small and T is sufficiently large (T > N), under the basic hypothesis that there is no cross-sectional dependence (H_0: $\rho_{ij} = \rho$

$_{ji}$=correlation(u_{it}, u_{jt}=0)), it corresponds to χ^2 distribution with N(N-1)/2 degrees of freedom.

In cases where N and T are large, the following version of this test can be used:

$$CD_{LM2} = \sqrt{\frac{1}{N(N-1)} \sum_{i=1}^{N-1} \sum_{j=i+1}^{N} \left(T\hat{\rho}_{ij}^2 - 1\right)} \qquad (3)$$

Under the basic hypothesis that there is no cross-sectional correlation, $CDLM_2$ is distributed normally asymptotically.

It is not possible to use these statistics when N is quite large compared to T. In this case, the cross-sectional dependence test proposed by Pesaran (2004) can be used. This test is based on the sum of the correlation coefficients between the cross-sectional residuals and is calculated as follows:

$$CD_{LM2} = \sqrt{\frac{2T}{N(N-1)} \sum_{i=1}^{N-1} \sum_{j=i+1}^{N} \hat{\rho}_{ij}} \qquad (4)$$

This test can be applied to both stationary and non-stationary dynamic and heterogeneous panels. It is strong against structural breaks and gives good results in small samples. In addition to the balanced panels, it is also possible to use this test in unbalanced panels (the case where sizes of samples of the cross sections are different). Modifications can be made to unbalanced panels in this test statistics.

The hypotheses for both BP (1980) and Pesaran (2004) tests are as follows:

H_0: There is no cross-sectional dependence.
H_1: There is cross-sectional dependence.

Since the time dimension in the panel (T = 28 years) is larger than the horizontal cross-sectional dimension (N = 6), BP (1980) CD_{LM1} test result is considered (Tab. 4.3). The null hypothesis cannot be rejected as the BP (1980) CD_{LM1} probability values for all variables are smaller than $\alpha = 0.05$, and there is cross-sectional dependence in the panel.

3.3 Testing homogeneity: PY (2008) delta test

In the next step of the analysis, PY (2008) test is used to investigate whether the slope coefficients are homogeneous in the cointegration equation. This is a more advanced version of the homogeneity test developed by Swamy (1970) and calculates two separate test statistics for large and small samples ($\tilde{\Delta}$ and $\tilde{\Delta}_{adj}$, respectively):

Tab. 4.3: Results of Cross-Sectional Dependence Test. Source: Author's own calculations.

	GDP		LBF		NETFDI		TRADE	
	Test Statistics	Probability Value	Test Statistics	Probability Value	Test Statistics	Probability Value	Test Statistics	Probability Value
Breusch-Pagan LM	34.046	0.0034	146.65	0.00	55.49	0.00	82.62	0.00
Pesaran scaled LM	2.381	0.0172	22.94	0.00	6.29	0.00	11.25	0.00
Bias-corrected scaled LM	2.270	0.0232	22.83	0.00	6.18	0.00	11.14	0.00
Pesaran CD	4.839	0.00	−0.35	0.72	5.70	0.00	2.49	0.0127

Tab. 4.4: Results of PY (2008) Homogeneity Test. Source: Author's own calculations.

	$\tilde{\Delta}_{adj}$	Probability Value
LBF → GDP	1.466	0.071
NFDI → GDP	−0.704	0.759
TRADE → GDP	5.573	0.000
GDP → LBF	1.868	0.031
GDP → NFDI	4.057	0.000
GDP → TRADE	7.563	0.000
PANEL	1.562	0.059

$$\tilde{\Delta} = \sqrt{N}\left(\frac{N^{-1}\tilde{S} - k}{2k}\right) \quad (5)$$

$$\tilde{\Delta}_{adj} = \sqrt{N}\left(\frac{N^{-1}\tilde{S} - k}{v(T,k)}\right) \quad (6)$$

Here, $\tilde{S}$ represents the Swamy test statistic, N is the number of horizontal cross-sectional observations, k is the number of independent variables and $v(T, k)$ is standard error. The hypotheses for this test are:

H_0: $\beta_i = \beta$ The slope coefficients are homogeneous, and
H_1: $\beta_i \neq \beta$ The slope coefficients are not homogeneous.

If the probability values of the test statistics are less than the selected significance level, the null hypothesis is rejected. PY (2008) homogeneity test results are given in Tab. 4.4. Only $\tilde{\Delta}_{adj}$ statistics are shown considering the sample size.

When the relationships between the variables are considered, it is seen that long-term cointegration coefficients may be homogeneous only for NFDI series, and the null hypothesis is rejected at 10 % significance level for all other relations. When all the variables in the panel are taken into consideration, it is seen that the panel shows heterogeneity at 10 % significance level. Therefore, it is meaningful to investigate individual causality relations for all countries.

3.4 Test of stationarity: Pesaran (2007) CIPS test

In the third step of econometric analysis, the CIPS test developed by Pesaran (2007) was used to test the stationarity of the series since there is horizontal cross-sectional dependence in the panel. This test is the improved form of the test developed by Im, Pesaran and Shin (IPS-2003) by taking into account the cross-sectional dependence of the series. The CIPS test extends the standard ADF regression, taking the first differences of individual series and the cross-sectional means of lagged values:

$$\Delta y_{it} = a_i + \rho_i y_{i,t-1} + b_i \bar{y}_{t-1} + \sum_{j=0}^{p} c_{ij} \Delta \bar{y}_{t-j} + \sum_{j=0}^{p} d_{ij} \Delta \bar{y}_{t-j} + e_{i,t} \tag{7}$$

This regression is called CADF. Here, y_t and Δy_t are expressed as follows:

$$\bar{y}_t = \frac{1}{N} \sum_{i=1}^{N} y_{it}, \, ve \Delta \bar{y}_t = \frac{1}{N} \sum_{i=1}^{N} \Delta y_{it} \tag{8}$$

According to Pesaran (2007), cross-sectional means of these two values are included in the model as a proxy of the unobservable common factor. The hypotheses of the test are as follows:

H_0: Each serial in the panel has a unit root.
H_1: At least one of the series in the panel is stationary.

The test statistic used to examine these hypotheses is the average of individual CADF tests. In addition, a modified (truncated) state of the CADF test is used to avoid the effects of contradictory results that may occur in small samples. Pesaran's (2007) CIPS test statistics, which is the extended version of IPS (2003) test statistics, can be calculated as follows:

$$\text{CIPS} = \frac{1}{N} \sum_{i=1}^{N} \text{CADF}_i \tag{9}$$

Even for large N, the distribution of CIPS statistics is not standard. Appropriate critical values can be obtained from Table II in Pesaran (2007). The value of the lag length that gives the smallest mean information criterion (IC) statistics is taken as CIPS test statistic. This statistic is compared with the critical values of Pesaran (2007) Table II (c). If the test statistic is less than the table value (or the absolute value is greater than the absolute value of the table statistics), the alternative hypothesis cannot be rejected. In order to avoid further data loss in the small sample, the maximum number of lags is taken as 4 and the CIPS test is applied to the series at level primarily. While the GDP, NFDI and TRADE series are determined as stationary in "constant and trend" model at level, the LBF series are found to contain unit roots. In the "constant only" model, GDP and NFDI series are stationary at level. After taking the first differences of the

Tab. 4.5: Pesaran (2007) CIPS Test Results. Source: Author's own calculations.

		CONSTANT		CONSTANT and TREND	
		Level	First Difference	Level	First Difference
GDP		−3.383*	−5.690*	−3.491*	−5.886*
LBF		−1.215	−3.371*	−1.816	−3.886*
NFDI		−2.741*	−4.524*	−2.784***	−4.448*
TRADE		−2.019	−4.819*	−3.175*	−4.698*
Critical Values	%1	−2.57		−3.10	
	%5	−2.33		−2.86	
	%10	−2.21		−2.73	

Note: Test statistics are truncated CIPS values. * and *** indicate 1 % and 10 % significance levels, respectively.

series and performing the CIPS test again, all series became stationary at the first differences at 1 % significance level in both models (Tab. 4.5).

3.3 Causality analysis: EK (2011) panel causality test

In order to analyze the causality relationship, panel causality test developed by EK (2011) is performed. This test is a simple causality test developed for heterogeneous mixed panels. The most advantageous property of this test is that it can be used in the presence of cross-sectional dependence and in the absence of a cointegration relationship between variables in heterogeneous panels (Altıner et al., 2018).

EK (2011) panel causality test is an extended version of the Toda-Yamamoto (1995) test, which is modified for panels. While modifications on the probability values of the ADF unit root tests are carried out in the Maddala-Wu (1999) test, a similar arrangement is carried out on the Toda-Yamamoto (1995) probability values in the EK panel causality test. In the first stage of the test, the following model is estimated:

$$y_{i,t} = \mu_i + A_{i1}y_{i,t-1} + \ldots + A_{ik}y_{i,t-k_i} + \sum_{j-k_i+1}^{k_i+dmax_i} A_{ij}y_{i,t-j} + e_{i,t} \quad (10)$$

Here, the appropriate lag length can be obtained by utilizing information criteria. While the null hypothesis states that there is no causality relationship in the panel, the alternative hypothesis is that there is Granger causality relationship between the variables of at least one series. The test statistics used are calculated as follows:

$$\lambda = -2\sum_{i=1}^{N} \ln(p_i) \qquad (11)$$

This test statistic is a Fisher test statistic and it is compatible with the χ^2 distribution with 2N degrees of freedom.

The results of EK (2011) panel causality test are shown in Tab. 4.6. According to the results, the null hypothesis that "there is no causality from LBF to GDP in any country in the whole panel" cannot be rejected. Similarly, the null hypothesis of "there is no Granger causality from GDP to LBF" cannot be rejected as well. While the null hypothesis cannot be accepted at 5 % significance level for the causality from GDP to NFDI, it cannot be rejected for the Granger causality from NFDI to GDP. On the other hand, the null hypothesis cannot be accepted for the Granger causality from TRADE to GDP at 5 % significance level. However, the null hypothesis that there is no causality relationship between GDP and TRADE cannot be rejected for the whole panel.

The individual results of the countries are in parallel with the homogeneity test. Causality from the ratio of young labor force participation (youth employment) to GDP growth rate cannot be rejected at the 5 % significance level for Egypt and at the 10 % significance level for Turkey. For all countries, there is no causality from the net FDI flow to the GDP growth rate. The causality from trade openness to GDP growth rate cannot be rejected at 5 % significance level for Egypt and at 1 % significance level for South Africa. On the other hand, the causality from the growth rate in South Africa to the ratio of young labor force participation cannot be rejected at 10 % significance level and the causality from the growth rate to the trade openness at 1 % significance. The causality from growth rate to net FDI cannot be rejected at 1 % significance level for Indonesia, at 5 % significance level for Turkey and at 10 % significance level for South Africa.

4 Conclusion

In this chapter, whether some characteristics the CIVETS defined on have created similar effects on the growth rates of their GDP in the long run is analyzed. According to PY (2008) homogeneity test results, there are differences between youth employment and GDP growth rates in the CIVETS countries. Here, it is expected that a young workforce will accelerate supply-push growth by being more productive and innovative, and demand-pull growth by increasing

Tab. 4.6: Results of EK (2011) Test. Source: Author's own calculations.

		LBF→GDP	NFDI→GDP	TRADE→GDP
Colombia		0.131 (1)	0.015 (1)	0.006 (1)
		[0.717]	[0.902]	[0.937]
Egypt		**9.773 (3)****	5.366 (3)	**9.644 (3)****
		[0.021]	[0.147]	**[0.022]**
Indonesia		0.081 (1)	0.193 (2)	0.003 (2)
		[0.776]	[0.908]	[0.999]
South Africa		4.435 (3)	0.124 (1)	**30.675 (3)***
		[0.218]	[0.724]	**[0.000]**
Turkey		**3.163 (1)*****	0.052 (1)	0.652 (1)
		[0.075]	[0.820]	[0.419]
Vietnam		0.012 (1)	0.046 (1)	4.429 (2)
		[0.911]	[0.830]	[0.109]
PANEL		17.339	5.648	**41.588****
CRITICAL	%1	35.556	34.619	63.205
VALUES	%5	26.857	26.254	30.169
	%10	22.625	22.613	23.251
		GDP →LBF	GDP→NFDI	GDP→TRADE
Colombia		0.351 (1)	0.944 (1)	**9.350 (1)***
		[0.554]	[0.331]	**[0.002]**
Egypt		3.529 (3)	5.172 (3)	0.240 (3)
		[0.317]	[0.160]	[0.971]
Indonesia		1.545 (1)	**11.584 (2)***	1.144 (2)
		[0.214]	**[0.003]**	[0.564]
South Africa		**7.494 (3)*****	**2.839 (1)*****	**7.478 (3)***
		[0.058]	**[0.092]**	**[0.058]**
Turkey		0.461 (1)	**5.741 (1)****	0.576 (1)
		[0.497]	**[0.017]**	[0.448]
Vietnam		0.707 (1)	0.758 (1)	0.455 (2)
		[0.401]	[0.384]	[0.796]
PANEL		15.497	**32.349****	21.166
CRITICAL	%1	38.409	33.397	86.500
VALUES	%5	28.145	25.363	32.400
	%10	23.279	21.595	24.524

Note: Statistics were obtained by 10000 iterations. The maximum lag length is taken as 3 and the Akaike Information Criteria is used to determine the appropriate lag length. Statistics are Modified Wald (mWald) statistics. Numbers in parentheses are the number of lags. Numbers in brackets are probability values. *, ** and *** indicate significance level at 1 %, 5 % and 10 %, respectively.

the aggregate demand. However, this relationship does not appear in a similar (homogenous) way in the countries investigated in this chapter. It is thought that this result might be due to differences in labor quality (human capital) education and experience accumulation (learning by doing). Indeed, the Granger causality from youth employment to GDP growth rate is found significant for only Egypt and Turkey. The causality from GDP growth rate to youth employment is only seen in South Africa. Thus, in South Africa, the youth employment increases as the economy grows, while growth accelerates as the youth employment increases in Egypt and Turkey.

Similarly, countries differ as to the long-term relationship between their shares of trade in GDP, which represent trade openness, and GDP growth rates. Countries with a diversified economic base should at least have specialized in producing some goods and have a comparative advantage in the trade of these goods. On the other hand, foreign trade is an important development tool in a diversified economy: While the production and exports of various goods and services increase the welfare of producers, imports of goods and services in order to meet the demands of both a growing young population and a diversified production economy would also have welfare-improving effects. Therefore, in a diversified economy, the trade openness and its effect on growth rates are expected to be significant. In this context, and according to the results of this study, it can be said that individual CIVETS countries are not diversified as to the expected level that their economies are defined on and foreign trade has no significant impact on the long-term growth of these countries. On the other hand, it is noteworthy that the causality from trade openness to growth rate and vice versa is significant for the primary goods exporters in the panel. The causality from trade openness to growth rate is significant only for Egypt. The causality from GDP growth rate to trade openness is found to be significant only for Colombia, while in South Africa there is bidirectional causality between these two variables. According to these results, it can be said that growth increases as trade deficit increases in Egypt, while growth induces more trade openness in Colombia, and growth and trade openness affect each other in South Africa.

Countries are similar to each other in terms of the relationship between the ratio of net foreign direct investment rates to GDP and GDP growth rates. The reason for this similarity is the lack of a significant relationship between the ratio of net FDI to GDP and the GDP growth rate in these six countries. Nevertheless, the finding of a significant causality from GDP growth rate to the ratio of net FDI to GDP in Indonesia, South Africa and Turkey indicates that flows of foreign direct investment accelerate in these countries when GDP growth rate goes up.

Looking at the results for the whole panel, unidirectional causality is found only from trade openness to GDP and from GDP to net FDI. The fact that there is no causality relationship between the ratio of young labor force to employment and GDP growth rates in the whole panel indicates that there is no long-term causality between young labor force and growth for the overall panel. Bidirectional causality is not found between the share of net FDI flows in GDP and GDP growth rate throughout the panel. Causality from GDP growth rate to net FDI flow is significant at 5 % significance level. A significant unidirectional causality relationship is also found at the 5 % significance level from trade openness and GDP. On the other hand, there was no Granger causality relationship from growth rate to trade openness.

Yet, economic growth and development in CIVETS countries are likely to be better understood with a more detailed set of data and different methods. An analysis by taking into account the composition of imports and exports, for example, using the export and import values of capital goods, intermediate goods and consumer goods, will be more accurate. Similarly, the different effects of educated and uneducated youth employment on growth should be considered for a better understanding of the relationship between youth employment and growth in the studied countries. On the other hand, since most of the investments in emerging economies are invested in the financial sector and national stock exchanges, investigating the impact of such monetary flows may also help to better understand the growth dynamics in these countries.

References

Agarwal, G. (2015). Foreign Direct Investment and Economic Growth in BRICS Economies: A Panel Data Analysis. *Journal of Economics, Business and Management*, 3(4), 421–424.

Almudhaf, F. (2014). Testing for Random Walk Behaviour in CIVETS Exchange Rates. *Applied Economics Letters*, 21(1), 60–63.

Altıner, A., Bozkurt, E., & Toktaş, Y. (2018). Küreselleşme ve Ekonomik Büyüme: Yükselen Piyasa Ekonomileri İçin Bir Uygulama. *Finans Politik & Ekonomik Yorumlar*, 639: 117–162.

Anonymous (2018a). Retrieved from https://www.investopedia.com/terms/b/brics.asp(06.12.2018).

Anonymous (2018b). Retrieved from https://pitt.libguides.com/emergingmarkets (10.10.2018).

Anonymous (2018c). Retrieved from https://www.investopedia.com/terms/c/civets.asp(14.10.2018).

Apergis, N. & Cooray, A. (2016). Old Wine in a New Bottle: Trade Openness and FDI Flows—Are the Emerging Economies Converging? *Contemporary Economic Policy*, 34(2), 336–351.

Basu, T.,Barik, D. & Arokiasamy, P. (2013). Demographic Determinants of Economic Growth in BRICS and selected Developed Countries, *XXVII IUSSP International Population Conference (IUSSP 2013)*, 2631 August 2013 at BEXCO, Busan, Republic of Korea.

Breusch, T. S. & Pagan, A. R. (1980). The Lagrange Multiplier Test and Its Applications to Model Specification in Econometrics. *The Review of Economic Studies*, 47(1),239-253.

Cohn, C. (2014). BRIC, MINT, CIVETS: Money Managers Are So Over Investing in Catchy Acronyms, Retrieved from https://www.businessinsider.com/bric-mint-civets-acronym-investing-2014-1 (06.12.2018).

Doğanay, M. A. & Değer, M. K. (2017). Yükselen Piyasa Ekonomilerinde Doğrudan Yabancı Yatırımlar ve İhracat İlişkisi: Panel Veri Eşbütünleşme Analizleri (19962014). *Çankırı Karatekin Üniversitesi İİBF Dergisi*, 7(2), 127–145. DOI: 10.18074/ckuiibfd.345332.

Dritsakis, N. & Stamatiou, P. (2016). Trade Openness and Economic Growth: A Panel Cointegration and Causality Analysis for the Newest EU Countries. *The Romanian Economic Journal*, XVIII(59), 45–60.

Dumitrescu, E. I., & Hurlin, C. (2012). Testing for Granger Non-Causality in Heterogeneous Panels. *Economic Modelling*, 29(4), 1450–1460.

Emirmahmutoglu, F. & Kose, N. (2011). Testing for Granger Causality in Heterogeneous Mixed Panels. *Economic Modelling*, 28(3), 870–876.

Guerra-Baron, A. & Mendez, A. (2015). A Comparative Study of Foreign Economic Policies: The CIVETS Countries, *LSE Global South Unit*. Working Paper No. 3/2015.

Im, K. S., Pesaran, M. H., & Shin, Y. (2003). Testing for Unit Roots in Heterogeneous Panels. *Journal of Econometrics*, 115: 53–74.

Jadhav, P. (2012). Determinants of Foreign Direct Investment in BRICS Economies: Analysis of Economic, Institutional and Political Factors. International Conference on Emerging Economies—Prospects and Challenges (ICEE- 2012). *Procedia—Social and Behavioral Sciences*, 37: 5–14.

Khanna, P. (2013). BRICS, VISTA, BROOMS—Just because it's an acronym doesn't mean it's an investment opportunity, Retrieved from https://qz.com/97085/the-brics-were-coined-using-superficial-indicators-these-are-the-real-frontier-markets/ (06.12.2018).

Konya, L. (2006). Exports and Growth: Granger Causality Analysis on OECD Countries with a Panel Data Approach. *Economic Modeling*, 23: 978–992.

Maddala, G. S. & Wu, S. (1999). A Comparative Study of Unit Root Tests with Panel Data and a New Simple Test. *Oxford Bulletin of Economics and Statistics*, 61(S1), 631–652.

Öncü, E. & Çelik, Ş. (2018). Doğrudan Yabancı Yatırımlar ve Ekonomik Büyüme İlişkisi: BRICT Ülkeleri Panel Nedensellik Analizi. *Uluslararası İktisadi ve İdari İncelemeler Dergisi*, (17. UİK Özel Sayısı), 403–414.

Özcan, C. C., Özmen, İ. & Özcan, G. (2018). Ticari Dışa Açıklık ve Ekonomik Büyüme Arasındaki Nedensellik İlişkisi: Yükselen Piyasa Ekonomileri. *Selçuk Üniversitesi Sosyal Bilimler Enstitüsü Dergisi*, 40: 60–73.

Peseran, M. H. (2004). General Diagnostic Tests for Cross Section Dependence in Panels. *IZA Discussion Paper*, 1240, August 2004.

Pesaran, M. H. (2007). A Simple Panel Unit Root Test in the Presence of Cross-Section Dependence. *Journal of Applied Econometrics*, 22: 265–312.

Pesaran, M. H. & Yamagata, T. (2008). Testing Slope Homogeneity in Large Panels. *Journal of Econometrics*, 142: 50–93.

Phillips, P. C. B. & Sul, D. (2007). Transition Modeling and Econometric Convergence Tests. *Econometrica*, 75: 1771–855.

Sahin, S. (2018). Foreign Direct Investment, International Trade and Financial Development in BRICS-T Countries: A Bootstrap Panel Causality Analysis. *Business and Economics Research Journal*, 9(2), 301–316.

Saleem, K., Al-Hares, O. & Ahmed, S. (2016). Financial Integration and Portfolio Diversification: Evidence from CIVETS Stock Markets. *Theoretical Economics Letters*, 6: 1304–1314.

Sezer, S. (2018). Dış Ticaret ve Doğrudan Yabancı Yatırımlar Arası İlişki: BRICS Ülkeleri ve Türkiye Üzerine Bir Analiz. *Turkish Studies*, 13(18), 1171–1189.

Sökmen, F. Ş., Accı, Y. & Taşar, İ. (2017). The Lack of Savings Might Be a Reason for Poverty? Panel Data Analysis of Feldstein-Horioka Puzzle in CIVETS Countries. In *Different Aspects of Poverty and Corruption*, edited by A. Yildiz, 97–114. Ankara: Gece Kitaplığı.

Swamy, V. & Narayanamurthy, V. (2018). What Drives the Capital Flows into BRICS Economies? *The World Economy*, 41: 519–549. Retrieved from https://doi.org/10.1111/twec.12606 (11.04.2019).

Şaşmaz, M. Ü. & Yayla, Y. E. (2018). Doğrudan Yabancı Sermaye Yatırımlarının Ekonomik Kalkınma Üzerindeki Etkisi: OECD Ülkeleri Örneği. *Hitit Üniversitesi Sosyal Bilimler Enstitüsü Dergisi*, 11(1), 359–374.

Toda, H. Y. & Yamamoto, T. (1995). Statistical Inference in Vector Autoregression with Possibly Integrated Processes. *Journal of Econometrics*, 66: 225–250.

Topallı, N. (2016). Doğrudan Sermaye Yatırımları, Ticari Dışa Açıklık ve Ekonomik Büyüme Arasındaki İlişki: Türkiye ve BRICS Ülkeleri Örneği. *Doğuş Üniversitesi Dergisi*, 17(1), 83–95.

Yi, Y., Qi, W. & Wu, D. (2013). Are CIVETS the Next BRICs? A Comparative Analysis from Scientometrics Perspective. *Scientometrics*, 94(2), 615–628.

Zeren, F. & Ari, A. (2013). Trade Openness and Economic Growth: A Panel Causality Test. *International Journal of Business and Social Science*, 4(9), 317–324.

Tuncer Gövdeli

5. Tourism, Oil Prices and Economic Growth in the Mediterranean Countries: Bootstrap Panel Granger Causality Analysis

Abstract The purpose of this chapter is to investigate the causal relationship of both the tourism revenues and crude oil prices and the economic growth of eight countries of the coastal Mediterranean (France, Croatia, Spain, Italy, Egypt, Slovenia, Turkey, and Greece). In the chapter, annual data of the period between 1995 and 2016 was used. The precondition of Konya (2006) causality test is cross-sectional dependence and heterogeneity of the slope coefficients. After the prerequisites of the causality test of Konya (2006) are met, the variables were analyzed at three different stages: (i) According to the results of the causality relationship between economic growth and tourism revenues, in Turkey, Egypt and Croatia "tourism-oriented economic growth hypothesis" is true. In Slovenia, the "economic growth-oriented tourism development hypothesis" is also valid. In Italy, there is bi-directional causality between economic growth and tourism revenues. (ii) The causality relationship between economic growth and crude oil prices was analyzed. Accordingly, in France, Egypt, Slovenia and Turkey, causality was determined from crude oil prices to economic growth. (iii) The relationship between tourism revenues and crude oil prices and economic growth has been analyzed. In Italy, tourism revenues and crude oil prices are the cause of economic growth.

Keywords: Bootstrap panel Granger causality, Economic growth, Tourism, Oil prices, Mediterranean countries

1 Introduction

The economies of the Mediterranean countries have a fragile structure against global crises. Tourism is an important sector in these countries. Therefore, the results of the crisis should be investigated and the instruments to improve the performance of the tourism sector should be investigated (Farčnik et al., 2015). Studies on how to increase the tourism revenues, which have made a significant contribution to the economies of the Mediterranean countries, have gained momentum in these countries. These countries, which enter the race in attracting tourists, also increase international competition, thus improving the quality.

Economic prosperity is linked to the growth of agricultural and manufacturing sectors as well as the flow of foreign capital. While the role of tourism in economic growth is often diminished, it is seen as a non-growing sector, and therefore attracts little attention from both economists and policy makers (Papatheodorou, 1999). There is no consensus on whether tourism promotes an economic activity or economic activity leads to the development of tourism on both theoretical and empirical grounds. This is because changes in economic and tourism conditions may modify the nature and magnitude of the relationship between these two series over time. However, its examination in a time-varying framework has been largely disregarded in the literature (Antonakakis et al., 2015; Tang and Tan, 2015).

Tourism has become the fourth largest export sector in the world after fuels, chemicals and foods (Tugcu, 2014; Balli et al., 2015). Tourism, which has a significant share in economic growth, affects many different sectors. Tourism has a direct impact on the development of traditional industries such as civil aviation, railway, motorway, trade, food and accommodation (Gövdeli, 2019). The impact of tourism on national economies is becoming increasingly important due to the growing size of the tourism market. In this context, the tourism-oriented growth hypothesis proposed by Balaguer and Cantavella-Jordá (2002) suggests that the expansion of international tourism activities boosts the economic growth. The tourism-driven growth hypothesis is derived from the widely known "export-driven growth hypothesis", which suggests that economic growth can be supported not only by the expansion of human resources and technology within the economy, but also by a rise in foreign exchange revenues (Ohlan 2017).

The number of international tourists in the world increased from 25 million in 1950 to 278 million in 1980, to 674 million in 2000 and to 1.235 million in 2016. At the same time, international tourism revenues earned by worldwide destinations rose from 2 billion dollars in 1950 to 104 billion dollars in 1980, 495 billion dollars in 2000 and 1.220 billion dollars in 2016. Tourism has grown faster than the world trade for the last five years and represents 7 % of the world exports of the goods and services. Tourism, which is the worldwide export category, ranks third after chemicals and fuels, followed by automotive products and food products. In many developing countries, tourism is the biggest exporting sector (UNWTO).

Tab. 5.1 shows the number of tourists in the top 10 countries according to the number of international tourists. France is the country with the highest number of tourists in the world. While the number of tourists in 2015 was 84.5 million, the number of tourists decreased by 2.2 % to 82.6 million in 2016 compared to the previous year. In 2016, the number of tourists in France is followed by the United States and Spain with 75.6 million tourists. The number of tourists in the USA (2015) is 77.5 million and the change in 2015–2016 period is −2.4 %. The increase in the number of tourists in Spain is quite significant. The number of

Tab. 5.1: Top 10 Countries by International Number of Tourists. Source: UNTWO, 2017.

COUNTRIES	(Millions $)		Change (%)	
	2015	2016	2015/2014	2016/2015
France	84,5	82,6	0,9	−2,2
USA	77,5	75,6	3,3	−2,4
Spain	68,5	75,6	5,5	10,3
China	56,9	59,3	2,3	4,2
Italy	50,7	52,4	4,4	3,2
Britain	34,4	35,8	5,6	4,0
Germany	35,0	35,6	6,0	1,7
Mexico	32,1	35,0	9,4	8,9
Thailand	29,9	32,6	20,6	8,9
Turkey	39,5	30,3	−0,8	−23,3

Note: Turkey's 2016 data is taken from the World Bank database.

Tab. 5.2: Top 10 Countries by International Tourism Revenue. Source: UNTWO, 2017.

COUNTRIES	(Billions $)		Change (%)	
	2015	2016	15/14	16/15
USA	205,4	205,9	7,0	0,3
Spain	56,5	60,3	−13,3	6,9
Thailand	44,9	49,9	16,9	11,0
China	45,0	44,4	2,1	−1,2
France	44,9	42,5	−22,9	−5,3
Italy	39,4	40,2	−13,3	2,0
England	45,5	39,6	−2,3	−12,9
Germany	36,9	37,4	−14,8	1,4
Hong Kong (China)	36,2	32,9	−5,8	−9,1
Australia	28,9	32,4	−8,2	12,3

tourists went up by 5,5 % in 2014–2015 period and an increase of 10.3 % was realized in 2015–2016 period. Four of the countries bordering the Mediterranean Sea (France, Spain, Italy and Turkey) are in the top 10 countries by the number of international tourists and the number of tourists visiting four countries in 2016 is 240.9 million people.

The tourism income of the top 10 countries according to international tourism income is given in Tab. 5.2. The tourism income of the USA is quite high compared to other countries. Tourism revenue, which was 205.4 billion dollars in

2015, increased by 0.3 % in 2016 to 205.9 billion dollars. There is instability in the tourism revenues of Spain. In 2015, tourism revenues of 56.5 billion dollars were realized, but decreased by 13.3 % compared to the previous year. In 2016, it increased by 6.9 % compared to 2015 and reached 60.3 billion dollars. Thailand is the third country with the highest tourism income in the world and the rise in tourism income is noteworthy. In 2015, there was 44.9 billion dollars in tourism revenue in Thailand, which rose by 11 % in 2016 to 49.9 billion dollars. Spain, France and Italy, which are among the top 10 countries in terms of international tourism income, accounted for 143 billion dollars in total tourism revenue in 2016.

The motivation for this chapter is to analyze empirically the casual relationship of both tourism revenues and crude oil prices with the economic growth in eight countries bordering the Mediterranean Sea (France, Croatia, Spain, Italy, Egypt, Slovenia, Turkey, and Greece). The difference of this study from other studies is to make an econometric analysis to find out if the tourism incomes and crude oil are the causality of the economic growth. Previous studies have often tested the causality relationship between tourism revenues and economic growth, or the causality relationship between crude oil prices and economic growth. The aim of this chapter is to analyze the causality of two variables (tourism revenues and crude oil prices) with economic growth and bring a different dimension to the literature.

In this context, the chapter consists of four parts. The first part is the introduction and theoretical information is given. The second part is the literature review and discusses the findings from previous studies. The third part explains the data set and methods in detail. The fourth part includes the interpretation of the empirical results. And in the conclusion section, results and political recommendations are given.

2 Literature

There are many studies examining the relationship between growth and tourism for different countries. One of the studies supporting the tourism-based economic growth hypothesis, Tang and Tan (2015), confirmed the tourism-focused growth hypothesis in Malaysia for the years between 1975 and 2011 using a multivariate model derived from Solow. Balaguer and Cantavella-Jorda (2002) investigated the period between 1975 and 1997 by applying Granger causality test using quarterly data. In this chapter, the tourism-driven growth hypothesis was valid for the Spanish economy, and concluded that tourism was the cause of economic growth. For Romania, Surugiu and Surugiu (2013) analyzed the time period between 1988 and 2009 by using the VECM Granger causality test. The findings confirmed the existence of a tourism-driven growth hypothesis. Huseyin and Kara (2011) examined Turkey by using the 1964 and 2006 annual time series data. According to the results of GDP, tourism revenues and exchange rate data using the VECM Granger causality test, tourism revenues are the cause of economic growth. Thus, in Turkey, "tourism-driven economic growth" was reached as the current result of that hypothesis. For Turkey, Savas et al. (2012) investigated tourism-driven growth hypothesis by using two variables (tourism income and tourism expenditure) for tourism development during the period of 1984M1–2008M3. In the results obtained, tourism is the cause of economic growth. Similarly, Terzi and Pata (2016) analyzed the relationship of the number of tourists and tourism revenue with growth in Turkey for the years between 1964 and 2014 and accordingly tourism-driven economic growth hypothesis was confirmed, while also confirming that tourism is the cause of economic growth.

In the literature, there are studies supporting the hypothesis of tourism development focused on economic growth. For Croatia, Payne and Mervar (2010) examined the relationship between tourism incomes, economic growth and real effective exchange rate by applying the Toda Yamamota causality approach for the quarters between 2000 and 2008. The findings support the hypothesis of tourism development focused on economic growth, and it is determined that economic growth is the cause of tourism incomes. For Morocco, Bouzahzah and Menyari (2013) analyzed the hypothesis of tourism development focusing on economic growth, using annual data from 1980 to 2010. Since there is one-way causality from economic growth to tourism, the hypothesis of economic growth-focused tourism development in Morocco has been confirmed. For Sri Lanka, Suresh and Senthilnathan (2014) investigated the causality relationship between economic growth and tourism revenues between 1977 and 2012. In the findings, due to the fact that economic growth is the cause of tourism revenues, the hypothesis of economic growth-driven tourism development has been confirmed. He and Zheng (2011) used China's tourism expenditures, GDP and exchange rate data for the period between 1990 and 2009. In the study, they analyzed the data by running VECM Granger causality test, and according to the results they determined a causal relationship between economic growth and tourism expenditures.

The last hypothesis combines the previous hypotheses, and provides information on the bi-directional causality between economic growth and tourism. Nowak et al. (2007), using the VECM Granger causality test, analyzed tourism revenues, GDP, industrial goods and machinery imports data. They found a bi-directional causality relationship between tourism incomes and economic growth for the 1960 – 2003 period. Cortes-Jimenez and Pulina (2010) examined the tourism income, GDP, physical and human capital series for Italy and Spain by using the VECM Granger causality test. The tourism-driven economic growth hypothesis has been confirmed for Italy; however, there is bi-directional causality between tourism revenues and economic growth for Spain. Demiroz and Ongan's (2005) work includes the 1980 and 2004 period for Turkey. According to the results of the VECM Granger causality test, they obtain a bi-directional causality relationship between economic growth and tourism revenues.

3 Data Set and Econometric Methodology

3.1 Data set

The data used in the empirical study is on the annual basis and covers the period between 1995 and 2016. Variables used for the eight countries with a Mediterranean coast (France, Croatia, Spain, Italy, Egypt, Slovenia, Turkey, Greece) are gross domestic product (GDP), tourism incomes (TG) and crude oil prices (PF). The data of GDP and TG were obtained from the World Bank database and PF data from the Energy Information Administration (EIA). GDP and TG (current US dollar), PF (dollar per barrel). All series were analyzed by taking their natural logarithms.

3.2 Econometric methodology

3.2.1 Horizontal cross-section dependence tests and homogeneity

In this study, the methods used to test the cross-sectional dependence of the panel data sets are Breusch-Pagan (1980) $CDLM_1$ test, Pesaran (2004) $CDLM_2$ test, Pesaran (2004) CDLM and Pesaran et al. (2008) Bias Adjusted LM_{adj} tests.

H_0: No cross-sectional dependency.
H_1: Horizontal cross-section dependence.

In Breusch-Pagan (1980) $CDLM_1$ test, Pesaran (2004) $CDLM_2$ test, Pesaran (2004) CDLM and Pesaran et al. (2008) and Bias Adjusted LM_{adj} test results, H_0 is rejected when the probability values are less than 0.05 and at the 5 % significance level, and the cross-sectional dependence between the units forming the panel is decided.

$$CDLM_1 = T \sum_{i=1}^{N-1} \sum_{j=i+1}^{N} \hat{\rho}_{ij}^2 \tag{1}$$

$\hat{\rho}_{ij}$: shows the estimates of cross-sectional correlations between residuals.

$$\hat{\rho}_{ij} = \hat{\rho}_{ji} = \frac{\sum_{t=1}^{T} \hat{\nu}_{it}\hat{\nu}_{jt}}{\left(\sum_{t=1}^{T} \hat{\nu}_{it}\right)^{1/2}\left(\sum_{t=1}^{T} \hat{\nu}_{jt}\right)^{1/2}} \quad (2)$$

There is no cross-sectional dependence under the H_0 hypothesis. Under the H_0 hypothesis, N is constant and $T \to \infty$. There is an asymptotic distribution of Chi-square with N (N-1)/2 degrees of freedom. The $CDLM_1$ test gives good results when the time dimension is greater than the horizontal cross-sectional dimension (T> N).

$$CDLM_2 = \left(\frac{1}{N(N-1)}\right)^{1/2} \sum_{i=1}^{N-1} \sum_{j=i+1}^{N} (T\hat{\rho}_{ij}^2 - 1) \quad (3)$$

Pesaran (2004) $CDLM_2$ statistics is normally distributed as standard in the case of $T \to \infty$ and $N \to M$ under hypothesis H_0. The $CDLM_2$ test provides good results when the time dimension is greater than the horizontal cross-sectional size (T> N).

$$CDLM = \left(\frac{2T}{N(N-1)}\right)^{1/2} \sum_{i=1}^{N-1} \sum_{j=i+1}^{N} \hat{\rho}_{ij}^2 \quad (4)$$

Pesaran (2004) CDLM statistic is normally distributed in the case of $T \to \infty$ and $N \to \infty$ under the H_0 hypothesis. The CDLM test is used when the cross-sectional size is greater than the time dimension (N> T).

$$LMadj = \left(\frac{2}{N(N-1)}\right)^{1/2} \sum_{i=1}^{N-1} \sum_{j=i+1}^{N} \hat{\rho}_{ij}^2 \frac{(T-K-1)\hat{\rho}_{ij} - \hat{\mu}_{Tij}}{v_{Tij}} \sim N(0,1) \quad (5)$$

In Equation 5, $\hat{\mu}_{Tij}$ represents the mean and v_{Tij} indicates the variance and the test statistics to be obtained will be asymptotically standard normal distributed. The LM_{adj} test is used when the cross-sectional dimension is greater than the time dimension (N> T) (Pesaran et al., 2008).

Swamy (1970) tested the homogeneity of slope coefficients in the cointegration equations. Pesaran and Yamagata (2008) have further developed the Swamy test and introduced it into the literature. In this test, a general cointegration equation in the form of

$$Y_{it} = \alpha + \beta_i X_{it} + \varepsilon_{it} \quad (6)$$

is tested whether the slope coefficients β_i differ between the horizontal sections. Test hypotheses are as follows:

$H_0 : \beta_i = \beta$ Slope coefficients are homogeneous.
$H_1 : \beta_i \neq \beta$ Slope coefficients are heterogeneous.

The required test statistics are generated by estimating the panel data first with the Ordinary Least Squares, then with the Weighted Fixed Effect. Pesaran and Yamagata (2008) have developed two different test statistics to test hypotheses:

$$\text{Forlargesamples} : \tilde{\Delta} = \sqrt{N} \frac{N^{-1}\tilde{S} - k}{\sqrt{2k}} \qquad (7)$$

$$\text{Forsmallsamples} : \tilde{\Delta}_{adj} = \sqrt{N} \frac{N^{-1}\tilde{S} - k}{\sqrt{Var(t-k)}} \qquad (8)$$

where N is the number of horizontal sections; S is the Swamy test statistic; k is the number of explanatory variables and Var(t,k) refers to the standard error. When the calculated probability values are less than 0.05, the H_0 hypothesis is rejected at 5 % significance level and H_1 hypothesis is accepted. Thus, it is decided that the cointegration coefficients are heterogeneous (Pesaran and Yamagata, 2008).

3.2.2 Konya causality test

Konya (2006) causality test is applied based on "Seemingly Unrelated Regression" (SUR) systems and country-specific bootstrap critical values and Wald tests. This test has two advantages: The first one assumes that the panel is not homogeneous. Thus, Granger causality can be tested separately for each country included in the panel. The second one, the simultaneous correlation is allowed between countries; hence, it allows using the additional information provided by the panel data. On the other hand, this application can be analyzed without the need for unit root and cointegration analyzes (Konya, 2006: 990–991).

The bootstrap panel causality model used in the bivariate model is given below:

$$y_{1,t} = \alpha_{1,1} + \sum_{l=1}^{mly_1} \beta_{1,1,l} y_{1,t-1} + \sum_{l=1}^{mlx_1} \varphi_{1,1,l} x_{1,t-1} + \mu_{1,1,t} \qquad (9)$$

$$y_{1,t} = \alpha_{1,2} + \sum_{l=1}^{mly_1} \beta_{1,2,l} y_{2,t-1} + \sum_{l=1}^{mlx_1} \varphi_{1,2,l} x_{2,t-1} + \mu_{1,2,t} \qquad (10)$$

⋮

$$y_{N,t} = \alpha_{1,N} + \sum_{l=1}^{mly_1} \beta_{1,N,l} y_{N,t-1} + \sum_{l=1}^{mlx_1} \varphi_{1,N,l} x_{N,t-1} + \mu_{1,N,t} \qquad (11)$$

and

$$x_{1,t} = \alpha_{2,1} + \sum_{l=1}^{mly_2} \beta_{2,1,l} y_{1,t-1} + \sum_{l=1}^{mlx_2} \varphi_{2,1,l} x_{1,t-1} + \mu_{2,1,t} \tag{12}$$

$$x_{1,t} = \alpha_{2,2} + \sum_{l=1}^{mly_2} \beta_{2,2,l} y_{2,t-1} + \sum_{l=1}^{mlx_2} \varphi_{2,2,l} x_{2,t-1} + \mu_{2,2,t} \tag{13}$$

⋮

$$x_{N,t} = \alpha_{2,N} + \sum_{l=1}^{mly_2} \beta_{2,N,l} y_{N,t-1} + \sum_{l=1}^{mlx_2} \varphi_{2,N,l} x_{N,t-1} + \mu_{2,N,t} \tag{14}$$

The bootstrap panel causality model used in the three variable models is given below:

$$y_{1,t} = \alpha_{1,1} + \sum_{l=1}^{mly_1} \beta_{1,1,l} y_{1,t-1} + \sum_{l=1}^{mlx_1} \varphi_{1,1,l} x_{1,t-1} + \sum_{l=1}^{mlz_1} \gamma_{1,1,l} z_{1,t-1} + \mu_{1,1,t} \tag{15}$$

⋮

$$y_{N,t} = \alpha_{1,N} + \sum_{l=1}^{mly_1} \beta_{1,N,l} y_{N,t-1} + \sum_{l=1}^{mlx_1} \varphi_{1,N,l} x_{N,t-1} + \sum_{l=1}^{mlz_1} \gamma_{1,N,l} z_{N,t-1} + \mu_{1,N,t} \tag{16}$$

and

$$x_{1,t} = \alpha_{2,1} + \sum_{l=1}^{mly_2} \beta_{2,1,l} y_{1,t-1} + \sum_{l=1}^{mlx_2} \varphi_{2,1,l} x_{1,t-1} + \sum_{l=1}^{mlz_2} \gamma_{2,1,l} z_{1,t-1} + \mu_{2,1,t} \tag{17}$$

⋮

$$x_{N,t} = \alpha_{2,N} + \sum_{l=1}^{mly_2} \beta_{2,N,l} y_{N,t-1} + \sum_{l=1}^{mlx_2} \varphi_{2,N,l} x_{N,t-1} + \sum_{l=1}^{mlz_2} \gamma_{2,N,l} z_{N,t-1} + \mu_{2,N,t} \tag{18}$$

and

$$z_{1,t} = \alpha_{3,1} + \sum_{l=1}^{mly_3} \beta_{3,1,l} y_{1,t-1} + \sum_{l=1}^{mlx_3} \varphi_{3,1,l} x_{1,t-1} + \sum_{l=1}^{mlz_3} \gamma_{3,1,l} z_{1,t-1} + \mu_{3,1,t} \tag{19}$$

⋮

$$z_{N,t} = \alpha_{3,N} + \sum_{l=1}^{mly_3} \beta_{3,N,l} y_{N,t-1} + \sum_{l=1}^{mlx_3} \varphi_{3,N,l} x_{N,t-1} + \sum_{l=1}^{mlz_3} \gamma_{3,N,l} z_{N,t-1} + \mu_{3,N,t}$$
(20)

where x, y, and z indicate natural logarithms of GDP, tourism income and crude oil prices respectively. N is the number of countries in the panel (i = 1,..., N); t represents the time period (t = 1,..., T) and l, the lag length.

Each equation belongs to a different country, so it is estimated by a different sample. Variables are the same in all equations, but observations are different. Each equation has predetermined variables and the possible link between individual regressions is in the cross-sectional dependence (Konya, 2006: 981). Granger causality can be found for each country. For example: (i) there is a one-way Granger causality relationship from lnGDP to lnTG while all $\varphi_{1,i}$ are not equal to zero and all $\beta_{2,i}$ are equal to zero. (ii) There is a one-way Granger causality relationship from lnTG to lnGDP when all $\beta_{2,i}$ are not equal to zero and all $\varphi_{1,i}$ are equal to zero. (iii) There is a bi-directional causality relationship between lnGDP and lnTG when neither $\varphi_{1,i}$ nor all $\beta_{2,i}$ are equal to zero. (iv) There is no causality between lnGDP and lnTG if all $\varphi_{1,i}$ and all $\beta_{2,i}$ are equal to zero.

4 Empirical Results

Konya (2006) causality test is a test, which can be used without unit root and cointegration test. As a precondition for this test, the cross-sectional dependence and the homogeneity of the slope coefficients should be tested.

The results of horizontal cross-section dependence and homogeneity tests are displayed in Tab. 5.3. When the Tab. 5.3 is examined, the null hypothesis 'no cross-sectional dependency' is rejected at the significance level of 1 %. Thus, the shock that may occur in any of the eight Mediterranean-coastal countries used in the study may affect the other country. The homogeneity of slope coefficients is shown in the table. Accordingly, the zero hypothesis 'slope coefficients are homogeneous' is rejected at the 1 % significance level, and the alternative hypothesis 'slope coefficients are heterogeneous' is accepted. The requirement of heterogeneity of slope coefficients under the cross-sectional dependence, which is the precondition for Konya (2006) causality test, is met.

According to the results of the causality between economic growth and tourism incomes of eight countries, which have coast to the Mediterranean Sea, tourism incomes are the cause of economic growth in Turkey at the 5 % significance

Tab. 5.3: Horizontal Cross-Section Dependence and Homogeneity Tests. Source: Author's own calculations.

Tests	Statistic	p-Value
$CDLM_1$	122,817*	0,000
$CDLM_2$	12,670*	0,000
CDLM	8,075*	0,000
LM_{adj}	10,374*	0,000
$\tilde{\Delta}$	6,153*	0,000
$\tilde{\Delta}_{adj}$	6,770*	0,000

Note: * indicates the level of significance at 1 %.

Tab. 5.4: Economic Growth—Tourism Income Causality Results. Source: Author's own calculations.

Countries	lnTG ↛ lnGDP				lnGDP ↛ lnTG			
	Statistical Value	Critical Values			Statistical Value	Critical Values		
		1 %	5 %	10 %		1 %	5 %	10 %
France	4,205	23,142	11,648	7,953	0,149	16,369	8,494	5,782
Croatia	9,447*	20,326	10,830	7,360	0,119	15,239	8,548	5,860
Spain	1,678	20,362	11,138	7,808	0,036	17,179	9,206	6,389
Italy	14,684**	18,506	9,762	6,855	11,035**	15,460	8,561	5,971
Egypt	21,930*	18,363	9,363	6,227	4,412	18,866	10,012	6,841
Slovenia	0,153	20,350	10,321	7,203	35,256*	17,166	9,725	6,584
Turkey	12,638**	22,604	9,958	6,455	1,535	18,841	9,449	6,332
Greece	0,765	24,599	12,658	8,358	0,175	14,838	8,398	5,669

Note: *, ** and *** signify the significance levels at 1 %, 5 % and 10 % respectively. Critical values were obtained with 10,000 bootstrap replications.

level. In Egypt and Croatia, tourism incomes are the cause of economic growth at the 1 % significance level. Therefore, the results show that in Turkey, Egypt and Croatia, 'tourism-driven economic growth' hypothesis is valid. Since the economic growth in Slovenia is the cause of tourism, the hypothesis of 'economic growth-oriented tourism development' is valid in this country. In Italy, there is bi-directional causality between economic growth and tourism revenues (Tab. 5.4).

Tab. 5.5 shows the causality results between economic growth and crude oil prices. According to the findings. In Egypt, oil prices are the cause of economic

Tab. 5.5: Economic Growth—Crude Oil Prices Causality Results. Source: Author's own calculations.

Countries	lnPF ↛ lnGDP				lnGDP ↛ lnPF			
	Statistical Value	Critical Values			Statistical Value	Critical Values		
		1 %	5 %	10 %		1 %	5 %	10 %
France	13,192**	13,666	7,676	5,216	0,537	24,978	13,597	9,507
Croatia	4,113	13,190	7,226	4,965	2,325	13,236	6,586	4,788
Spain	0,087	13,741	7,422	5,090	4,375	12,292	6,180	4,416
Italy	3,935	13,533	7,346	5,026	1,536	11,893	6,094	4,335
Egypt	30,060*	14,514	7,082	4,670	0,293	7,667	3,559	2,412
Slovenia	6,096***	13,121	7,473	5,127	2,113	10,791	5,492	3,812
Turkey	6,432***	12,055	6,658	4,478	0,977	2,573	0,988	0,529
Greece	0,775	12,258	6,926	4,859	3,783	12,268	6,213	4,455

Note: *, ** and *** signify the significance levels at 1 %, 5 % and 10 % respectively. Critical values were obtained with 10,000 bootstrap replications.

growth at 1 % significance level. Egypt is the country, which has more oil reserves than the seven other countries in the Mediterranean. In France, Slovenia and Turkey, according to the results of causality; oil prices seem to be the cause of economic growth at 10 % significance level. In the eight countries along the Mediterranean, no causality was found between economic growth and oil prices.

Tab. 5.6 shows the causality results of both tourism incomes and crude oil prices with economic growth. In the results obtained, in Italy, it can be observed that tourism revenues and crude oil prices are the causes for economic growth at significance level of 5 %.

5 Conclusion

In this chapter, the relationship of both tourism revenues and crude oil prices with economic growth was analyzed for the eight countries (France, Croatia, Spain, Italy, Egypt, Slovenia, Turkey, and Greece), which have a Mediterranean coast and which have a certain tourism potential, by applying Konya (2006) panel causality test. Accordingly, in these countries, tourism is the cause of economic growth. In Slovenia, economic growth is the cause of tourism. For this reason, 'tourism-driven economic growth' hypothesis is valid in Slovenia. In Italy, there is bi-directional causality between tourism incomes and economic growth. Italy is the only country among the eight countries in which the causality relationship of tourism incomes and crude oil prices with economic growth is found.

Tab. 5.6: Tourism Revenues and Crude Oil Prices—Economic Growth Causality Results. Source: Author's own calculations.

Countries	lnTG and lnPF ↛lnGDP			
	Statistical Value	Critical Values		
		1 %	5 %	10 %
France	0,089	19,706	10,407	7,027
Croatia	0,344	15,714	8,360	5,720
Spain	4,907	18,798	10,215	7,200
Italy	10,708**	19,076	10,586	7,049
Egypt	0,263	17,490	9,310	6,059
Slovenia	0,007	17,685	9,359	6,474
Turkey	1,856	18,412	9,188	6,114
Greece	1,016	17,463	9,461	6,295

Note: *, ** and *** signify the significance levels at 1 %, 5 % and 10 % respectively. Critical values were obtained with 10,000 bootstrap replications.

The contribution of this study to the literature is to investigate the effect of tourism and oil prices on economic growth in the countries, which have coasts in the Mediterranean Sea. Both variables play an active role in economic growth. Thus, the changes in oil prices and tourism revenues can be examined together and the decisions of policy makers can become more consistent. In addition, minimizing the negative impact caused by the sudden changes in oil prices should be aimed at these oil- importing countries. For this reason, the fact that tourism revenues in these countries play a balancing role against the fluctuations in oil prices should be clearly emphasized in the policies to be implemented.

The economic growth of these eight countries, which have a high tourism potential due to their Mediterranean coasts, will increase by the investments in tourism. For this, the private sector and policy makers need to act together. Tourism, which has a real impact on the economic development of these countries, is an opportunity mechanism. From a global and theoretical point of view, the relationship between tourism and economic growth is one of the factors that determine the growth of tourism as a tool for economic development.

In these eight countries, which are foreign-dependent in terms of oil consumption and which have coast to the Mediterranean, oil prices are very important for the development and development of the country. Fluctuations in oil prices can directly affect the economy of these countries. If the countries design their policies correctly, achieving their goals on subjects such as wealth, competitiveness, and economic growth will be easier. For this reason, policy makers should implement new policies and allocate more funds to promote tourism by developing existing policies.

References

Antonakakis, N., Dragouni, M., & Filis, G. (2015). How Strong is the Linkage Between Tourism and Economic Growth in Europe? *Economic Modelling*, 44, 142–155.

Balaguer, J. & Cantavella-Jorda, M. (2002). Tourism as a Long-Run Economic Growth Factor: The Spanish Case. *Applied economics*, 34(7), 877–884.

Balli, F., Curry, J., & Balli, H. O. (2015). Inter-Regional Spillover Effects in New Zealand International Tourism Demand. *Tourism Geographies*, 17(2), 262–278.

Bouzahzah, M., & El Menyari, Y. (2013). International Tourism and Economic Growth: The Case of Morocco and Tunisia. *The Journal of North African Studies*, 18(4), 592–607.

Breusch, T. S. & Pagan, A. R. (1980). The Lagrange Multiplier Test and Its Applications to Model Specification in Econometrics. *The Review of Economic Studies*, 47(1), 239–253.

Cortes-Jimenez, I. & Pulina, M. (2010). Inbound Tourism and Long-Run Economic Growth. *Current Issues in Tourism*, 13(1), 61–74.

Demiroz, D. M. & Ongan, S. (2005). The Contribution of Tourism to the Long-Run Turkish Economic Growth. *Ekonomický časopis*, 9, 880–894.

Farčnik, D., Kuščer, K., & Trobec, D. (2015). Indebtedness of the Tourism Sector in Mediterranean Countries. *Tourism economics*, 21(1), 141–157.

He, L. H. & Zheng, X. G. (2011). Empirical Analysis on the Relationship Between Tourism Development and Economic Growth in Sichuan. *Journal of Agricultural Science*, 3(1), 212.

Husein, J. & Kara, S. M. (2011). Research Note: Re-Examining the Tourism-Led Growth Hypothesis for Turkey. *Tourism Economics*, 17(4), 917–924.

Gövdeli, T. (2019). Investigating the Relationship on CO_2, Tourism, Economic Growth and Trade Openness in Turkey. *Yönetim ve Ekonomi: Celal Bayar Üniversitesi İktisadi ve İdari Bilimler Fakültesi Dergisi*, 26(1), 321–331.

Kónya, L. (2006). Exports and Growth: Granger Causality Analysis on OECD Countries with a panel data approach. *Economic Modelling*, 23(6), 978–992.

Nowak, J. J., Sahli, M., & Cortés-Jiménez, I. (2007). Tourism, Capital Good Imports and Economic Growth: Theory and Evidence for Spain. *Tourism Economics*, 13(4), 515–536.

Ohlan, R. (2017). The Relationship Between Tourism, Financial Development and Economic Growth in India. *Future Business Journal*, 3(1), 9–22.

Papatheodorou, A. (1999). The Demand for International Tourism in the Mediterranean Region. *Applied Economics*, 31(5), 619–630.

Payne, J. E. & Mervar, A. (2010). Research Note: The Tourism–Growth Nexus in Croatia. *Tourism Economics*, 16(4), 1089–1094.

Pesaran, M. H. (2004). General Diagnostic Tests for Cross Section Dependence in Panels. Cambridge Working Papers in Economics no. 435. University of Cambridge.

Pesaran, M. H., Ullah, A., & Yamagata, T. (2008). A Bias-Adjusted LM Test of Error Cross-Section Independence. *The Econometrics Journal*, 11(1), 105–127.

Pesaran, M. H. & Yamagata, T. (2008). Testing Slope Homogeneity in Large Panels. *Journal of Econometrics*, 142(1), 50–93.

Savaş, B., Beşkaya, A., & Şamiloğlu, F. (2012). Analyzing the Impact of International Tourism on Economic Growth in Turkey. *Uluslararası Yönetim İktisat ve İşletme Dergisi*, 6(12), 121–136.

Suresh, J. & Senthilnathan, S. (2014). Relationship Between Tourism and Economic Growth in Sri Lanka. *Published as the 7th chapter of a book titled "Economic Issues in Sri Lanka" compiled by Dr. S. Vijayakumar*, 115–132.

Surugiu, C. & Surugiu, M. R. (2013). Is the Tourism Sector Supportive of Economic Growth? Empirical Evidence on Romanian Tourism. *Tourism Economics*, 19(1), 115–132.

Swamy, P. A. (1970). Efficient Inference in a Random Coefficient Regression Model. *Econometrica: Journal of the Econometric Society*, 38(2), 311–323.

Tang, C. F. & Tan, E. C. (2015). Does Tourism Effectively Stimulate Malaysia's Economic Growth? *Tourism Management*, 46, 158–163.

Terzi, H. & PATA, U. K. (2016). Türkiye'nin İktisadi Büyümesinde Turizm Sektörünün Katkısı. *Erciyes Üniversitesi İktisadi ve İdari Bilimler Fakültesi Dergisi*, 48, 45–64.

The World Tourism Organization—UNWTO (2017), "Tourism Highlights 2017" https://www.e-unwto.org/doi/pdf/10.18111/9789284419029 (16.07.2018).

Tugcu, C. T. (2014). Tourism and Economic Growth Nexus Revisited: A Panel Causality Analysis for the Case of the Mediterranean Region. *Tourism Management*, 42, 207–212.

Habibe Günsel Doğrul, Mediha Mine Çelikkol, and İlhan Korkmaz

6. The Analysis of the Relationship Between Creative Class, Financial Development and Regional Innovativeness in Turkey

Abstract Today, innovation is an important element of competitiveness for businesses, regions and countries. Spatial and social proximities between the actors of economic system contribute to regional innovation by facilitating knowledge spillover. The aim of this study is to examine the effect of creative class workers, financial development and spatial interaction on regional innovation and it is found that considered factors have an impact on regional innovation. Measuring the human capital with creative class workers other than conventional measure of human capital and including spatial proximity as a factor affecting innovation make this study different from other studies conducted in Turkey. It is expected that the chapter will shed light on politicians to assess the existing provincial innovation strategies in Turkey and contribute to the related literature.

Keywords: Creative class, Skilled labour force, Innovation, Financial development, Spatial analysis

1 Introduction

Today's economies are transformed into economies with the increasing weight of knowledge-based activities in the production process. In these economies, one of the driving forces of sustainable economic growth is considered as innovation. Innovation is a process in which ideas are transformed into a new product and service, new production or distribution method, new institutional structures or new management approaches. The region where people live creates effects that reveal or prevent this process due to the spatial sharing between actors in the economic system. However, the fact that innovation has both concrete and abstract elements complicates it to make a clear assessment of the innovation performance of the regions (Camagni and Capello, 2013; Mas-Verdu, Ortiz-Miranda, Garcia-Alvarez-Coque, 2016).

The purpose of this chapter is to investigate the impact of "creative class", which is thought to affect the regional innovation performance and which has an increasing importance in the recent years, "financial development" and "spatial dependence" factors on the "innovation performance of provinces in Turkey".

In the second section of the chapter, the literature emphasizing the importance of regional dimensions of innovation is reviewed. In the third section, the data and method of the research are explained. In the fourth section, the results of the analysis are discussed, and in the conclusion section the findings are evaluated, and recommendations are given.

2 Regional Dimensions of Innovation

From a regional perspective, it is possible to accept innovation as the "driving force of economic growth", because it leads to a high level of production and prosperity. In terms of their capacity for knowledge and innovation, and their ability to make use of existing ideas and technologies, the differences between regions are indicative of the importance of the "region factor" in the process of innovation (Krakowiak-Bal and Ziemiańczyk, 2017). The unique characteristics of a region and its ability to create a suitable environment for information transfer affect regional innovation.

In the literature, it is seen that the impact of innovation on regional development and competitiveness level has been examined extensively (Cooke et al., 2011; Asheim and Gertler, 2005; Autio, Kenney, Mustar, Siegel and Wright, 2014; Roig-Tierno, Alcazar and Riberio-Navarrete.,2015; Rodriguez-Pose and Villarreal Peralta, 2015). In the studies, it has been observed that innovation is not evenly distributed among regions, innovation tends to be spatially over time, and even regions with similar innovation capacity have different economic growth levels.

Lucas (1988) emphasizes that the concentration of creative and talented people is particularly important in terms of innovation. Additionally, he proposes that innovators, practitioners, and financial supporters should communicate both at work and outside the workplace in order to disseminate ideas and take them quickly.

2.1 Regional innovation and creative class

Creativity was added by the studies of Florida (2002, 2011, and 2012) to the definition of capital, which is considered by economists according to physical, monetary, natural, human and social factors. These studies, which have initiated important debates on explaining regional development, are important for emphasizing the important connections between creativity, innovation, competitiveness and the role that cities play as supporting environments (Bradford, 2004).

Florida's (2002) 'creative class' includes scientists and engineers, academics, poets and novelists, film and show artists, actors, designers, architects, and also thought leaders of modern society. In addition to all of these professions, there are employees in knowledge-intensive industries such as financial professionals, law, healthcare and business management who work in high-tech sectors under the name of creative professionals. These employees create new products and processes and participate in the creative problem-solving process and increase the economic performance of the companies. Innovation, economic growth and prosperity occur in places attracting the most creative individuals.

The related literature suggests that highly qualified creative employees act as mediators of information circulation and innovation by bringing knowledge, ideas and practices to the company through a series of mechanisms. Qualified employees participate in non-corporate networks and communities of practice (Amin and Cohendet, 2004; Bathelt, Malmberg and Maskell, 2004; Storper and Venables, 2004). Non-firm networks may include interactions with local or non-local actors involved in similar activities, as well as social interactions with highly qualified employees in different professions. Through interaction, creative employees help local production and contribute to the creation of a global line that brings new knowledge, ideas and practices to the company and the region (Bathelt et al., 2004). These employees move their knowledge and practice experiences to companies and projects on a national and/or international scale, improving knowledge exchange and learning.

2.2 Regional innovation and financial development

In a developed financial system, it is thought that the transfer of foreign resources to the financing of business activities in a faster and lower cost will affect the ability of enterprises to adopt and develop innovation. The development of financial technology, and the increase in the institutions in the system, the financial development that comes with the existence of legal regulations, will facilitate the employment of relatively more risky enterprises to benefit from the banking sector compared to others who adopt new ideas and practices, but will allow opportunities for innovation-oriented works.

In the literature, there are studies that have concluded that financial development is an important component of the innovation infrastructure of the region and positively affects the innovation performance (Girma, Gong and Görg, 2008; Hanley, Liu and Vaona, 2011; Maskus, Neumann and Seidel, 2012; Meierrieks, 2014; Tee, Low, Kew and Ghazali, 2014; Laeven, Levine and Michalopoulos, 2015; Zhao, 2016, Bal, Iscan, Serin and Kara, 2017; Kocak, 2018). In addition to these studies, there are some studies that advocate a negative effect between financial development and innovation (Chu, Cozzi, Fan, Pan and Zhang, 2016). Many of these studies are carried out at the country level. The study by Hanley, Liu and Vaona (2011) was carried out for 30 Chinese provinces except Tibet. This study argues that the financial depth of a region has a significant positive impact on regional innovation performance, and that these effects are larger in size than in more complex innovations such as utility models and patents on minor innovations.

3 Data Set and Method

3.1 Variables used in research

Based on the literature mentioned in the first section, the level of innovation of the province is used as the dependent variable for this study; while the independent variables are determined as creative class workers, financial development and the share of the industrial sector in the total production of the province were used (Tab. 6.1). Descriptive statistics for the variables are presented in Tab. 6.2.

Tab. 6.1: Selected Variables. Source: Authors' own construction.

Variable Name	Symbol	Definition	Source
Innovation	INNOVA	Innovation Index (2014)	URAK (2016)
Creative Class	CREA	The share of creative class workers in the province in the province's labor force (15–64 age group) (2013)	Doğrul et al. (2016)
Financial Development	FINDEV	Financial Development Index	Gül and Çelik (2015)
Industry Share	INDSHR	The share of the industrial sector in the province's GDP (2014)	TSI

Tab 6.2: Descriptive Statistics. Source: Authors' own calculations

Variable	Mean	Std.Dev.	Min.	Max.
INNOVA	10,16	14,97	0,02	93,34
CREA	0,06	0,017	0,03	0,11
FINDEV	0,89	2,83	0,01	24,02
INDSHR	0,25	0,09	0,09	0,50
Observation	81	81	81	81

3.2 Method of research

Since the sample of the study shows a structure neighbor to each other with boundaries, and because the innovation, which is the dependent variable in the model, is a variable that can lead to spillovers between observations, Spatial Statistics and Spatial Econometrics techniques have been used as analysis method. Exploratory Spatial Data Analysis (ESDA) was used to determine the spatial clusters related to the variables used in the study while spatial interactions between variables were tested by Spatial Lag model. ESDA is a set of techniques that aim to describe and visualize spatial distributions, and to identify spatial outliers, clusters and problematic regions. It consists of the measurement of global and local spatial autocorrelations (Le Gallo and Ertur, 2003: 3).

Spatial autocorrelation can be defined as the tendency of high values in a location to coexist with high values in close locations or the tendency of low values in a location to coexist with low values in close locations (positive spatial autocorrelation). In contrast, when high values at a location are surrounded by nearby low values, or vice versa, negative spatial autocorrelation is present in the form of spatial outliers (Anselin, Cohen, Cook, Gorr and Tita, 2000). In order to measure spatial autocorrelation, a spatial weight matrix is created first. The spatial weight matrix describes the relationship between neighbors and is expressed numerically in order to calculate the spatial association between them (Lee and Wong, 2001). In general, the spatial weight matrix is calculated according to the boundaries between the units (Rook and queen[1]) or the distance between each other (k-nearest and distance) (Anselin, 1988).

Spatial autocorrelation analysis includes calculation of both global (test for clustering) and local (test for clusters) Moran's I statistics and visualization with Moran scatter plots (Anselin, Syabri and Kho, 2006). Moran's I statistics does not allow to understand the regional structure of spatial autocorrelation because it is a global statistic (Le Gallo and Ertur, 2003). The Local Indicators of Spatial Association (LISA) analysis should be performed to see the local characteristics of spatial dependency more clearly. The LISA analysis basically helps identify

---

1 Rook and queen matrices used in the definition of neighborhood based on borderline are named after the movement directions of rook and queen figures in the chess game (Anselin, 1988). While rook matrix includes cardinal points such as eastern, western, northern, and southern; queen matrix contains intercardinal points such as north-east, north-west, South-east, South-west.

significant local spatial clusters and identifies local instabilities as spatial outliers (Anselin, 1995). The LISA statistics are calculated separately for each location of the sample and provide a measure of the extent to which the arrangement of values around a specific location deviates from spatial randomness (Anselin, 1995; Anselin et al., 2000).

3.3 Econometric model used in research

Although many spatial econometric models are derived in the literature, spatial lag and spatial error models are seen as the foundations of spatial econometric models. The determination tests to be given in the Findings section indicate that the Spatial Lag Model should be used in this study. The Spatial Lag Model is created by adding the spatial lag term, denoted by ρ, to the right side of the equation. The term "spatial lag" added to the model indicates the weighted average value of WY in neighboring regions (Anselin, 2003).

$$Y = \rho WY + X\beta + \varepsilon \qquad (1)$$

Above in Equation 1, Y and X represents the dependent variable and the independent variable respectively; and W indicates the n × n dimensional spatial weight matrix. ρ is the autoregressive parameter that measures the effect of the dependent variables in the neighboring positions on the dependent variable in the respective position and is mostly accepted as $|\rho| < 1$ (Zeren, 2011: 24).

The econometric model established in this study is as follows:

$LNINNOVA_i = \alpha + \rho WLNINNOVA_i + \beta_1 LNCREA_i + \beta_2 LNFINDEV_i + \beta 3 LNINDSHR_i + \varepsilon_i$

The natural logarithm of all variables in the model is taken. The variable expressed by WLINNOVA is the multiplication of innovation with the weight matrix.

4 Analysis Findings

In this part of the study, ESDA findings on the level of innovation, creative class share and financial development level of the provinces, which are the variables that are the focal point of the research, will be included. Then, the estimation results of spatial econometric model will be explained.

4.1 Moran scatter plots

Moran's I statistic was calculated for the focus variables due to the structure according to the boundaries of cities in Turkey, 'Queen' weight matrix is taken as baseline. As Moran's I statistics approach to 1, the degree of positive spatial autocorrelation increases. As it approaches to −1, the degree of negative spatial autocorrelation increases. In addition, spatial autocorrelation is visualized for the entire sample with Moran scatter plots.

Fig. 6.1 shows the Moran scatter plots of innovation, creative class and financial development variables from left to right. Each graph is composed of four quadrants and shows how spatial autocorrelation works for all observations. The first quadrant (High-High—HH) at the top right of the graph shows high-value units surrounded by high-value neighbors. The second quadrant (Low-High—LH) at the top left illustrates low-value units surrounded by high-value neighbors. The third quadrant (Low-Low—LL) at the bottom left displays low value units surrounded by low-value neighbors. Finally, the fourth quadrant in the lower right (High-Low-HL) depicts the high-value units surrounded by low-value neighbors. Quadrants 1 and 3 indicate positive spatial autocorrelation, while quadrants 2 and 4 indicate negative spatial autocorrelation (Le Gallo and Ertur, 2003: 6, Dall'erba, 2005: 131).

In this scatter plot, the degree of spatial autocorrelation is given by Moran's I statistics. The Moran's I statistics of the innovative and creative class variables are 0.52 and 0.58, respectively, indicating strong positive spatial autocorrelation.

The Moran's I statistic, calculated for financial development, shows positive spatial autocorrelation, albeit weakly with 0.26. This situation shows the tendency of high-value locations to coexist with high-value locations, while also showing the tendency of low-value locations to coexist with low-value locations on the geography examined. The findings show that there is strong regional segregation in the geography studied for these variables.

4.2 LISA analysis findings

In the LISA analysis, it was determined that spatial autocorrelation measured by the Moran's I statistic creates significant spatial clusters in the local aspect. Places tinted with red within the maps (Figs. 6.2, 6.3 and 6.4) created by the LISA analysis represent the "HH Cluster Type" consisting of provinces which are above the average for Turkey and which are surrounded by high-value neighbors for the said variable. Places tinted with blue represent the "LL Cluster Type" consisting of provinces which are below the average for Turkey and which are surrounded by low-value neighbors. Places tinted with light blue represent the "LH Cluster Type" consisting of provinces which are below the average for Turkey and which are surrounded by high-value neighbors. Places tinted by pink represent the "HL Cluster Type" consisting of provinces which are above the average for Turkey and which are surrounded by low-value neighbors.

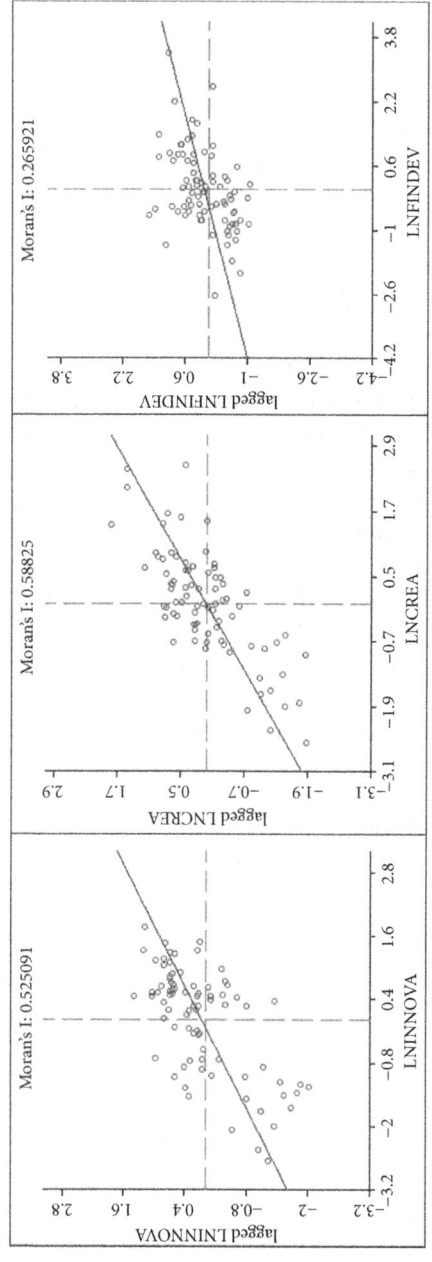

Fig. 6.1: Moran Scatter Plots Regarding to Innovation, Creative Class and Financial Development Level. Source: Authors' own construction

Creative Class, Development and Innovativeness 105

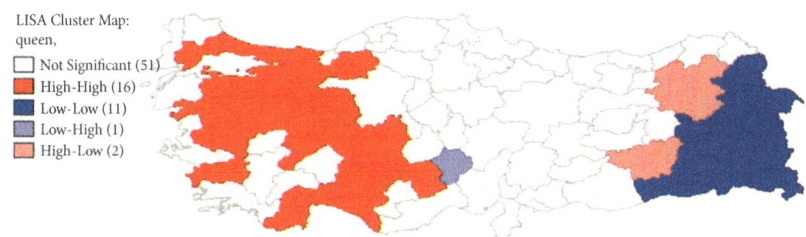

Fig. 6.2: LISA Map for Innovation Level. Source: Authors' own construction

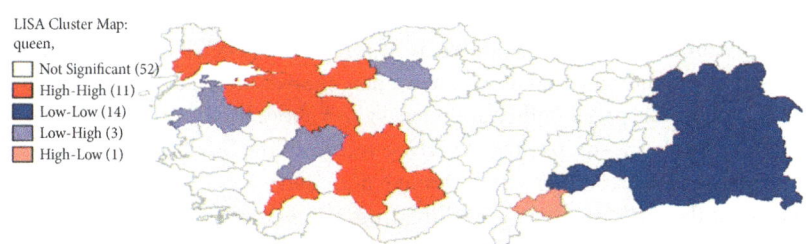

Fig. 6.3: LISA Map for Creative Class. Source: Authors' own construction

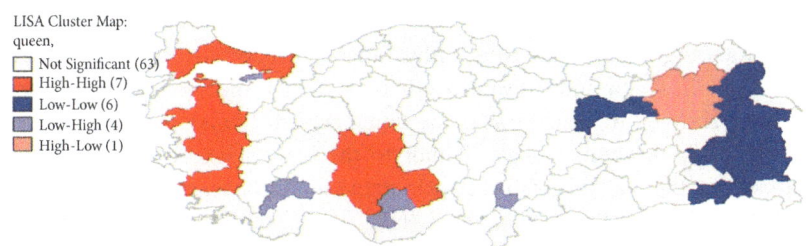

Fig. 6.4: LISA Map for Financial Development. Source: Authors' own construction

When the LISA map for the level of innovation (Fig. 6.2) is examined, 'regional development' differences, which are one of Turkey's main problems, are also observed at innovativeness level. It is understood that the provinces of Tekirdağ, İstanbul, Kocaeli, Yalova, Sakarya, Bilecik, Bursa, Balıkesir, Bolu, Kütahya, Eskişehir, Afyon, Manisa, Aydın, Antalya and Konya form HH type clusters in terms of innovation level. It can be thought that the provinces within the HH cluster have a positive effect on each other in spatial terms and benefit from the positive spillover effects that may arise from the spatial neighborhood. On the map, it is seen that the province of Niğde has an LH appearance although it is neighbor to the HH cluster. This situation shows that Niğde province is below the average in Turkey and cannot take advantage of the spillover effects from the cluster, even though it is in a spatial relationship with high-value provinces.

On the other hand, it is observed that Kars, Agri, Igdir, Van, Mus, Bitlis, Bingol, Siirt, Batman, Mardin, Sirnak and Hakkari provinces (all of which are located in the east part of Turkey) create aLL cluster. It is thought that the provinces located in the LL cluster affect each other negatively from a spatial point of view, and the negative spillover effects are dominant for this region. The liberation of a province located within the LL cluster is very difficult due to the negative propagation effects. However, in terms of innovation level, it can be observed that Erzurum and Diyarbakir, which are neighbor to a LL cluster, establishing the HL type of a spatial outlier and have values above Turkey average in terms of innovation level—despite the negative effect from the LL cluster. The fact that they are in spatial relations with low-value units can adversely affect these cities in time. If the spillover effect prevailing from these provinces is dominant, it is possible for some provinces within the LL cluster to join in an LH structure.

When the LISA map for variable indicating the share of creative class (Fig. 6.3) is examined, it is seen that Turkey has a clear disparity between the east and west. While Tekirdag, Istanbul, Kocaeli, Yalova, Sakarya, Bursa, Bilecik, Bolu, Eskisehir, Konya and Burdur provinces form a HH cluster in terms of creative class, Kars, Ağrı, Iğdır, Van, Muş, Bitlis, Bingöl, Siirt, Batman, Mardin, Sirnak, Hakkari, Diyarbakir and Adiyaman provinces exhibit the LL cluster structure. Creative class LISA map shows some differences in terms of spatial outliers. Despite their positions being neighbor to the high-value areas with creative class values, the share of creative class in Balıkesir, Afyon and Çankırı provinces (which are in spatial relationship with 'HH cluster') remains below the average in Turkey. Another spatial outlier is the HL structure exhibited by Gaziantep. Gaziantep, despite of the negative spillover effects from neighboring LL clusters, exhibits a creative class values above the average of Turkey.

When both maps are evaluated together, it is observed that the spatial clusters related to the creative class in the provinces with the LISA analysis corresponded to the spatial clusters of the level of innovation in accordance with the theoretical expectation. In addition, there are provinces in the HH cluster at the level of innovation, but are not in the HH cluster in terms of the creative class. The provinces of Balıkesir, Kütahya, Manisa, Afyon and Antalya are located in the HH cluster at the level of innovation but are not included in the cluster in terms of the share of the creative class. One of the reasons that lead to this situation is probably the factors

Tab. 6.3: LISA Clusters by Variables. Source: Authors' own construction.

	H-H	L-L	L-H	H-L
LNINNOVA	Tekirdağ, İstanbul, Kocaeli, Yalova, Sakarya, Bilecik, Bursa, Balıkesir, Bolu, Kütahya, Eskişehir, Afyon, Manisa, Aydın, Antalya, Konya.	Kars, Ağrı, Iğdır, Van, Muş, Bitlis, Bingöl, Batman, Siirt, Mardin, Şırnak, Hakkari.	Niğde.	Erzurum, Diyarbakır.
LNCREA	Tekirdağ, İstanbul, Kocaeli, Yalova, Sakarya, Bilecik, Bursa, Bolu, Eskişehir, Konya, Burdur.	Kars, Ağrı, Iğdır, Van, Muş, Bitlis, Bingöl, Batman, Siirt, Mardin, Şırnak, Hakkari. Diyarbakır.	Balıkesir, Afyon, Çankırı.	Gaziantep.
LNFINDEV	Tekirdağ, İstanbul, Kocaeli, Balıkesir, Manisa, Aydın, Konya.	Erzincan, Kars, Ağrı, Van, Şırnak, Bitlis.	Yalova, Osmaniye, Karaman, Burdur.	Erzurum.

affecting innovations in these provinces beside the creative class could be more influential. According to the creative class theory of Florida (2002, 2012), if the share of creative class workers in a region is high, the level of innovation is expected to be high. However, the creative class is not the only determinant of innovation.

Fig. 6.4 shows the LISA map of the level of financial development, which is another critical factor for innovation. HH cluster-forming provinces are Tekirdağ, İstanbul, Kocaeli, Balıkesir, Manisa, Aydın and Konya. Yalova, Burdur, Karaman and Osmaniye provinces, which exhibit a LH type spatial outliers, are low-value cities surrounded by high-value neighbors in terms of financial development. Kars, Ağrı, Van, Şırnak and Bitlis provinces, which are in LL clusters in other maps, are also observed in the LL cluster in terms of financial development. In addition to this, Erzincan province, which is not included in the LL cluster in other maps, is in the LL cluster in terms of financial development. Erzurum, as well as the level of innovation, has an HL appearance in financial development.

When considering the level of financial development and innovation together, it is observed that LL clusters generally overlap while HH clusters overlap only partially. It is noteworthy that Sakarya, Bilecik, Bursa, Bolu, Eskisehir, Kutahya, Afyon and Antalya provinces, which are included in the HH cluster in terms of innovation level, are not included in this cluster in financial development.

Tab. 6.3 presents the distribution of provinces included in the LISA clusters according to the variables.

4.3 Spatial econometric analysis findings

In the first stage, the econometric model established in the study was estimated according to the Ordinary Least Squares (OLS) method. Besides, Lagrange Multiplier (LM) determination tests were also applied to determine the existence of spatial effect in the model. Finally, the spatial econometric model for the type of spatial effect indicated by the tests was estimated.

According to the results of the OLS regression, R^2 (adjusted) value, which indicates the explanatory power of the independent variable, is quite high with a value 0.71. The coefficients of the independent variables in the model are positive and the probability values are statistically significant at 1 % significance level. In OLS regression analysis, the effects that may arise from the spatial dependence between the observations cannot be captured. Ignoring spatial effects can lead to a biased evaluation of the results. Therefore, LM determination tests were performed to see if there were spatial effects. The test that is statistically significant from the tests in the lower part of Tab. 6.4 is the LM lag test. This indicates the presence of an impact from the spatial lag of the dependent variable (Tab. 6.4).

Tab. 6.4: OLS Regression Results. Source: Authors' own calculations.

Dependent Variable: LNINNOVA	OLS Regression	
	Coefficients	Probability Values
Constant	10.76 (1.36)	0.00
LNCREA	2.55 (0.52)	0.00
LNFINDEV	0.68 (0.11)	0.00
LNINDSHR	0.88 (0.34)	0.01
R^2	0.73	
Corrected R^2	0.71	
Jarque-Bera	4.16	0.12
Breusch-Pagan	6.77	0.07
Moran's I	1.99	0.04
LM (lag)	6.88	0.00
Firm LM (lag)	4.57	0.03
LM (error)	2.31	0.12
Firm LM (error)	6.88	0.94

Note: In-parenthesis values in the coefficients column indicate standard deviations.

Tab. 6.5: Results of Spatial Lag Model. Source: Authors' own calculations.

Dependent Variable: LNINNOVA	Spatial Lag Model (Queen Weight Matrix)	
	Coefficients	Probability Values
Constant	8.31 (1.54)	0.00
WLNINNOVA (Rho)	0.21 (0.09)	0.01
LNCREA	1.86 (0.53)	0.00
LNFINDEV	0.66 (0.10)	0.00
LNINDSHR	0.80 (0.32)	0.01
R^2	0.75	
LR Test	5.99	0.01

Note: In-parenthesis values in the coefficients column indicate standard deviations.

As the LM lag test shows, the Spatial Lag Model is estimated in the study. According to the results in Tab. 6.5, the sign of the Rho coefficient is positive and the probability value is significant at 1 % level of significant. This shows that one of the factors affecting the innovation level of a province is the level of innovation of the neighboring provinces. Since the spatial effects are taken into account with the Spatial Lag Model, it is healthier to interpret the results. Accordingly, the intensity of the creative class in a province, the level of financial development of the province and the weight of the industrial sector have positive effects on the innovation level of the province in question, and it is statistically significant. In addition, 0.21 of the Rho coefficient indicates that a 1 % increase in the level of innovation in the neighboring provinces could lead to a 0.47 % increase in the innovation level of the province in question.

5 Conclusion and Recommendations

In this chapter, as compared with other chapters in the book, the effects of creative class workers, financial development and spatial proximity on the innovation performance of the provinces in Turkey are investigated by using Spatial Statistics and Spatial Econometric Techniques. When the literature on the subject is examined, it is understood that innovative activities are realized more in the places where the creative class employees are concentrated, the industrial sectors are developed and the financial development is high. While creative class workers meet the high-quality human capital needs that the region needs for innovation, the financial development of the region facilitates innovative individuals and firms to engage in innovative activities or increase their physical capital access to increase their existing innovative activities.

In the geography that examined for the variables included in the study, spatial autocorrelation has been shown to exhibit strong regional segregation. LISA analysis was used to determine the regions where spatial autocorrelation forms significant spatial clusters locally. According to the results of the analysis, it is understood that for all three variables, Tekirdağ, Istanbul, Kocaeli, Balıkesir, Manisa, Aydın and Konya provinces formed HH type clusters. It has been determined that the provinces within this cluster affect each other positively from the spatial perspective and they benefit from the positive spillover effects that may arise from the spatial neighborhood. It is observed that Kars, Ağrı, Van, Bitlis and Şırnak provinces formed LL clusters for each of the three variables, that they affect each other negatively in a spatial manner, and that the negative spillover effects are dominant for this region.

According to the spatial econometric analysis of the findings in Turkey (provincial level), the presence of creative class workers, financial development and the weight of the industrial sector positively affect innovation. Increasing the financial development, attracting the creative class employees, contributes to the acceleration of the economic activities in the province and paving the way for innovation processes. The results seem to be in line with the literature on creative class theory and innovation studies.

The spatial analysis conducted within the framework of the established model also shows that there is a spatial dependence on the level of innovation in the provinces. In other words, one of the factors affecting the innovation level of a province is the level of innovation of the neighboring provinces. In this aspect, the study can shed light to the decision makers about evaluation of existing innovation strategies in provinces of Turkey and is expected to contribute to the literature on the subject.

References

Amin, A. & Cohendet, P. (2004). *Architectures of Knowledge: Firms, Capabilities, and Communities*. Oxford: Oxford University Press.

Anselin, L. (1995). Local Indicator of Spatial Association–LISA. *Geographical Analysis*, 27(2), 93–115.

Anselin, L. (1988). *Spatial Econometrics: Methods and Models*. Hollanda: Kluwer Academic Publishers.

Anselin, L., Cohen, J., Cook, D., Gorr, W. & Tita, G. (2000). Spatial Analyses of Crim. *Criminal Justice*, 4(2), 213–262.

Anselin, L. (2003). Spatial Externalities. *International Regional Science Review*, 26(2), 147–152.

Asheim, B. T. & Gertler, M. S. (2005). The Geography of Innovation: Regional Innovation Systems. In *The Oxford Handbook of Innovation*, edited by J. Fagerberg, D. C. Mowery and R. R. Nelson, 291–317. New York: Oxford University Press.

Anselin, L., Syabri, I. & Kho, Y. (2006). Geo Da: An Introduction to Spatial Data Analysis. *Geographical Analysis*, 38(1), 5–22.

Autio, E., Kenney, M., Mustar, P., Siegel, D. & Wright, M. (2014). Entrepreneurial Innovation: The Importance of Context. *Research Policy*, 43(7), 1097–1108.

Bal, H., İşcan, E., Serin, D., & Kara, D. (2017). Finansal Büyüme ve İnovasyon İlişkisi: OECD. *International Conference on Eurasian Economies (İstanbul 1012 July 2017)*, 546–549. Retrieved from https://avekon.org/papers/1942.pdf (11.08.2019)

Bathelt, H., Malmberg, A., & Maskell, P. (2004). Clusters and Knowledge: Local Buzz, Global Pipelines and the Process of Knowledge Creation. *Progress in Human Geography*, 28(1), 31–56.

Bradford, N. (2004). *Creative Cities Structured Policy Dialogue Backgrounder*. Ottawa: Canadian Policy Research Networks. Retrieved from http://www.urbancenter.utoronto.ca/pdfs/elibrary/CPRN_Creative-Cities-Backgr.pdf (12.08.2019)

Camagni, R. & Capello, R.(2013). Regional Competitiveness and Territorial Capital: A Conceptual Approach and Empirical Evidence from the European Union. *Regional Studies*, 47(9), 1383–1402.

Chu, A. C., Cozzi, G., Fan, H., Pan, S. & Zhang, M. (2016). Do Stronger Patents Stimulate or Stifle Innovation? The Crucial Role of Financial Development. *Munich Personal RePEc Archive*, 1–30. Retrieved from https://mpra.ub.uni-muenchen.de/73630/1/MPRA_paper_73630.pdf (12.08.2019)

Cooke, P., Asheim, B., Boschma, R., Martin, R., Schwartz, D. & Tödtling, F., eds. (2011). *Handbook of Regional Innovation and Growth*. Cheltenham, UK: Edward Elgar.

Dall'erba, S. (2005). Distribution of Regional Income and Regional Funds in Europe 1989–1999: An Exploratory Spatial Data Analysis. *The Annals of Regional Science*, 39(1), 121–148.

Doğrul, H. G., Çelikkol, M. M., & Murat, N. (2016). İllerin Ekonomik Gelişmişliği Üzerinde Yaratıcı Sınıfın Etkisi: Türkiye Örneği. *Business & Economics Research Journal*, 7(4), ss.79–95.

Florida, R. (2002). *The Rise of the Creative Class*. New York: Basic Books.

Florida, R. (2011). *Yaratıcı Sınıf Adres Değiştiriyor* (translated by Zeynep Kökkaya Chalar). İstanbul: Kapital Medya.

Florida, R. (2012). *The Rise of the Creative Class*, revised paperback edition. New York: Basic Books.

Girma, S., Gong, Y., & Görg, H. (2008). Foreign Direct Investment, Access to Finance, and Innovation Activity in Chinese Enterprises. *The World Bank Economic Review*, 22(2), 367-382.

Gül, E. & Çevik, B. (2015). *2013 Verileriyle Türkiye'de İllerin Gelişmişlik Düzeyi Araştırması*, Türkiye İş Bankası İktisadi Araştırmalar Bölümü, İstanbul.

Hanley, A., Liu, W., & Vaona, A. (2011). Financial Development and Innovation in China: Evidence from the Provincial Data. Kiel Working Paper: 1673, 33.

Koçak, E. (2018). Finansal Gelişme ve Yenilik (İnovasyon): Türkiye Üzerine Ampirik Bir Araştırma. *Kapadokya Akademik Bakış*, 2(1), 12-28.

Krakowiak-Bal, A. & Ziemiańczyk, U. (2017). Factors Influencing the Level of Regional Innovation-Qualitative Comparative Analysis. *Barometr Regionalny*, 4(50), 7-14.

Laeven, L., Levine, R. & Michalopoulos, S. (2015). Financial Innovation and Endogenous Growth. *Journal of Financial Intermediation*, 24(1), 1-24.

Le Gallo, J. & Ertur, C. (2003). Exploratory Spatial Data Analysis of the Distribution of Regional Per Capita GDP in Europe 1980-1995. *Papers in Regional Science*, 82(2), 175-201.

Lee, J. & Wong, D. (2001). *Statistical Analysis with ArcView GIS*. New York: John Wiley & Sons.

Lucas, R. E. (1988). On the Mechanics of Economic Development. *Journal of Monetary Economics*, 22, 3-42.

Mas-Verdu, F., Ortiz-Miranda, D., & Garcia-Alvarez-Coque, J. M. (2016). Examining Organizational Innovations in Dierent Regional Settings. *Journal of Business Research*, 69(11), 5324-5329.

Maskus, K. E., Neumann, R., & Seidel, T. (2012). How National and International Financial Development Affect Industrial R&D. *European Economic Review*, 56(1), 72-83.

Meierrieks, D. (2014). Financial Development and Innovation: Is There Evidence of a Schumpeterian Finance Innovation Nexus? *Annals of Economics and Finance*, 15(2), 343-363.

Rodriguez-Pose, A. & Villarreal Peralta, E. M. (2015). Innovation and Regional Growth in Mexico: 20002010. *Growth and Change*, 46(2), 172-195.

Roig-Tierno, N., Alcazar, J., & Ribeiro-Navarrete, S. (2015). Use of Infrastructures to Support Innovative Entrepreneurship and Business Growth. *Journal of Business Research*, 68(11), 2290–2294.

Storper, M. & Venables, A. J. (2004). Buzz: Face-to-face Contact and the Urban Economy. *Journal of Economic Geography*, 4, 351–370.

Tee, L., Low, S., Kew, S. & Ghazali, N. (2014). Financial Development and Innovation Activity: Evidence from Selected East Asian Countries. *Prague Economic Papers*, 2(2), 162–180.

Turkish Statistical Institute – TSI, Retrieved from http://www.tuik.gov.tr (21.01.2019).

Uluslararası Rekabet Araştırmaları Kurumu—URAK (2016). İllerarası Rekabetçilik Endeksi 20132014. İstanbul. Retrieved from www.urak.org (21.01.2019).

Zeren, F. (2011). Mekânsal Etkileşim Analizi. *Ekonometri ve İstatistik*, 12, 18–39.

Zhao, W. (2016). Financial Development and Regional Innovation Output Growth: Based on Empirical Analysis of Provincial Panel Data in China. *Modern Economy*, 7, 10–19.

SECTION 2: Labor Markets and Unemployment

Umut Akduğan and Seyhun Doğan

7. Youth Unemployment: An Empirical Analysis

Abstract The implementation of policies towards the solution of "youth unemployment", which is seen as one of the unemployment types and has become a structural problem in most countries, stands out in the fight against unemployment in general. Especially with the impact of economic crises, not only in developing countries, but also in many developed countries, the problem of youth unemployment has reached significant levels. On the other hand, weak and irregular economic recovery causes the young labor market to worsen. In most of the countries, the most affected by population from economic problems is the young labor force. However, while in some countries, the increase in the youth unemployment rate has been limited, in some other countries, it has increased rapidly. Due to the fact that the factors affecting the youth labor market differ from country to country, it is very important to determine the factors affecting youth unemployment. In this chapter, the problem of youth unemployment was discussed, and the effects of some macroeconomic variables on youth unemployment were analyzed using the panel vector autoregression (VAR) approach in the context of a sample of selected countries.

Keywords: Youth unemployment, Macroeconomic variables, Panel VAR, Impulse-response

1 Introduction

The implementation of policies to solve youth unemployment, which has become a structural problem in many countries, generally becomes prominent in the fight against unemployment. At this point, it is essential to examine the dynamics of the youth labor market. The International Labor Organization (ILO) uses the 15–24 age range, which the United Nations accept, in the definition of "young individuals". Hence, the youth unemployment rate refers to the proportion of people who are unemployed and seeking employment in the age group of 15–24. Also, depending on the demographic structure of countries, different age categories are used in some particular cases. For example, ILO's school-to-work transition questionnaires include the definition of extended youth between the ages of 15–29 (ILO, 2018: 6).

With the impact of global economic crises, the youth unemployment problem has also reached serious levels in developing countries as well as in developed countries. For example, considering the 2008 Global Crisis, the economic downturn and other negativities that emerged immediately after this crisis had a significant impact on the youth labor market. Due to the crisis, labor demand decreased and unemployment increased, so the crisis in financial markets was reflected in the real economy. The weakness and irregularity of the economic recovery led to a worsening of the youth labor market in the post-crisis period and the majority of the young people who were not in education and training remained unemployed. On the other hand, the debt crisis, which had an impact on the eurozone in the period 2010–2014, and the subsequent savings measures had serious effects on the youth labor markets in Europe (Bruno, Marelli and Signorelli, 2016: 249). Therefore, the developments show that the young labor force is more sensitive to changes in economic conditions than adults. However, the increase in youth unemployment caused by macroeconomic shocks was limited in some countries, whereas it was quite rapid in some countries where the economic instability lasts long time. The differences in the factors affecting the youth labor market play a role in this situation.

In this study, firstly the youth unemployment problem and its causes are discussed. Then, the factors affecting youth unemployment are summarized in the literature survey. Besides, the results of the empirical studies in the literature on macroeconomic determinants of youth unemployment are included. Later, an empirical analysis is made on the determinants of youth unemployment. The main purpose of this study is to analyze the macroeconomic determinants of youth unemployment and to make policy alternatives by generalizing the results under certain assumptions. Unlike the studies in the literature, considering the period of 1990–2017 examined, in this study, macroeconomic determinants of youth unemployment are investigated from a broad perspective. In this respect, the effects of macroeconomic variables selected by the panel VAR approach on the youth unemployment were analyzed and the findings obtained were evaluated by using the panel data set of twenty OECD countries and the 1990–2017 period.

2 Youth Unemployment and Its Causes

Along with the integration of young people into the labor market, the creation of new employment areas for young people and the reduction of youth unemployment have been an important objective of economic administrations in the world, with the aim of approaching full employment. Against the economic crises and economic instability that have affected many countries in the recent past, it has been apparent that the concept of youth employment is quite fragile. It can be seen that the youth unemployment rate has risen dramatically due to these negativities (ILO, 2010).

There are many studies on the factors affecting youth unemployment in the literature. Some common conclusions have been reached overall, and the determinants of youth unemployment have been studied in specific groups. In this context, the first group includes demographic factors affecting youth unemployment. In many studies, empirical evidence has been reached regarding the effect of the changes in the age composition of the population and the labor force on the position of youth in the labor market and the youth unemployment rate (Gomez-Salvador and Leiner-Killinger, 2008: 18). These studies test the hypothesis that increasing the ratio between the young population and the adult population will negatively affect the expectations of the young people in the labor market and increase the youth unemployment rate unless youth labor force and adult workforce are perfect substitutes. This situation is expressed as "cohort effect" and explains the effect of "cohort size" on the youth unemployment, which is calculated as [(population 15–24)/(population 25–54)]. When the literature on this issue is summarized, the increase in the proportion of youth labor force in the total population in many countries has had a negative impact on employment and wages among young people (Korenmark and Neumark, 1997: 4–8). Because in the case that the increase of the young population rate is greater than the increase of total population rate, the young labor supply will increase rapidly. In the labor market, the increase in youth labor supply more than the demand for youth labor force will create an increase in the youth unemployment rate. On the other hand, population density and migration flows are among the other demographic factors affecting youth unemployment and overall unemployment.

Another factor affecting youth unemployment is the reduction of the elasticity of the labor market as a result of the legal regulations and policies implemented towards the labor market. Variables such as labor taxes, unemployment benefits, collective labor agreements, temporary employment contracts, unionization rate, minimum wages, and active labor market programmes (ALMP), which concern the level of unemployment, are the institutional variables affecting youth unemployment. In addition, several results have been reached in many studies in the literature that employment protection legislations (EPL) affect youth unemployment. However, in general, by the results of empirical studies, it was concluded that employment protection legislations and lay-off legislation affect the working cycle and the duration of unemployment, which is more important for the youth labor force than the adult workforce (Brada, Marelli and Signorelli, 2014: 559).

One of the main variables related to youth unemployment is the school-to-work transition (STWT) process (Quintini, Martin and Martin, 2007). The negativities that arise during this process or the application disruptions resulting from the system increase the lack of experience of the young people and decrease the employment rate. At this point, Brunello, Garibaldi, Wasmer and Bassanini (2007) stated that the dual education system, in which apprenticeship or internship training plays an important role besides formal education, can be an effective way to reduce youth unemployment. Another reason for the increase in youth unemployment is the mismatch between the skills required by the labor market and the information acquired through education. Due to this discrepancy, the number of qualified staff in the labor market is low and thus, the employment rate decreases (Brada et al., 2014: 560). In this context, education is a very important variable and its impacts on the youth labor market are more evident, especially in periods of economic stagnation.

Finally, the general economic situation and some macroeconomic variables also affect youth unemployment. It is known that the inclination of the youth labor force to be negatively affected by economic fluctuations is higher than that of adults (Scarpetta, Sonnet and Manfredi, 2010; Choudry, Marelli and Signorelli, 2012; O'Higgins, 2012; Bernal-Verdugo, Furceri and Guillaume, 2012). The decrease in labor demand due to macroeconomic shocks and also the deterioration in the labor market adversely affect all segments of society. However, it is stated that the risk of unemployed for young individuals is generally still higher than that of adults. There are many factors that cause this situation: the inadequacy of education, the inexperience of the young workforce and the lower quality of some of them, inadequate efforts in job search, weak employment contracts and the situations where the characteristics demanded by employers do not correspond to the characteristics of young people seeking for jobs. That also negatively affects the employment of young people (Tomić, 2016: 7; Brada et al., 2014: 558). Bell and Blanchflower (2011) concluded that, after the 2008 Global Crisis, the most affected by the economic conditions was the youth labor force with the lowest level of education. The high rate of frictional employment and part-time employment within the youth labor force also increases youth unemployment alongside the total unemployment rate in the long run (O'Higgins, 2015: 1). In addition, the concentration of youth labor force in sectors sensitive to economic fluctuations, such as the construction sector (Scarpetta et al., 2010: 14), and the dismissal practices of firms which involves dismissing the most recently recruited young employees with the least experience in times of crisis (Caroleo, Ciociano and Destefanis, 2017: 43) also affects the youth unemployment rate. Additionally, the long-term effects of macroeconomic shocks on labor demand, they also have effects on labor market-related institutions and policies (Marelli, Choudhry and Signorelli, 2013: 70). Therefore, the impact of economic conditions on youth unemployment is manifested in a variety of ways.

3 Literature Research

In many empirical studies investigating the macroeconomic determinants of youth unemployment, economic growth or real GDP and inflation rate variables are used as independent variables or as control variables. In addition, there are also studies using real interest rate, labor productivity, domestic savings, domestic investment rate, foreign direct investment, and trade openness variables.

As a result of the analysis conducted by Cvecic and Sokolic (2018) using the 2005–2014 period data of 27 European Union countries; they found that the effect of GDP and inflation on the youth unemployment rate, Not in Education, Employment, or Training (NEET) and the total unemployment rate were statistically significant and negative. Tomić's (2016) analysis using the panel data for the 28 member countries of the EU in the period of 2002–2014 found that the effect of economic growth on the youth unemployment rate was negative. Caporale and Gil-Alana (2014) examined the macroeconomic determinants of youth unemployment for the period of 1980–2005 and 15 EU countries in the context of GDP and inflation variables. The results show that youth unemployment has a significant and negative relationship with two basic macroeconomic variables in the long run. Bayrak and Tatlı (2018) examined the main macroeconomic variables affecting youth unemployment for 31 OECD countries by running panel data models for the 2000–2015 period data. The results show that economic growth, inflation and domestic saving rate have adverse effects on youth unemployment; while labor productivity positively affects youth unemployment. Pastore and Giuliani (2015) used the panel data for the period of 1970–2013 for 21 selected countries and found that the youth population rate was positive for youth unemployment. They also found that economic growth reduced youth unemployment. Gomez-Salvador and Leiner-Killinger (2008), in the analysis of the panel data of the period of 1985–2004, consisting of 20 units selected from the EU and OECD countries, detected that the youth population rate had a positive effect on youth unemployment. Also, another finding is that economic growth, which represents economic conditions, negatively affects youth unemployment, that is, when the economic situation worsens, the youth unemployment rate increases. Jimeno and Rodriguez-Palenzuela (2002) examined the effects of young population, labor market institutions and macroeconomic shocks on youth unemployment in the 1960–1996 period and 19 OECD member countries with panel data models. In addition, they found that the young population rate is an important variable affecting youth unemployment and that the effects of institutional factors and macroeconomic variables on different age groups differ. Marelli et al. (2013) investigated the impact of institutions and policies on youth unemployment and total unemployment for the 27 OECD countries with a high-income level for the period of 1980-2009. However, some macroeconomic variables were included in the model as a control variable. Resultantly, they have found that economic growth and inflation have a negative impact on both youth unemployment and total unemployment.

Some macroeconomic factors have been used as explanatory variables in studies that examine the effects of financial crises on youth unemployment. For example, Bruno, Choudhry, Marelli and Signorelli (2017) examined the effects of financial crises on youth unemployment for the 27 OECD countries in the period 1981-2009. The results show that economic growth and inflation have a negative impact on youth unemployment in the long run. In a study of Choudhry et al. (2012), a sample of 75 units of EU countries and OECD countries for the period 1980-2005 examined the determinants of youth unemployment based on the economic crisis. They found that the effects of economic growth, inflation, foreign direct investment and trade openness on youth unemployment rate were statistically significant and negative.

Anyanwu (2013) investigated the macroeconomic variables affecting youth employment for 48 African countries in the period 1991-2009. Results show that there is a significant and negative relationship between youth employment and domestic investment rate, public expenditures, inflation and per capita income. Moreover, they have found that foreign direct investment and trade openness variables affect youth employment as indicators of globalization. Wajid and Kalim (2013) found a negative correlation between youth unemployment and economic growth in the analysis of the Pakistani economy for the period 1973-2010 and found that inflation increased unemployment and the effect of trade openness was statistically insignificant. In their studies, Gunaydin and Cetin (2015) analyzed the main macroeconomic determinants of youth unemployment with data belonging to Turkey's economy for the 1988-2013 period. Their results show that real income per capita, trade openness and foreign direct investment have adverse effects on youth unemployment; and in the long run, shows that there is a causal relationship between inflation and youth unemployment.

4 An Empirical Analysis: Macroeconomic Determinants of Youth Unemployment

In this study, panel VAR approach was used to investigate the determinants of youth unemployment within the scope of the selected variables by taking into account the literature. Accordingly, the VAR model was estimated and the results of the impact-response functions were evaluated. In the analysis of macroeconomic variables affecting youth unemployment, the focus has been on the impulse-response functions that show the response of a variable to the shock in another variable. Because of the impulse-response functions, by keeping the effects of other variables constant, the effect of the shock in a single time period and a single variable on the respective variable can be determined.

On the other hand, in the VAR approach, all variables are considered endogenously and all relationships between them are estimated. However, the VAR approach can also be used to isolate the effect of a particular variable on other key factors. In this direction, considering the purpose of the study, only the results and comments of the models, where the youth unemployment rate is dependent variable, will be included.

4.1 Data and variables

In the empirical analysis part of the study, 1990–2017 period panel data set is used for the 20 member countries of the OECD (Australia, Austria, Canada, Denmark, France, Germany, Greece, Ireland, Italy, Republic of Korea, Mexico, Netherlands, Norway, Portugal, Spain, Sweden, Switzerland, Turkey, United Kingdom, United States). In the context of selected countries, macroeconomic variables were chosen as the determinants of youth unemployment by taking into consideration the relevant literature. In this sense, the dependent variable is youth unemployment rate (YUP). The youth unemployment rate variable shows the proportion of unemployed in the 15–24 age group expressed as a percentage of the youth labor force. This definition of youth unemployment is the most common indicator of youth unemployment (see OECD, Eurostat). The explanatory variables chosen as the factors affecting youth unemployment are economic growth (GDP), gross domestic savings (GDS), inflation (INF), foreign direct investment (FDI), trade openness (TRADE) and percentage change of young population (POP).

The real GDP growth rate (GDP) has been selected in terms of showing the general situation of the economy. Together with the increase in production capacity, high GDP growth rates are expected to reduce the total unemployment rate along with youth unemployment (see Section 3). Besides, GDP is very important in terms of the fact that youth unemployment is more sensitive to fluctuations in economic growth (Bruno, Choudhry, Marelli and Signorelli, 2014; Bruno et al., 2017).

Among the macroeconomic variables that are important in explaining the total unemployment, besides the GDP or output gap variables, inflation rate, domestic savings, trade openness, real interest rate, productivity increase, and foreign direct investments are the variables. The impact of some of these variables on employment is more pronounced. For example, it is generally accepted that savings are financing investments, and thus, they support economic growth and employment. According to most economic theories, investments are not independent of savings and savings are a key determinant of economic growth. Therefore, the GDS variable, which is measured as a percentage of GDP, is included in the model. On the other hand, the negative impact of inflation on unemployment may arise if the actual price level exceeds the expected prices. In such a case, in the wage negotiation process, real wages can be determined at a level lower than expected and as a result, employment may be increased (Marelli et al., 2013: 69). Therefore, it is expected that the inflation (INF) variable will have a negative impact on youth unemployment along with total unemployment in terms of affecting the wage determination policy and costs of firms.

In open economies, the effects of some of the advantages of globalization on the labor market are expected to be positive (Anyanwu, 2013: 116). In order to determine the effect of such factors on youth unemployment, foreign direct investment (FDI) and trade openness (TRADE) variables, which are measured as a percentage of GDP, are included in the model. Foreign direct investment and international trade have an increasing effect on employment in terms of creating new employment areas. For this reason, one of the reasons for policymakers to attract more foreign direct investment in both developed and developing countries is to create new jobs in their economies (Javorcik, 2013: 1). TRADE variable refers to the "terms of trade" indicator, which is measured as a percentage of imports and exports in GDP. For example, in economies where export-oriented manufacturing sectors are labor intensive, foreign trade is expected to reduce unemployment. On the other hand, it can be expected that a change in the relative prices of imports would increase the costs and overall wage pressures in industries where imported inputs are used more in production. This may lead to an increase in unemployment (Bassanini and Duval, 2006: 11).

Finally, in addition to macroeconomic factors, the "percentage change of young population (POP)" variable is also included in the model to examine the effect of population structure on youth unemployment. By a more explicit description, the POP variable shows the rate of change (%) in the number of individuals aged 15–24 compared to the previous year. When the studies in the literature are examined, in general, the increase in the youth labor force is seen to increase youth unemployment (Pastore and Giuliani, 2015; Gomez-Salvador and Leiner-Killinger, 2008; Jimeno and Rodriguez-Palenzuela, 2002; Korenmark and Neumark, 1997).

4.2 Empirical methodology: Panel data VAR

The panel VAR method combines the traditional VAR approach that internally incorporates all the variables in the system with a panel data approach that can be used in heterogeneous observations. The general format of VAR is as follows (Grossmann, Love and Orlov, 2014: 11):

$$z_{it} = \Gamma_0 + \Gamma_1 z_{it-1} + f_i + d_t + e_{it}$$

In this model, the z_{it} vector contains YUP, GDP, GDS, INF, FDI, TRADE and POP. Since the autoregressive structure of the model permits the lagged values of all internal variables to enter the model, the dynamic effect of macroeconomic factors on youth unemployment can be estimated. The term f_i, which allows a fixed effect to be included in the model, contains all factors that cannot be observed at the country level and do not change over time. Another advantage of the panel VAR model is that it is possible to include a macroeconomic shock that can affect all countries by means of the term d_t which shows common time effects (Grossmann et al., 2014: 12).

The difference of the panel VAR models from the traditional VAR models used for time series is the unit effects in the model. In the implementation of the panel VAR model, the assumption that all parameters are homogeneous, that is, no unit effect is a very limited assumption and generally cannot be achieved. One way to overcome this constraint on the parameters is to allow individual heterogeneity in the variables by adding the fixed effects indicated by f_i to the model (Love and Zicchino, 2006: 195). Therefore, since the panel VAR model is a dynamic model in terms of structure, estimators that allow unit variables to be correlated with unit effects and independent variables are preferred in estimating the model. Anderson and Hsiao (1981) suggest that the lagged values of the dependent variables should be used as an instrumental variable after the first difference transformation. However, the correlation between the lagged dependent variable and error terms may cause the OLS estimator to be inconsistent (Ciarreta and Zarraga, 2010: 3794). Arellano and Bond (1991) state that the Generalized Method of Moments (GMM) estimator using the adapted instrumental variable matrix is more effective because of the possibility of variance and autocorrelation problems that change in the error terms of the first difference model. On the other hand, the lagged structure of the model and the use of the first difference transformation for the GMM estimation will result in data loss. At this point, Arellano and Bover (1995) recommends "orthogonal deviation", where the difference of the mean of all possible future values of the variable is taken instead of the first differences. Because of this transformation, lagged regressors can be used as a variable, and coefficients can be estimated by GMM (Love and Zicchino, 2006: 195). The VAR model, which was created in this study, was estimated with GMM using forward orthogonal deviations. Then, the orthogonalized impulse-response functions (Orthogonalized IRF) were generated using the estimated VAR model.

Impulse-response functions explain the response of a variable to changes occurring in another variable in the model, while keeping the effect of all other shocks equals to zero. Besides, since the variance-covariance matrix of the error terms is not generally diagonal, it is necessary to separate the residues orthogonal to determine the shocks associated with one of the variables in the system (Love and Zicchino, 2006: 194). In this context, specifically, the focus is on the orthogonalized IRF, which reacts to the orthogonal shock in the other explanatory variables (i.e., GDP, GDS, INF, FDI, TRADE, POP). In this way, the effect of a single shock on the YUP variable was determined by orthogonalizing the response while keeping all the other shocks constant.

4.3 Results

In the analysis of the determinants of youth unemployment, first of all, the stationarity of the variables in the model was tested by performing the first and second generation panel unit root tests. All of the results are shown in Tab. 7.1. Levin, Lin and Chu (2002) and Im, Pesaran and Shin (2003) tests, which are the first generation panel unit root tests were applied; and for all variables, null hypothesis in the form of "panels contain unit roots" was rejected at 1 % significance level. Therefore, according to the results of both tests, all variables were found stationary at the level. On the other hand, it is stated that the first generation unit root tests do not take into account the correlation between the units and if the series is cross-sectional, the null hypothesis can be rejected in a wrong way (O'Connel, 1998: 1–19). Therefore, second-generation panel unit root tests that take inter-unit correlation into account are recommended. In this context, for all the variables, whether or not there is an inter-unit correlation was examined by Pesaran (2004) CD test, the null hypothesis expressing cross-sectional independence for each variable was rejected at the significance level of 1 % (see Tab. 7.1). In this respect, the stationary of the series of variables was examined by Pesaran (2003) panel unit root test (CADF), and Pesaran (2007) panel unit root test (CIPS). The results show that the level of GDS and TRADE variables is unit rooted, and the other variables are stationary. The first differences of GDS and TRADE variables are stationary (see Tab. 7.1). Considering that there is a

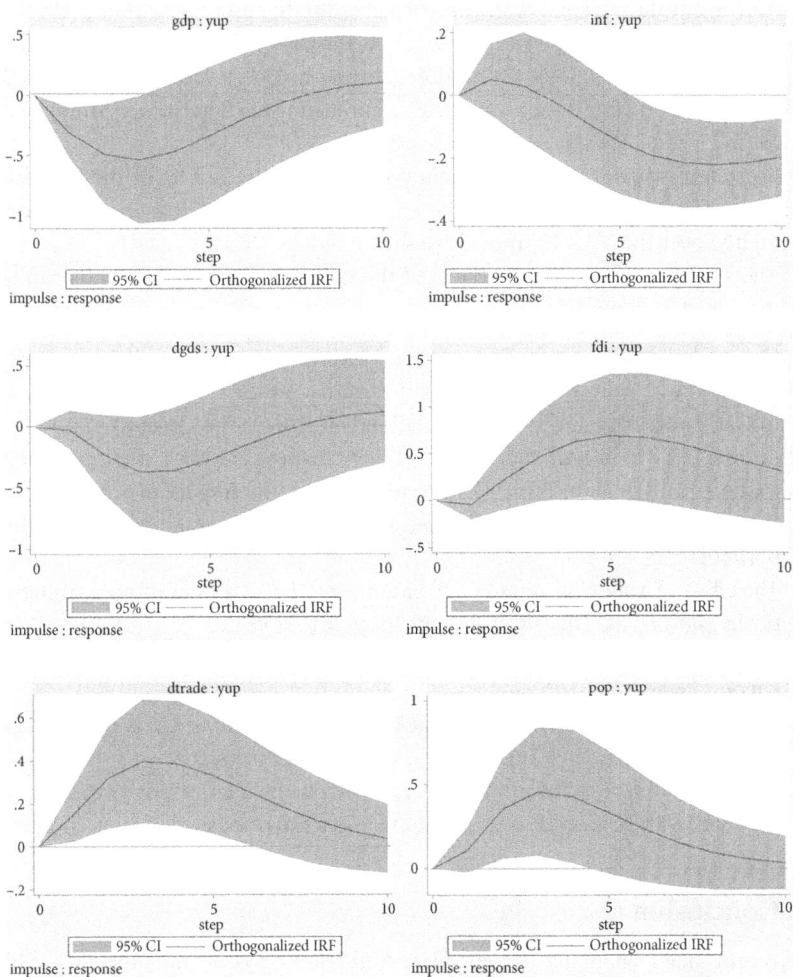

Fig. 7.1: Impulse-Response Functions. Source: Authors' own construction

correlation between the units, the results of CADF and CIPS tests will be taken into consideration when creating the VAR model.

In the continuation of the analysis, the lag length of the VAR model, which includes the stationary forms of the variables, was determined as two according to the MAIC information criterion and the VAR (2) model was estimated. The estimation results of the model are shown in Tab. 7.2. Furthermore, it can be said that all eigenvalues related to the model are within the unit circle, thus providing the stationarity conditions of the model. At this stage, firstly Granger Causality Test was applied over VAR (2) model and the results are presented in Tab. 7.3. According to the findings, a causality relationship was determined from GDP, GDS, INF, FDI, TRADE, and POP variables to YUP.

As stated earlier, the chapter mainly focused on the results of the impulse-response functions. Accordingly, the graphs of the impulse-response functions estimated over the VAR (2) model are shown in Fig. 7.1.

The statistically significant results of the impulse-response functions can be summarized as follows:

- Generally, it is seen that the reaction of youth unemployment to a change in economic growth and inflation is negative, and its reaction to a change in trade openness and youth population is positive. The effect of a shock on domestic savings and foreign direct investment on youth unemployment is statistically insignificant. On the other hand, the impact of inflation and the change in the youth population on youth unemployment appears with a particular lag.
- The effect of a shock given to GDP variable on the YUP is negative and statistically significant. This effect disappears after four years.
- The effect of a shock given to INF variable on the YUP is negative, and this effect is lagged.
- The effect of a shock given to the TRADE variable on the YUP is positive and statistically significant. This effect disappears after six years.
- The effect of a shock given to POP variable on the YUP is positive, and this effect is delayed. Later, it seems that the impact disappears within five years.

5 Conclusion

In recent years, in many developed and developing countries, together with EU countries, youth unemployment has become one of the main economic problems. While many measures are being taken to combat with this problem, the results of the implemented policies are open to discussion. Because of the problems in the youth labor market vary according to the internal dynamics of each country. Nevertheless, in studies conducted on the subject, differences between countries are often disregarded, and the causes of youth unemployment are addressed holistically.

Considering the primary empirical results of the study, economic growth, inflation, and trade openness, as well as the percentage change of young population, are the most important determinants of youth unemployment. The relationship between economic growth and inflation and the rate of youth unemployment is in the opposite direction, as expected and consistent with the literature. In particular, the link between GDP and youth unemployment rate confirms the sensitivity of youth unemployment to general macroeconomic conditions (Marelli et al., 2013; Caporale and Gil-Alana, 2014). In addition to macroeconomic variables, the education system, labor market policies, institutions, and regulations affect the total unemployment rates along with youth unemployment. However, it is supported by the studies that employment performance is mostly dependent on output level and economic growth, especially in economies with low-income levels.

Another finding obtained from the analysis is the positive effect of the young population's change rate on youth unemployment. That is, the increase in young population boosts the youth unemployment rate. This is an expected result and is consistent with the literature, as the young population growth increases the supply of young labor and has a negative impact on wages (see Section 3).

One interesting result is the positive effect of trade openness on youth unemployment. According to the findings, the increase in foreign trade volume increases youth unemployment rates. This finding does not support the view that increasing exports and imports towards foreign markets may increase employment areas due to imports. In this context, it can be said that the change in trade openness is mainly due to imports in some of the countries examined. In addition, it can be concluded that export-oriented sectors perform more capital-intensive production rather than labor-intensive production.

The conclusion that can be drawn from the findings of the study is clear: Sustainable, stable economic growth should be realized, and an effective incentive policy should be implemented in this direction because the young population growth has many positive contributions that require neglecting the negative impact on youth unemployment. Therefore, the most important goal that needs to be realized to employ the growing young population is to increase the production capacity and output level. The slowdown in economic growth as a result of stabilization programs implemented (especially in the face of economic crises and involving solid measures) is likely to have serious adverse effects on employment. Additionally, it is essential to support economic measures in the reduction of total unemployment, especially in the youth unemployment rate, by structural reforms related to the labor market. The risks involved in the realization of structural reforms should be managed effectively.

References

Anderson, T. W. & Hsiao, C. (1981). Estimation of dynamic models with error components. *Journal of the American Statistical Association*, 76(375), 598–606.

Anyanwu, J. C. (2013). Characteristics and macroeconomic determinants of youth employment in Africa. *African Development Review*, 25(2), 107–129.

Arellano, M. & Bond, S. (1991). Some tests of specification for panel data: Monte Carlo evidence and an application to employment equations. *Review of Economic Studies*, 58(2), 277–297.

Arellano, M. & Bover, O. (1995). Another look at the instrumental variable estimation of error-components models. *Journal of Econometrics*, 68(1), 29–51.

Bassanini, A. & Duval, R. (2006). The determinants of unemployment across OECD countries: Reassessing the role of policies and institutions. *OECD Economic Studies*, 42, 7–86.

Bayrak, R. & Tatlı, H. (2018). The determinants of youth unemployment: A panel data analysis of OECD countries. *The European Journal of Comparative Economics*, 15(2), 231–248.

Bell, D. N. F. & Blanchflower, D. G. (2011). Young people and the Great Recession. *Oxford Review of Economic Policy*, 27(2), 241–267.

Bernal-Verdugo, L. E., Furceri, D., & Guillaume, D. M. (2012). Crises, labor market policy, and unemployment. IMF Working Paper, No.12/65, International Monetary Fund.

Brada, J. C., Marelli, E., & Signorelli, M. (2014). Introduction: Young people and the labor market: Key determinants and new evidence. *Comparative Economic Studies*, 56(4), 556–566. doi:10.1057/ces.2014.30.

Brunello, G., Garibaldi, P., Wasmer, E., & Bassanini, A. (eds.). (2007). *Education and Training in Europe*. New York: Oxford University Press.

Bruno, G. S. F., Choudhry, M. T., Marelli, E., & Signorelli, M. (2017). The short- and long-run impacts of financial crises on youth unemployment in OECD countries. *Applied Economics*, 49(34), 3372–3394. doi:10.1080/00036846.2016.1259753.

Bruno, G. S. F., Choudhry, M. T., Marelli, E., & Signorelli, M. (2014). Youth unemployment: Key determinants and the impact of crises. In *Disadvantaged Workers-Empirical Evidence and Labour Policies*, edited by M. A. Malo and D. Sciulli, 121–148. Switzerland: Springer, Cham.

Bruno, G. S. F., Marelli, E., & Signorelli, M. (2016). The regional impact of the crisis on young people in different EU countries. In *Youth and the Crisis-Unemployment, Education and Health in Europe*, edited by G. Coppola and N. O'Higgins, 249–271. Abingdon, Oxon: Routledge.

Caporale, G. M. & Gil-Alana, L. (2014). Youth unemployment in Europe: Persistence and macroeconomic determinants. *Comparative Economic Studies*, 56(4), 581–591. doi:10.1057/ces.2014.29.

Caroleo, F. E., Ciociano, E., & Destefanis, S. (2017). Youth labour-market performance, institutions and VET systems: A cross-country analysis. *Italian Economic Journal*, 3(1), 39–69. doi:10.1007/s40797-016-0045-8.

Choudry, M. T., Marelli, E., & Signorelli, M. (2012). Youth unemployment rate and impact of financial crises. International Journal of Manpower, 33(1), 76–95. doi:10.1108/01437721211212538.

Ciarreta, A. & Zarraga, A. (2010). Economic growth-electricity consumption causality in 12 European countries: A dynamic panel data approach. *Energy Policy*, 28, 3790–3796.

Cvecic, I. & Sokolic, D. (2018). Impact of public expenditure in labour market policies and other selected factors on youth unemployment. *Economic Research-Ekonomska Istraživanja*, doi:10.1080/1331677X.2018.1480970.

Gomez-Salvador, R. & Leiner-Killinger, N. (2008). An analysis of youth unemployment in the Euro area. *ECB Occasional Paper Series*, 89, 1–43.

Grossmann, A., Love, I., & Orlov, A. G. (2014). The dynamics of exchange rate volatility: A panel VAR approach. *Journal of International Financial Markets, Institutions & Money*, 33, 1–27.

Günaydın, D. & Çetin, M. (2015). Genç işsizliğin temel makroekonomik belirleyicileri: Ampirik bir analiz. *Pamukkale Üniversitesi Sosyal Bilimler Enstitüsü Dergisi*, 22, 17–34.

International Labor Organization (ILO). (2018). *Guide on measuring decent jobs for youth-Monitoring, evaluation and learning in labour market programmes. Note 2: Concepts and definitions of employment indicators relevant for young people*. Geneva: International Labour Office.

ILO. (2010). *Global employment trends for youth. Special issue on the impact of global economic crisis on youth*. Geneva: International Labour Organization.

Im, K. S., Pesaran, M. H. & Shin, Y. (2003). Testing for unit roots in heterogeneous panels. *Journal of Econometrics*, 115(1), 53–74.

Javorcik, B. (2013). Does FDI bring good jobs to host countries? World Development Report 2013 Background Paper. Washington, DC: World Bank.

Jimeno, J. F. & Rodriguez-Palenzuela, D. (2002). Youth unemployment in the OECD: Demographic shifts, labour market institutions and macroeconomic shocks. ECB Working Paper, No. 155. Frankfurt: European Central Bank (ECB).

Korenman, S. & Neumark, D. (1997). Cohort crowding and youth labor markets: A cross national analysis. NBER Working Paper, No. 6031.

Levin, A. & Lin, C. & Chu, J. (2002). Unit root tests in panel data: Asymptotic and finite-sample properties. *Journal of Econometrics, Elsevier,* 108(1), 1–24.

Love, I. & Zicchino, L. (2006). Financial development and dynamic investment behavior: Evidence from panel VAR. *The Quarterly Review of Economics and Finance,* 46, 190–210.

Marelli, E., Choudry, M., & Signorelli, M. (2013). Youth and total unemployment rate: The impact of policies and institutions. *Rivista Internazionale di Scienze Sociali,* 121(1), 63–86.

O'Connell, P. (1998). The overvaluation of purchasing power parity. *Journal of International Economics,* 44, 1–19.

O'Higgins, N. (2015). Youth unemployment. IZA Policy Papers, No. 103. Bonn: Institute for Study of Labor (IZA).

O'Higgins, N. (2012). This time it's different? Youth labour markets during "The Great Recession". *Comparative Economic Studies,* 54, 395–412.

Pastore, F. & Giuliani, L. (2015). The determinants of youth unemployment. A panel data analysis. Discussion Papers-CRISEI. Italy: University of Naples "Parthenope".

Pesaran, M. H. (2003). A simple panel unit root test in the presence of cross section dependence. *Cambridge Working Papers in Economics 0346,* Faculty of Economics, University of Cambridge. Retrieved from http://www.econ.cam.ac.uk/research-files/repec/cam/pdf/cwpe0346.pdf (14.08.2019).

Pesaran, M. H. (2004). General diagnostic tests for cross section dependence in panels. *Cambridge Working Papers in Economics* No. 435, University of Cambridge, and CESifo Working Paper Series No. 1229.

Pesaran, M. H. (2007). A simple panel unit root test in the presence of cross-section dependence. *Journal of Applied Econometrics,* 22, 265–312.

Quintini, G., Martin, J. P., & Martin, S. (2007). The changing nature of the school-to-work transition process in OECD countries. Discussion Paper No.2582. Bonn: IZA.

Scarpetta, S., Sonnet, A., & Manfredi, T. (2010). Rising youth unemployment during the crisis: How to prevent negative long-term consequences on a generation? OECD Social, Employment and Migration Working Papers, No. 106. Paris: OECD Publishing. doi:10.1787/5kmh79zb2mmv-en.

Tomić, I. (2016). What drives youth unemployment in Europe? EIZ Working Papers, 1601. Zagreb: The Institute of Economics.

Wajid, A. & Kalil, R. (2013). The impact of inflation and economic growth on unemployment: Time series evidence from Pakistan. *Proceedings of 3rd International Conference on Business Management.*

Appendices:

Tab. 7.1: Results of Panel Unit Root Tests. Source: Authors' own calculations.

Variables	First Generation Panel Unit Root Tests		Pesaran (2004) CD test	Second Generation Panel Unit Root Tests		
	IPS (2003)	LLC (2002)		CADF		CIPS
	Stat. (p-value)	Stat. (p-value)	Stat. (p-value)	t-bar	Z [t-bar] (p-value)	Stat.
YUP	−4,49* (0,00)	−5,10* (0,00)	16,00* (0,00)	−2,24**	−2,28* (0,01)	−2,14***
GDP	−11,09* (0,00)	−10,9* (0,00)	34,09* (0,00)	−3,42*	−7,7* (0,00)	−3,62-
GDS	−1,96* (0,024)	−2,63* (0,004)	11,52* (0,00)	−1,59	0,734 (0,77)	−1,61
INF	−9,46* (0,00)	−8,78* (0,00)	40,10* (0,00)	−3,16*	−6,53* (0,00)	−2,94
FDI	−8,42* (0,00)	−8,94* (0,00)	19,05* (0,00)	−2,45*	−3,25* (0,001)	−3,76
TRADE	−4,02* (0,00)	−4,89* (0,00)	50,25* (0,00)	−2,46	−0,73 (0,23)	−1,87
POP	−8,19* (0,00)	−5,26* (0,00)	4,13* (0,00)	−2,57*	−3,81 (0,00)	−3,19*
Δ GDS	-	-	-	−3,41*	−7,65* (0,00)	−4,35*
Δ TRADE	-	-	-	−3,16*	−6,49 (0,00)	−3,86*

Explanation: The model containing the "constant & trend" for the TRADE variable is estimated. For the other variables, the model containing the "constant" is estimated. Critical values for t-statistic in CADF test and CIPS test statistic are as follows: −2,38 (1 %), −2,20 (5 %) and −2,11 (10 %) in fixed model; −2,88 (1 %), −2,72 (5 %) and −2,63 (10 %) in the model with constant&trend. * indicates significance at 1 % level. ** indicates significance at 5 % level. *** indicates significance at 10 % level.

Tab. 7.2: Estimation of Panel VAR Model. Source: Authors' own calculations.

Equation Variable: YUP_t		
	Coef.	Z (P-value)
YUP_{t-1}	1,24*	17,41 (0,000)
YUP_{t-2}	−0,40*	−6,39 (0,000)
GDP_{t-1}	−0,14*	−2,71 (0,007)
GDP_{t-2}	0,03	0,88 (0,377)
ΔGDS_{t-1}	−0,04	−0,66 (0,507)
ΔGDS_{t-2}	−0,20*	−3,08 (0,002)
INF_{t-1}	0,002	0,08 (0,935)
INF_{t-2}	−0,02	−0,97 (0,332)
FDI_{t-1}	−0,007	−0,54 (0,586)
FDI_{t-2}	0,05*	4,20 (0,000)
$\Delta TRADE_{t-1}$	0,04**	2,33 (0,020)
$\Delta TRADE_{t-2}$	0,03***	1,67 (0,095)
POP_{t-1}	0,03***	1,67 (0,096)
POP_{t-2}	0,05*	3,19 (0,001)
N obs	440	
N countries	20	

Note: VAR model is estimated by GMM. Final GMM Criterion Q(b) = 0,289. * indicates significance at 1 % level. ** indicates significance at 5 % level. *** indicates significance at 10 % level.

Tab. 7.3: Results of Panel Granger Causality Test. Source: Authors' own calculations.

Equation Variable: YUP	χ^2	df	Prob.
GDP	11,31*	2	0,004
ΔGDS	9,53*	2	0,009
INF	17,33*	2	0,000
FDI	18,91*	2	0,000
$\Delta TRADE$	7,26**	2	0,027
POP	10,97*	2	0,004

Note: * indicates significance at 1 % level. ** indicates significance at 5 % level.

Mehmet Güçlü

8. Industrial Revolution and Labor Market

Abstract The aim of this chapter is to analyze the possible effects of the fourth industrial revolution on the labor market in light of the historical examples of earlier industrial revolutions. The increasing and spreading speed of recent technological advances led to a serious concern in the labor market. Accordingly, technology will significantly reduce the need for human labor, and hence a mass technological unemployment. Historical examples have shown that technology, while extinguishing some jobs, leads to the emergence of many new jobs that did not exist before. Therefore, although there were periodic fluctuations in unemployment, there was no increase in unemployment in the long run due to technology. We foresee that the fourth industrial revolution cannot create a mass technological unemployment by following a similar path. However, while positive results can be observed in the labor markets of countries preparing for technological transformation in this process, the labor markets of countries that do not prepare themselves for transformation will have problems such as skill mismatch, inequality and polarization.

Keywords: Industrial revolution, Technological progress, Labor market, Employment, Unemployment

1 Introduction

As of the second half of the 18th century, there have been various periods in which extraordinary technological advances have been observed, called "the industrial revolution", and each time there have been concerns that these technological advances will cause many people to be unemployed. However, in the long run, these concerns have not been correct. Recent technological developments (automation, robotics, digitalization, artificial intelligence, big data, cloud technology, etc.) were named as "the fourth industrial revolution" and revived the concerns about the future of the labor force. In fact, the most extreme example of these concerns is the claim by the famous physicist Stephen Hawking that "artificial intelligence is going to replace human in the future" (Sulleyman, 2017). Although there are such extreme discourses, there is no consensus on the effects of the new industrial revolution. Some claim that this time is different from the previous ones (Schwab, 2016; Bonciu, 2017), and some argue that the results of this would not be much different from the previous ones (Autor, 2015a; Bessen, 2015; McKinsey, 2017a).

The aim of this chapter is to analyze the possible effects of the fourth industrial revolution on the labor market. Because, with the new industrial revolution, some mismatches are likely to emerge in the labor market. The possible effects of these mismatches on labor markets are the research topic of this study. *The first mismatch* will arise with regard to labor demand. With the decline in demand for some jobs along with the technological transformation, the demand for labor in these areas is expected to decrease, while at the same time new jobs will emerge, and the demand for labor will increase in these new areas: Who will win and who will lose in this situation? And how much will the net impact on labor demand be? *The second mismatch* is related to the nature of the demanded labor. As a result of the increase in automation, there will also be a mismatch between the nature of the labor that is no longer needed (in other words the capabilities of the to-be-unemployed) and the nature of the labor that will be needed for new jobs: How will this qualification mismatch affect the labor market? Who will win, who will lose?

In order to answer these questions above, firstly in the second section, in the light of historical examples, the effect of the technological advances in the past on employment is revealed. The third section discusses whether the effect of the fourth industrial revolution on labor markets will be different from the previous ones or not. In the following section, it is tried to determine who will benefit in the labor market and who will suffer from the recent technological advances. Finally, general evaluations are included.

2 A Historical View on the Relationship Between Industrial Revolution and Employment

In the last 200 years, significant advances have been observed in the technology in four different periods, which have been named as four different industrial revolutions (Bruckner, LaFleur and Pitterle, 2017). *The first industrial revolution* began in the late 18th century with the use of water and steam power in industry. It has emerged in the textile industry in the UK. This revolution is defined as *the mechanization of manufacturing*. The driving force of the revolution was the steam engine. *The second industrial revolution* is the name given to the period of *rapid industrialization* that started in the second half of the 19th century and continued until the beginning of the 20th century. The most significant features of this revolution are assembly lines, mass production and electrical energy. The central role of electricity in this revolution has led to the revolution being called the *electric revolution*. Also during this period, railways were built, the use of iron and steel was the subject of a great deal of use, and the use of telegraph and oil has begun. *The third industrial revolution* is defined as the *automation revolution* in which electronics (transistors and microprocessors) and computers were used in production, which started in the second quarter of the 20th century and continued until the end of it. *The fourth industrial revolution* appeared in the early years of the 21st century, and still continues. This revolution is defined by *cyber-physical systems* which include artificial intelligence, machine learning, big data, cloud technology, and internet of things.

While the first industrial revolution initially emerged in a single country and in a single sector, subsequent revolutions began to manifest in multiple countries and in many sectors. Since the first industrial revolution, all technological advances have increased production capacity, reduced costs and increased productivity in economies. Each time, the question of whether machines will replace human labor has created concerns. The first reflection of this concern was found in the famous Luddite movement, in the early 19th century, when British textile workers protested automation by destroying automated looms (PwC, 2018). The effects of machines on human labor have not only worried the workers but also the economists. D. Ricardo, a classical economist, added a chapter on the third edition of his book, Principles of Political Economy, titled, "Machinery". Here, he reconsidered how the invention of machines influenced labor. Ricardo stated that in the past he accepted that the use of a machine in any branch of production was a good thing, but later, the substitution of human labor by machinery changed the idea that it was mostly against the working class (Mokyr, Vickers and Ziebarth, 2015). J. M. Keynes (1930) pointed out that during the Great Depression, human beings are able to solve the long-term scarcity problem, but they are worried about a new disease that has not been heard yet, but will be heard a lot in the coming years. He said that the name of this disease is *technological unemployment*. W. Leontief (1952) argued that labor would become increasingly important, that more and more machines would replace workers, and that new industries would not be able to employ anyone looking for work.

This concern created by the technology has found its place in the famous *TIME* magazine in 1961. In an article titled "The Automation Jobless", the following are mentioned (TIME, 1961): The number of jobs lost due to machines is only part of the problem. What is more important is that automation is an obstacle to creating new jobs that will ensure the employment of labor. Automation has also handled office work and started to eliminate them. In the past, new industries have created jobs for more workers than the number of unemployed workers, but this is not the case in today's new industries.

These concerns were not only limited to workers and academics, but also by the country's governments. For example, in 1964, US President Lyndon B. Johnson had established "the National Commission on Technology, Automation and Economic Progress" in the deep recession between the years 1960–1961, to investigate the unemployment concern that was thought to be caused by technological progress. He wanted to investigate the possibility of 'automation eliminating the jobs' and how to deal with it (Bowen, 1966).

Since the first industrial revolution, economies of countries have experienced dramatic structural changes. For example, in the period covering the first three industrial revolutions (from 1800 to 2000), the share of agriculture in the total employment in the United States fell from 74.7 % to 2.4 %. Similarly, 55.8 % of the workforce in the USA was employed in the manufacturing industry in 1850, while in 2000 it was 14.7 % (Carter, 2003). Although structural transformations are largely realized in other countries in the same manner, there are considerable differences in the rate of transformation. For example, the share of agriculture in total employment in China has fallen from 60 % to 32 % in just 25 years (from 1990 to 2015). In the same period, the share of agriculture in employment in Japan was decreased from 31 % to 3.5 %, while the share of manufacturing industry in employment dropped from 32 % to 3.5 %. In this process, with the transition of workers to other sectors and other jobs in large proportions, an increase has been observed in the ratio of employment to the population. Workers who were unemployed due to technology were employed in new industries and jobs, thus preventing a significant increase in unemployment. In the course of all these transformations, while some of the jobs have disappeared, new jobs have emerged which were not foreseen in those days. In this process, the share of employment in the population has generally increased (McKinsey, 2017a: 34).

The fact that some tasks can be done by computers or robots together with technology does not mean that all tasks of work will be eliminated. In this regard, Bessen (2015)'s example of Automated Teller Machines (ATMs) is a guiding one: ATMs were first introduced in the United States and other developed countries in the 1970s. These machines were able to fulfill the most routine tasks such as accepting deposits, which had being done by bank tellers. Banks have expanded the availability and the number of ATMs throughout the years to 400,000 by the year 2010. With this increase in the use of ATMs, you may think that the banks tellers would be laid off. However, surprisingly, the number of bank tellers has not decreased over time. According to Bessen (2015), there are two reasons for this: The first is that ATMs reduce costs associated with operating a bank branch. The reduced costs led banks to open new branches in an attempt to increase their market share. Thus, the need for fewer employees in a branch reduced the demand for bank tellers, but with new branches, the demand increased yet again. Second, while ATM's managed to automate certain tasks; other tasks they couldn't automate became more valuable. The banks' desire to increase their market shares led them to "relationship banking" practices. In this case, the expected skill of a bank teller has also changed. The ability of the teller to handle cash became less important, while customer relations have gained importance. This led to the expansion of the customer relationship banking teams.

Technological innovations can create new demands and new industries. Printing is an example of this. When the newspaper *Times of London* began to use a steam-powered printing press, which was invented by Friedrich Koenig in 1814, it had reached five times the speed of its previous presses (1,100 pages per hour). Of course, this situation caused a riot among the printers in the newspaper and the workers could only be assured that they would not be dismissed from their jobs. As of 1820, printing presses were able to print 2,000 pages per hour. This number was doubled by 1828. Then, more advanced printing machines were invented. In 1860, the most advanced machines could print 30,000 pages per hour. Developments in the use of electricity and photomechanical processes allowed the images to be reproduced and in 1890, a four-page newspaper could be produced by 90,000 prints per hour. This flow of innovation has been the driving force for the rapid growth of the journalism industry in the US and Europe. Millions of jobs were created in printing, journalism and many other sectors (McKinsey, 2017b: 30–31).

Thanks to the production technique he developed, Henry Ford has managed to reduce the installation period of a car from 12 hours to 1 hour 33 minutes. Under normal conditions, this is expected to lead to a decrease of about 90 % in employment. However, mass production has led to a considerable decline in car production costs, leading to more people buying cars than ever before. As a result, the increase in labor demand resulting from the increase in car demand was higher than the decline in labor demand from Ford's production technique (Berger, 2016: 14).

According to Bessen (2015), the industrial revolution did not create mass technological unemployment. For example, during the 19th century, weaving looms automated 98 % of the work to be done by workers (labor) in fabric production. This has led to a decrease in labor costs, a reduction in fabric prices, thus a significant increase in the demand for fabric, and demand for more weavers. Of course, the need for the remaining non-automated skills of weavers has also increased.

To put it briefly, technological advances and automation in the last two centuries have not eliminated the need for human labor. Although there have been fluctuations in unemployment over the course of time, no increase has been observed in the long term. Even though the transition period and speed of this period are different from the previous ones, we predict that the balance of the labor market will be similar to the previous ones.

3 Is the Industrial Revolution of the 21st Century Different from the Previous Ones?

In recent years, many researches were carried out on the potential effects of the fourth industrial revolution, and various predictions were made. Frey and Osborne (2017) suggested that 47 % of total employment in the United States is at high risk because of automation for over the next 10 or 20 years. Frey and Osborne (2016) also claimed that in OECD countries, 57 % of jobs (69 % in India and 77 % in China) on average are susceptible to automation. According to Chang and Hyunh (2016), in the next 10 or 20 years, approximately 56 % of jobs in the five ASEAN countries (Cambodia, Indonesia, the Philippines, Thailand and Vietnam) have automation risks. World Bank (2016: 126) stated that approximately 67 % of all jobs in developing countries are susceptible to automation. PwC's (2018) research for Singapore, Russia and 27 OECD countries expresses that the country with the highest automation risk by 2030 is Slovakia (44 %), while Finland and South Korea have the lowest risk by 22 %. McKinsey (2017b) predicts that 60 % of all jobs at least 30 % can be technically automated. WEF (2016) conducted research for 15 developed countries and emerging markets, claiming that 7.1 million jobs will be lost due to automation and technological advances between 2015 and 2020, and only 2 million new jobs will emerge. All of these studies tell us that in the future, almost 50 % of jobs are facing the risk of automation.

While interpreting these estimations, one should be careful. Because these estimates do not necessarily indicate the certainty for future automations to occur; rather, they indicate the *possibility* of automation (to occur). ILO (2017) claims that the difference between the concepts of 'can' and 'will' will be high, especially in developing countries, given the high financial costs associated with the adoption and implementation of advanced technologies. Furthermore, the fact that a specific task within a job can be performed by machines will not eliminate the entire job. In this case, only workers will need to adapt to the new work environment in which they would work with robots.

If the numerical estimations above were made at the beginning of earlier industrial revolutions, similar predictions could be made for those days as well. As we explained in the second section, with the industrial revolutions, there were expectations that the machines would replace human labor in large part and

Tab. 8.1: The Characteristics of Occupational Tasks. Source: Autor and Dorn (2009) and Acemoglu and Autor (2010).

Tasks	Cognitive	Manual
Routine	1 Bookkeepers Proof readers Clerks	2 Machine Operators Cashiers Typist
Nonroutine	3 Researchers Teachers Managers	4 Cleaners Hairdressers Street Vendors

that people would be faced with mass technological unemployment. But this has not happened. According to Autor (2015a), there are two main reasons for this. The first *is the overestimation of the substitution relationship between machines and labor.* The second is to disregard *the complementarity relationship between machines and labor.* The substitution feature negatively affects labor demand; the complementarity feature positively affects labor demand. Acemoglu and Restrepo (2017) describe the first one as displacement effect and the other as productivity effect. Technology increases the profits by increasing the productivity of labor, which leads to more production and labor demand. To determine the relationship between substitution and complementarity between human labor and machinery, Autor et al. (2003) and Acemoglu and Autor (2010) separated the tasks as *routine* and *non-routine*. They also subjected them to a second distinction of *cognitive* or *manual* (see Tab. 8.1). By making such a classification, the authors attempted to frame the interaction between technological advances and labor. *Cognitive tasks* are defined as problem-solving, creativity, or complex interpersonal interactions. *Manual tasks* are the tasks that require adaptation to the situation and/or environment, visual identification and/or language recognition, and in-person interaction (Autor and Dorn, 2009). *Routine tasks* are the tasks that can be accomplished by following explicit programmed rules. *Nonroutine tasks* are non-routine tasks that cannot be defined as computer code.

The machines are successful in tasks with clear and codable rules. Even so, if these are also cognitive tasks, they can perform these tasks in full. For example, some of these may be processing payrolls, bookkeeping or doing arithmetic (cell 1 in Tab. 8.1). They can also perform manual tasks such as conducting a train or a cashier job, but here they are slightly more limited (cell 2 in Tab. 8.1). These routine tasks can be easily automated. In contrast, it is more difficult to automate non-routine tasks such as research, maintaining personal relationships, and designing new products (cell 3). This is even more difficult in manual tasks such as cleaning, which requires considerable dexterity, personal care and providing security services (cell 4). Technology is a productivity enhancer for jobs in cell 3 of Tab. 8.1, while it is not directly effective in jobs in cell 4. In short, the easiest tasks to automate are routine and cognitive tasks, while the most difficult tasks to automate are non-routine and manual ones. Autor and Dorn (2009) claimed that the job growth graph was U-shaped (in other words, in time, amount of jobs would decrease and then increase). They mentioned two elements that led to this: The first is the replacement of labor by technology in the professions where routine tasks are intense (the element that leads to the reduction of number of jobs in the first stage). The other is the growth of professions in which non-routine tasks are intense (which leads to an increase in number of jobs in the future).

Recent technological developments have even rendered tasks that were previously described as non-routine tasks routine. For example, even the task of 'driving in city traffic', which was defined as a non-routine about 15 years ago by Autor et al. (2003), can be automated today by being converted to a computer code. But computers are still not capable of performing many things humans can do. Humans still have various advantages over computers: The first is *the depth and breadth of human perception*. The perceptions of computers/robots are still not at a level to compete with human perception. Robots can see well than humans, but they can't understand what they see. Imagine that a computer is programmed to identify a chair. An engineer thought that an object must have legs, arms, and a back, so that it could be called a "chair", and thus encoded it. A computer running with this code can identify a chair with high accuracy. If a chair with no back and arms is to be defined, the programmer may do so, but in this case, the computer's perception will expand, including many objects which are not chairs (such as small tables). The computer's correct identification rate will be reduced. But an elementary school child can describe it without error. It can be said that this can be overcome by machine learning, but the machine learning is not able to do so flawlessly either. Moreover, machines need a lot of energy and effort for this. For example, Google's X Lab project needs to use 16,000 processors, solely to identify cat images on YouTube (Autor, 2015a). Another advantage of humans is their *creativity*; computers are still far behind human creativity. Because, the psychological processes underlying human creativity and the coding of it is quite difficult. Finally, one more feature that makes humans superior to computers is *the social intelligence*. It does not seem possible for computers to reach social intelligence of humans, who have elaborate skills such as negotiation, persuasion, and so on. Although some of these features can be done by computers at basic level, they are not at the level that humans have. Frey and Osborne (2017) argue that it is unlikely that the tasks that require complex perception, creative intelligence and social intelligence to be achieved by computers in the near future.

With artificial intelligence, computers/robots are now thought to be able to do anything, and there is a serious concern that these can replace labor in every field. This concern has to do with the overestimation of two things: the first one, as previously said, is the overestimation of technology's substitution to labor, and the omission of the complementarity. The second is the exaggerated thinking that artificial intelligence is capable of doing anything. Kaku (2011) has described the fluctuating history of artificial intelligence since the 1950s: when computers emerged in the 1950s, scientists were thrilled by the idea of computers having miraculous skills: It was considered that in a short period of time, robots would be in every household cooking and cleaning the house. Finally, it's been observed that these first robots could only do simple tasks. After this realization, intelligence studies slowed and even reached a standstill, before peaking again in the 1980s. This time, the Pentagon's smart truck project has come up: This truck is intended to reach behind the enemy lines, conduct exploration and reconnaissance. But at that time, this project has failed and artificial intelligence studies have stopped once again. In 1997, IBM's "Deep Blue" artificial intelligence system was considered to be a historical breakthrough when it defeated chess champion Gary Kasparov. However, this success of Deep Blue, which is capable of performing 11 billion processes per second, emphasized the inadequacy of artificial intelligence studies rather than expediting artificial intelligence research. Because it was as clear as the day that Deep Blue cannot think like a human would: Deep Blue was perfected to play chess, but it could only take zero on the IQ. Twenty years later, nowadays computers are quite advanced and artificial intelligence researchers are still hopeful about the future. For example, the robot named STAIR at the artificial intelligence center at Stanford University is now faster than a human in the image recognition test. If you would want it to remove the orange on the table, it can analyze all the objects on the table, compare them with the thousands of images in its memory, and then it will detect the orange, and remove it. But STAIR is limited in terms of the number of objects it can recognize, and it'd be paralyzed if it was to walk out into the street and try recognizing random objects. Scientists use the latest technology to model the entire brain as close to reality as possible. With a computer called *Dawn* with 147,456 processors and 150,000 gigabytes of capacity, scientists were able to simulate 40 % of the mouse brain in 2006 (while the mouse brain contains 2 million neurons, the human brain contains 100 billion neurons). It should not be disregarded here that only the operation of neurons in the mouse could be simulated, not the behavior of the mouse. Up to now, only 1 % of the human brain cortex (the cerebral cortex) could have been simulated with *Dawn*. Besides, the simulation could be performed at a speed of about 1 in 600 of the human brain. For this operation, *Dawn* consumes 1 million watts of power and requires 6,675 tons of air-conditioning equipment, releasing chilled air at 76,500 m^3 per minute. To model a whole human brain you would have to multiply it by 1,000. This means that this super-computer's power consumption could reach 1 billion watts, which is equal to the total power available from a nuclear power reactor. But surprisingly, a human brain uses only 20 watts of power. In short, there are still serious problems of power, heat and money to create a single computer that will reach the capacity of thinking and working of a single human brain.

4 Who Will Win? Who Will Lose?

Technological change, of course, will not affect everyone equally. Brynjolfsson and McAfee (2014: 11) expressed this as:

> *"There's never been a better time to be a worker with special skills or the right education, because these people can use technology to create and capture value. But there's never been a worse time to be a worker with only 'ordinary' skills and abilities to offer, because computers, robots, and other digital technologies are acquiring these skills and abilities at an extraordinary rate".*

The impact of technological advances on the labor market on different levels of job groups causes job polarization. This polarization is particularly observed in developed countries in recent years. Employment is increasing in high-skilled jobs and low-skilled jobs, while it decreases in middle-skilled jobs (Autor, Katz and Kearney, 2006; Goos and Manning, 2007, Acemoglu and Autor, 2010; Goos, Manning and Salomons, 2014). Job polarization can bring a wage polarization (Boehm, 2014 and Autor, 2015b). The difference between the remuneration of the remaining job can become more evident when the middle-skilled jobs are eliminated.

Tab. 8.2: Expected Impacts of Technological Change on Employment and Earnings. Source: World Bank (2016).

Occupational Class	Type of occupation (by skills intensity)	Employment	Earnings
Low-skilled Jobs	Non-routine manual	Positive	Negative
Medium-skilled Jobs	Routine cognitive and manual	Negative	Negative
High-skilled Job	Non-routine cognitive	Positive	Positive

The possible impacts of technological advances on employment and earnings of various business groups are summarized in Tab. 8.2. The Table includes three occupation groups. Non-routine manual jobs (cleaners, hairdressers, etc.), routine cognitive jobs or routine manual jobs as middle-skilled jobs (office workers, cashiers, machine operators, etc.) and non-routine cognitive jobs are classified as highly skilled jobs (managers, researchers, etc.). The higher the skill requirements for a job, the more difficult it will be for new employees to enter this market. Therefore, the high demand for these workers will turn into higher wages. Because of high entry barriers, workers in high-skilled jobs will see the award of their higher efficiency as higher earnings. However, if the retraining of the worker for a new job is easy or the skills required by the job are low, competition will increase and there will be a downward pressure on wages. In addition, middle-skilled jobs will disappear in time and these workers will also lead to low-skilled existing jobs and competition will increase. For these reasons, earnings in these jobs will decrease over time. In short, in the coming period, employment is expected to increase in low-skilled jobs despite low wages. It is foreseen that both

Tab. 8.3: Comparing Skills Demand, 2018 vs. 2022 (top 10). Source: WEF (2018).

Today, 2018	Trending, 2022	Declining, 2022
Analytical thinking and innovation	Analytical thinking and innovation	Manual dexterity, endurance and precision
Complex problem-solving	Active learning and learning strategies	Memory, verbal, auditory and spatial abilities
Critical thinking and analysis	Creativity, originality and initiative	Management of financial, material resources
Active learning and learning strategies	Technology design and programming	Technology installation and maintenance
Creativity, originality and initiative	Critical thinking and analysis	Reading, writing, math and active listening
Attention to detail, trustworthiness	Complex problem-solving	Management of personnel
Emotional intelligence	Leadership and social influence	Quality control and safety awareness
Reasoning, problem-solving and ideation	Emotional intelligence	Coordination and time management
Leadership and social influence	Reasoning, problem-solving and ideation	Visual, auditory and speech abilities
Coordination and time management	Systems analysis and evaluation	Technology use, monitoring and control

employment and wages will be reduced in middle-skilled jobs. In high-skilled jobs, both employment and wages are estimated to rise. In other words, the ones who will gain from technological advances will be non-routine cognitive jobs (World Bank, 2016: 133–134).

In its report titled "Future of Job 2018", WEF (2018) has published a list of the top 10 skills that will decrease and increase the demand by 2022 (see Tab. 8.3). In this list, it can be seen that the skills with decreasing demand are routine-cognitive skills, while those with increasing demand are non-routine cognitive skills.

According to the same report, it is predicted that until 2022 some jobs would not be needed due to technological advances and some new jobs would be emerged. Here it is seen that as the middle-skilled jobs disappear, high-skilled jobs are maintained, even new high-skilled jobs are emerging. In the near future, it is envisaged that new high-skilled jobs will emerge, such as artificial intelligence and machine-learning specialists, Digital Transformation Specialists, innovation specialists, user experience and human-machine interaction designers, and so on. Examples of the stable, new and redundant job roles can be seen in Tab. 8.4.

"How quickly will technological advances eliminate existing jobs" or "How quickly will new jobs emerge" are some of the related questions that concern the minds the most. Although the speed and scope of automation varies from country to country, there are five key elements that will determine this (McKinsey, 2017b): The first is technical feasibility. Technological advances require scientific research, but there is also a need for engineering solutions or applied research to adopt these advances. There is a significant lag between the emergence of a technology and the release of a product using this technology. For example, there were about 20 years between Orville and Wilbur Wright leading the way to fly through an airplane that they produced in 1903, and the beginning of commercial aviation in the US. Similarly, there was a 15–20 year period between German engineers to obtain a patent for a four-cycle engine in the 1870s and commercial car production. Today, technological developments are remarkable in automation, physical hardware and robotics, artificial intelligence and software in many other areas. However, most of the innovations we see are still being developed: from driverless cars to digital personal assistants (Apple Siri or Google Assistant, etc.); and most of them still have significant shortcomings. This means that there is still a lot of work to be done by scientists and engineers. Secondly, it is related to the development of technology-based products and the cost of their implementation. Producing technology (even for software-based virtual solutions) requires high-cost investments. At the same time, the use of these technologies also requires high amounts of capital. The use

Tab. 8.4: Examples of Stable, New and Redundant Roles (All Industries). Source: WEF (2018).

Stable Roles	New Roles	Redundant Roles
Managing Directors and Chief Executives	Data Analysts and Scientists*	Data Entry Clerks
General and Operations Managers*	AI and Machine Learning Specialists	Accounting, Bookkeeping and Payroll Clerks
Software and Applications Developers and Analysts*	General and Operations Managers*	Administrative and Executive Secretaries
Data Analysts and Scientists*	Big Data Specialists	Assembly and Factory Workers
Sales and Marketing Professionals*	Digital Transformation Specialists	Client Information and Customer Service Workers*
Sales Representatives, Wholesale and Manufacturing, Technical and Scientific Products	Sales and Marketing Professionals*	Business Services and Administration Managers
	New Technology Specialists	Accountants and Auditors
Human Resources Specialists	Organizational Development Specialists*	Material-Recording and Stock-Keeping Clerks
Financial and Investment Advisers	Software and Applications Developers and Analysts*	General and Operations Managers*
Database and Network Professionals	Information Technology Services	Postal Service Clerks
Supply Chain and Logistics Specialists	Process Automation Specialists	Financial Analysts
Risk Management Specialists	Innovation Professionals	Cashiers and Ticket Clerks
Information Security Analysts*	Information Security Analysts*	Mechanics and Machinery Repairers
Management and Organization Analysts	Ecommerce and Social Media Specialists	Telemarketers
Electrotechnology Engineers	User Experience and Human-Machine Interaction Designers	Electronics and Telecommunications Installers
Organizational Development Specialists*	Training and Development Specialists	Bank Tellers and Related Clerks
Chemical Processing Plant Operators	Robotics Specialists and Engineers	Car, Van and Motorcycle Drivers
University and Higher Education Teachers	People and Culture Specialists	Sales and Purchasing Agents and Brokers

(continued on next page)

Tab. 8.4: (continued)

Stable Roles	New Roles	Redundant Roles
Compliance Officers	Client Information and Customer Service Workers*	Door-To-Door Sales Workers, News and Street Vendors, and Related Workers
Energy and Petroleum Engineers	Service and Solutions Designers	
Robotics Specialists and Engineers	Digital Marketing and Strategy Specialists	Statistical, Finance and Insurance Clerks
Petroleum and Natural Gas Refining Plant Operators		Lawyers

Note: Roles marked with * appear across multiple columns. This reflects the fact that they might be seeing stable or declining demand across one industry but be in demand in another.

of industrial robots in production cost millions of dollars. Although it may seem cheaper to use software-based solutions, there are also significant costs such as setting up data centers and building networks. The use of a software (or hardware) in a company can sometimes be much more costly than the software itself. These costs can be exemplified as customizing the software, changing some processes according to the organization of the company, and training the personnel. All these costs will be seen as an important element that determines when and where the technology will be used. Third main element is the dynamics of the labor market. The supply, demand and wages in the labor market are dynamic and all can change over time. For example, the high or low cost of labor in the face of automation can slow down or accelerate the transition to automation. The fourth is the economic benefits of automation. Potential economic benefits from automation are not limited to a reduction in labor cost. In the transition to automation, it will be decisive whether technology provides benefits to the company as well as efficiency, quality and safety. The fifth is about regulatory and social acceptance. There are some regulatory and social barriers to the implementation of automation. For example, a mistake made by robots can bring about important legal regulations. Who would be responsible if a decision given by artificial intelligence-operated autonomous vehicles would harm a person? This can initiate discussions of security and responsibility. At the end of the discussions, legal restrictions can be made to what robots can or can not do. One of the serious social barriers to automation is the confidentiality of the data. Due to the confidentiality of the data, some areas may be allowed limited automation. Another social barrier is the increase in social and economic pressures on the basis of society if many people lose their jobs and not find new jobs. This may lead to legal regulations that restrict the use of automation, or employers may refuse automation for humanitarian reasons. In addition to all these, discomfort and personal preferences for increased technology use can slow down the speed of automation to some areas.

Briefly, it is possible for automation to lead to high unemployment, but only if the transformation is done without sufficient time, and without preparation. According to McKinsey (2017a), if sufficient economic growth and investment take place, new jobs will be created to compensate for lost jobs in the future. The main challenge is to provide the necessary skills and support for the transition of these workers to new jobs. Countries that are not able to plan this transition well may face depressed wages and rising unemployment rates in the future.

5 Conculusion

When the subject is analyzed historically, it is clear that the historical industrial revolutions did not create mass unemployment. Although technological advances have eliminated some professions over time, it is certain that they have not eliminated the jobs. The pace of the recent technological advances and the breadth of the sphere of influence have started a debate on whether the fourth industrial revolution will differ from the others. In particular, even the extreme fears have come to the fore, stating that automation can make human labor completely obsolete. Neither the past, nor the present evidence supports this fear. Although technology replaces labor in some areas, it becomes a complement to labor in many fields and increases the employment within them. However, it should not be overlooked that some measures should be taken in order to ensure that technological progress does not lead to an increase in unemployment in the economies. In order to adapt the labor skills to new jobs, the necessary trainings should be provided by both public and private sector. These skill-building trainings should not be in a specific period and/or one-time; they should be continuous, and presented at specific intervals throughout the life cycle. Training programs should be planned separately according to low-, middle- and high-skilled jobs and should be kept up to date. New mechanisms should be established to ensure that those who lose their jobs in the transition phase find new jobs quickly, and these people should be provided with income support. Otherwise, significant reductions in total demand and social unrest may be observed. Education systems of countries should be at the level required by technological progress. Countries that cannot keep up with these advances will not be the producers of these technologies, nor will they be able to use and develop them. As a result, they may lose their competitiveness and cause their economies to decline. In order to create new jobs that can replace the technology-abolished jobs, new forms of work must be encouraged by the public. In countries where labor markets are not prepared for this transformation, significant problems regarding employment and unemployment will be observed. If this transformation is managed wisely, a better job and a better quality of life can be achieved for everyone. Otherwise, there will be an increase in skill mismatch, inequality and polarization. All these processes will also differ from developed and developing countries.

References

Acemoglu, D. & Autor, D. (2010). *Skills, Tasks and Technologies: Implications for Employment and Earnings*, NBER working Paper 16082.

Acemoglu, D. & Restrepo, P. (2017). *Robots and Jobs: Evidence from US Labor Markets*. NBER Working Paper, No: 23285.

Autor, D. & Dorn, D. (2009). This Job Is 'Getting Old': Measuring Changes in Job Opportunities Using Occupational Age Structure. *American Economic Review*, 99(2): 45–51.

Autor, D. H. (2015a). Why Are There Still So Many Jobs? The History and Future of Workplace Automation. *Journal of Economic Perspectives*, 29(3): 3–30.

Autor, D. H. (2015b). Polanyi's Paradox and the Shape of Employment Growth. Federal Reserve Bank of St. Louis: Economic Policy Proceedings, Reevaluating Labor Market Dynamics: 129–177. Retrieved from http://economics.mit.edu/files/9835 (14.08.2019).

Autor, D. H., Katz, L. F., & Kearney, M. S. (2006). The Polarization of the U.S. Labor Market. *American Economic Review*, 96(2): 189–194.

Autor, D. H., Levy, F., & Murnane, R. J. (2003). The Skill Content of Recent Technological Change: An Empirical Exploration. *The Quarterly Journal of Economics*, 118(4): 1279–1333.

Berger, R. (2016). The Industrie 4.0 Transition Quantified. *Think: Act Magazine*, Munich.

Bessen, J. (2015). Toil and Technology. *Finance and Development*, 52(1): 16–19.

Böhm, M. (2014). The Wage Effects of Job Polarization: Evidence from the Allocation of Talents. *Annual Conference 2014: Evidence-based Economic Policy, German Economic Association*, Hamburg.

Bonciu, F. (2017). Evaluation of the Impact of the 4th Industrial Revolution on the Labor Market. *Romanian Economic and Business Review*, 12(2): 7–16.

Bowen, H. R. (Chairman). (1966). *Report of the National Commission on Technology, Automation, and Economic Progress: Volume I*. Washington, DC: U.S. Government Printing Office.

Bruckner, M., LaFleur, M., & Pitterle, I. (2017). The Impact of the Technological Revolution on Labour Markets and Income Distribution. *Development Policy Seminar on the Frontier Issues*, United Nations—Department of Economic and Social Affairs.

Brynjolfsson, E. & McAfee, A. (2014). *The Second Machine Age: Work, Progress, and Prosperity in a Time of Brilliant Technologies*. New York: W.W. Norton.

Carter, S. B. (2003). *Labor Force for Historical Statistics of the United States*, millennial edition. Retrieved from https://economics.ucr.edu/papers/papers04/04-03.pdf (12.08.2019)

Chang, J. & Huynh, P. (2016). *ASEAN in Transformation: The Future of Jobs at Risk of Automation, Bureau for Employers' Activities*. ILO Regional Office for Asia and the Pacific Working Paper No. 9. Bangkok.

Frey, C. B. & Osborne, M. (2016). *Technology at Work v2.0: The Future Is Not What It Used to Be*. Oxford Martin School and Citi GPS, Oxford, England.

Frey, C. B., & Osborne, M. A. (2017). The Future of Employment: How Susceptible Are Jobs to Computerisation. *Technological Forecasting and Social Change*, 114: 254–280.

Goos, M. & Manning, A. (2007). Lousy and Lovely Jobs: The Rising Polarization of Work in Britain. *Review of Economics and Statistics*, 89(1): 118–133.

Goos, M., Manning, A., & Salomons, A. (2014). Explaining Job Polarization: Routine-Biased Technological Change and Offshoring. *American Economic Review*, 104(8): 2509–2526.

ILO. (2017). *Inception Report for the Global Commission on the Future of Work*. Geneva: International Labor Organization.

Kaku, M. (2011). *Physics of the Future: How Science Will Shape Human Destiny and Our Daily Lives by the Year 2100*. New York: Doubleday.

Keynes, J. M. (1930). *Economic Possibilities for Our Grandchildren*. Essays in Persuasion: 321–332. London: Palgrave Macmillan.

Leontief, W. (1952). Machines and Man. *Scientific American*, 187(3): 150–164.

McKinsey. (2017a). *Jobs Lost, Jobs Gained: Workforce Transitions in a Time of Automation*. MicKinsey Global Institute.

McKinsey. (2017b). *A Future That Works: Automation, Employment, and Productivity*. MicKinsey Global Institute.

Mokyr, J., Vickers, C., & Ziebarth, N. L. (2015). The History of Technological Anxiety and the Future of Economic Growth: Is This Time Different. *Journal of Economic Perspectives*, 29(3): 31–50.

PwC. (2018). *Will Robots Really Steal Our Jobs?: An International Analysis of the Potential Long Term Impact of Automation*. PricewaterhouseCoopers.

Schwab, K. (2016). *The Fourth Industrial Revolution*. Geneva: World Economic Forum.

Sulleyman, A. (2017, November2). Stephen Hawking Warns Artificial Intelligence May Replace Humans Altogether. *Independent*, Retrieved from https://www.independent.co.uk/life-style/gadgets-and-tech/news/stephen-hawking-artificial-intelligence-fears-ai-will-replace-humans-virus-life-a8034341.html (10.02.2019).

TIME. (1961, February 24). The Automation Jobless. *TIME Magazine*.Retrieved from http://content.time.com/time/subscriber/article/0,33009,828815-1,00.html (12.01.2019).

WEF. (2016). The Future of Jobs, Employment, Skills and Workforce Strategy for the Fourth Industrial Revolution. World Economic Forum, Retrieved from http://www3.weforum.org/docs/WEF_Future_of_Jobs.pdf (12.06.2019).

WEF. (2018). Future of Jobs Survey 2018, World Economic Forum, Retrieved from http://www3.weforum.org/docs/WEF_Future_of_Jobs_2018.pdf (02.04.2019).

World Bank. (2016). World Development Report 2016: Digital Dividends, Washington, DC. Retrieved from https://openknowledge.worldbank.org/handle/10986/23347 (10.04.2019).

Kemal Eker, Görkem Bahtiyar, and Hasan Bakır

9. Systemic Policies Towards Attracting Skilled Labor: An Investigation on Turkey

Abstract Immigration has become one of the most prominent issues of the 21st century. For this reason, it is unthinkable in part of governments to be indifferent towards this phenomenon. In the current era which has witnessed the rise of knowledge societies with the ever-increasing role of all kinds of knowledge in our lives, it has been more important than ever to attract the labor whose knowledge endowment is high. This kind of labor is termed 'high-skilled labor'. In this respect, it has been seen that some countries evaluate migration applications with various evaluation criteria. The most striking meaning of this is that, attracting skilled labor has now become a systematic process. Among 'point systems' of Australia, Canada and New Zealand and others, the Blue Card system of the EU and Turkey's plan to implement a Turquoise Card are worth a thorough academic consideration. Turkey now presents its systematic demand towards the supply of skilled immigrants which gains more and more importance with this plan. However, it is worthwhile to note that these point systems also have some deficiencies. Cultural factors, for instance, cannot be represented fully in those systems. The aim of this chapter is to investigate the possible advantages of Turkey in the pursuit of attracting skilled labor and how it can distinguish itself from other, especially developed countries which also try to attract skilled immigrants. A part of this investigation will surely be on the subject of the Turquoise card plan.

Keywords: Skilled labor, Immigration, Migration policy, Point systems, Turquoise card

1 Introduction

Today, more and more people are on the move than ever because of economic and social factors. These factors are the main reasons for immigration. The number of international immigrants reached 257.7 million people in 2017. This number higher by 85.1 million people from its 2000 levels than the number of immigrants in 2000, which was 172.6 million people (United Nations, 2017a: 1). The number of international immigrants worldwide reached 105.2 million people from 1990 to 2017. Most of this increase occurred during the period 2005–2017. Between 1990 and 2017, developed regions gained 63.6 million international migrants, and developing regions gained 41.5 million international migrants during the same period. While 89.6 million immigrants went to high-income countries to live and show their skills, only 2.4 million immigrants preferred to go to low-income countries. Middle-income countries gained 12.9 million immigrants in the same period (United Nations, 2017b: 1). Many developed countries such as the US, Australia, Canada and Germany have benefited from immigration. Similarly, Germany's great economic success after World War II resulted from low-skilled immigrants who came from Italy, Spain, Portugal and Greece along with Turkish and Yugoslavian immigrants. During the recent decades, however, immigration policies in the advanced countries focused on high-skilled workers instead of low-skilled ones. It could be easily understood why developed countries shifted their attentions to attract high-skilled labors. In the past, artisans and craftsmen were the elite of the labor force. Likewise, high-skilled workers have become popular at the current level of economic development in advanced countries (Martinez-Herrera, 2008:1–2; Chiswick, 2011: 1–2). In this context, concepts such as brain gain, brain drain as well as social capital transfer have become all the more important (see, Perruchoud & Redpath-Cross, 2013). Therefore, many countries have recently implemented skills-based migration programs.

There are two basic systems used to evaluate immigration. One of them is the individual worker/job evaluation system. This system is used by the US. The system is based on matching a specific immigrant with a specific job offer. Another system is the Point System. This system was first used by Canada and by Australia later. This system aims to evaluate the candidate. The candidate is evaluated by some criteria and receives points from each criterion. After that, the country decides whether to offer the candidate a job or not (Martinez-Herrera, 2008: 1–2). Countries such as Canada and Australia implement point tests to evaluate immigration applications. These tests include points assigned to each prospective immigrant according to his/her level of education, work experience and language skills (Yale-Loehr & Hoashi-Erhardt, 2001). When applicants pass the point test, it means that they provide the minimum standards in the area of health and good character (Tani, 2014: 3). It is obvious that with this tests the level of human capital of immigrants is identified quantitatively.

Points are given to applicants with high level of education or vocational training thus immigrants can be employed in the host country without any training cost. In addition to this, points are awarded to young applicants because immigrants contribute to public finance in the host country by paying income tax and these young immigrants does not need huge amounts of welfare payments. Furthermore, these points are also given to immigrants who speak the language of the host country, which reduces training costs and provides a quicker economic and social integration (Tani, 2014: 3). Therefore, this system provides some advantages such as objectivity and impartiality for candidates and host countries. For instance, candidates know which countries may accept them by looking at the evaluation system. On the other hand, the host country might always accept immigrants with certain skills (Martinez-Herrera, 2008: 5). However, these systems might entail deficiencies. For instance, research shows that migrant flow increases with factors such as common culture, common language, (Leblang et al., 2009). The language factor is taken into account in those systems but factors such as cultural and historical as well as political proximity are not. Thus, Section II of this chapter will provide some explanation over point systems which have been implemented in Canada, Australia and New Zealand as instruments of migration policy. Section III will deal with adapting these models to Turkey. Final remarks and policy implications are in Section IV.

2 Point Systems as Instruments of Migration Policy

According to Tani (2014: 1), "*A point system using measurable criteria selects economically desirable immigration applicants and result in the orderly management of population growth, which can reassure the native population that immigration is being properly managed*". On the other hand, this system cannot "*fix short term skilled labor shortages in a timely manner or prevent poor labor market outcomes for immigrants*" (Tani, 2014: 1). With this system, some countries are able to appoint scores in areas of age, education and language skills to evaluate their immigrants potentially (Tani, 2014: 2). A point system was firstly used by Canada in 1967. Then, Australia introduced a point system later and they are followed by New Zealand afterwards (Tani, 2014: 2). These three countries developed their point systems according to their labor market needs. For instance, while Canada is interested in the "*human capital of prospective immigrants*", Australia focuses on "*immediate employability*" and New Zealand focuses on employability (Tani, 2014: 3). These point systems used in these countries will be analyzed in this chapter and later on, Turkey's immigration policy and its potentials.

2.1 Canadian point system

Canada became the first country to implement an immigration points system in 1967 (Martinez and Herrera, 2008). Canada evaluates immigrant applications according to a point system known as the Comprehensive Ranking System (CRS)[1]

---

1   The details of the CRS are present on the Government of Canada website (https://www.canada.ca/en/immigration-refugees-citizenship/services/immigrate-canada/express-entry/eligibility/criteria-comprehensive-ranking-system/grid.html#a1 – 19.03.2019.

(Government of Canada, 2019). Factors related to age, education, language proficiency and work experience are included in the system. The points are calculated differently for potential immigrants with a spouse and without a spouse. If the spouse has skills that are included in the ranking system, the point of the applicant also varies. If candidates reach the minimum score which is required, their family members receive permanent residency in Canada. However, one of the disadvantages of the system is that they need to wait at least four years to obtain Canadian citizenship. Besides, candidates spend a great amount of time for the application process (Martinez-Herrera, 2008: 3–4). Canada wants to select high-skilled immigrants who would potentially contribute to Canada's productivity. Formal education, language proficiency and work experience are very important for Canada's point system (Tani, 2014: 3).

2.2 Calculation method of the CRS

2.2.1 Core human capital factors

Age: The calculation of the age factor in the CRS begins with 17 years of age which is assigned 0 points. Maximum points are assigned to applicants whose age range from 20 to 29 which is 100 points for applicants with a spouse or common law partner and 110 points for applicants without a spouse or common law partner. Points then start to decline from 30 years of age and become 0 for 45 years of age and more. *Level of Education:* The system assigns 0 points for applicant who does not have a high school diploma. Maximum points are given to PhD levels which are 140 for applicants with a spouse or common law partner and 150 points for applicants without a spouse or common law partner. *Language Proficiency:* Language proficiency points are assigned based on the applicant's CLB (Canadian Language Benchmark) score. An applicant gets 0 points for a CLB score less than 4. Maximum point is 128 for applicants with a spouse or common law partner and 136 for applicants without a spouse or common law partner are given for scores equal or more than 10. Since Canada has two official languages (English and French), applicants also could get language points from the other official language, given he/she has a CLB score equal or higher than 4. Maximum point an applicant could get from the second official language is 22 for applicants with a spouse or common law partner and 24 for applicants without a spouse or common law partner, and the CLB score must be 9 on more. *Work Experience (In Canada):* Applicants without a spouse or common law partner, who had worked in Canada for at least 5 years get a maximum 80 points whereas same type of applicants who have a spouse or common law partner get 70 points. Applicants with no work experience in Canada naturally get 0 points.

2.2.2 Spouse or common law partner factors

Spouse or common law factors affect the applicants' score in a way which would equalize it to the point one would get without a spouse or common law partner, provided that the spouse or the common law partner has a high level of human capital. For instance, if the spouse of one applicant has a 5 years or more work experience in Canada, then the applicant gets additional 10 points which equalizes his/her point to the maximum points an applicant without a spouse or common law partner could get from the same section. Same thing applies to points acquired from the level of education. In fact, points from official languages proficiency becomes higher for an applicant with a spouse or common law partner who gets a higher score from this part of the system if the spouse or common law partner has a CLB score of 9 or more. But the system does not assign any points for the age of the spouse or common law partner.

2.2.3 Skill transferability factors

Applicants get points from the combined effects of education, language proficiency and Canadian or foreign work experience. Applicants get maximum 100 points from this component of the system.

2.2.4 Additional points

Under this subsection, applicants get points if they had a post secondary education in Canada or have a valid job arrangement in Canada or if they have a sister or brother residing in Canada. In addition to these, an applicant gets 600 points, which is the maximum point one can get under this subsection, if he/she was nominated by one of the provinces in Canada.

2.3 The Australian point system

In 1973, Australia implemented its immigration points model six years later than Canada (Martinez and Herrera, 2008). Australian Point System[2] includes four main categories: skills, work experience, age and language (Australian Government, 2019). A candidate must score at least one point in each section, and immigrants who do not speak English will not be accepted in contrast to Canada (Martinez-Herrera, 2008: 4). Candidates who get enough points obtain permanent residency. In Australia, citizenship can be obtained after two years of

---

[2] The details of the Australian system are present at the Australian government department of home affairs website, https://immi.homeaffairs.gov.au.

consecutive lawful residence. Besides, Australia also gives awarding points for the educational achievements or experience of the candidate's spouse (Martinez-Herrera, 2008: 4–5).

2.3.1 Age

Beginning from 18 to 24 years, which get 25 points, points from the age component become maximum with 30 points for ages 25–32. Applicants get points until they are 44 years old. Forty-five and older are not included in the system.

2.3.2 English language ability

Points from the language ability diverge according to three categories for non-native English-speaking applicants: competent, proficient and superior English. These categories in turn are decided according to points an applicant gets from language proficiency tests such as International English Language Testing System (IELTS), Test of English as a Foreign Language (TOEFL), Occupational English Test (OET), Pearson Test of English (PTE) Academic or Cambridge English: Advanced (CAE).

2.3.3 Skilled employment

Applicants get points for their skilled work experience. Work experience could be inside or outside Australia. The assessment of whether a job is skilled or not is done according to Australian and New Zealand Standard Classification of Occupations (ANZSCO) standards and various occupational assessment authorities such as the Australian Computing Society (ACS) and Australian Institute of Management (AIM).

2.3.4 Qualifications

This subcategory includes the applicant's educational background. Education could be combined with relevant awards to give a maximum 30 points from this component of the Australian point system.

2.3.5 Other factors

Applicants can get additional 5 points if they had a proficiency on one of the accredited community languages. One can also get 5 points if he/she had lived or studied in a regional or low population growth area of Australia. This subsection also considers an applicant's partner's skills, similar to Spouse or Common Law Partner Skills in the Canadian Rating System (CRS).

2.3.6 Nomination

This subsection is similar to the provincial nomination in the CRS. The applicant gets up to 5 points if he/she is nominated by the state or a territorial government. To get additional 5 points from partner skills, the applicant's partner's age must be under 50; the partner must have at least a competent level English, must have been nominated in the same skilled type occupation as the applicant and must be assessed by the relevant authority as suitable for the aforementioned skilled occupation.

2.4 New Zealand point system

New Zealand's point system[3] supports skilled workers who have worked in New Zealand before (New Zealand Immigration, 2019). On the other hand, applicants must get a minimum score in formal English test of English proficiency. These formal qualifications are evaluated when applicants are selected (Tani, 2014: 4).

2.4.1 Skilled employment

An applicant gets points if he/she is currently working in a skilled employment, 60 points for employment longer than a year and 50 points for less. An applicant also gets additional points if he/she is employed in an area of absolute skills shortage (10 points) or in a region outside Auckland (30 points). An applicant can also get 20 points from his/her partner's skilled employment. Occupations that are in the area of skills shortage include occupations such as forest scientist, anesthetist, multimedia specialist, chef.[4] Applicants also gain points for their previous work experience in skilled sectors.

2.4.2 Recognized qualification

Under this section, an applicant gets points for his/her level of education, of which the maximum 70 points are given to the Ph.D. level. An applicant also gets additional points for post-graduate studies in New Zealand. Similar to the Canadian and Australian systems, applicants also could get points for their

---

[3] The details of the system can be found on the official New Zealand Immigration Department website, and the Self Assessment Guide for Residence in New Zealand-INZ 1003 published by it.

[4] Those lists are available at the http://skillshortages.immigration.govt.nz/assets/uploads/long-term-skill-shortage-list.pdf (15.03.2019).

partner's levels of education. The system also appoints points for age (max. 30 points ages from 20 to 39).

3 Turquoise Card: The Brand New Turkish Point System

Turkey has been known as a transit country, where immigrants crossed on their way to Europe. In the 1960s and 1970s, Turkey sent parts of its labor force to European countries such as Germany, Austria, Belgium, Holland, France and Sweden through agreements signed between corresponding governments (Kırışçı, 2007: 91). With the latest developments, the question of whether Turkey is transforming from a transit country to a target country gained attention. If so, this means a paradigm shift which Turkey needs to adapt to and implement policies and institutional arrangements towards being a target country. Turkey's attracting immigrants from its neighborhood with its growing economic power is mentioned by other authors (see, for example, Sirkeci & Martin, 2014), but the effect of the recent global financial and accompanying problems experienced by the European Union may also be important in the concept of Turkey's transition from a transit country to a target country. Facing these changes, Turkey moved towards her own point system with the Turquoise Card System. The Turquoise Card Regulation was published on March 14, 2017, in the Turkish Official Gazette (Turkuvaz Kart Yönetmeliği, 2017): The Turquoise Card application aims to attract high-skilled labor and high-tech investment in general. The Turquoise Card gives its holder sine die working and residence permit and living permit for the holder's family. The fifth article of the Turquoise Card Regulation states that the card is to be given to foreigners in the following categories: (a) foreigners who are identified as high-skilled labor due to his/her level of education, wage, experience, contribution to the advancement of science and technology and so on; (b) foreigners who are identified as a high-skilled investor due to the size of the investment, exports and employment it generates, contribution to the advancement of science and technology; (c) scientists or researchers who contribute to the advancement of science and technology or who pursue research on an international level which is deemed strategic for Turkish national interests; (d) foreigners who have success in cultural, artistic or sports activities on international level; (e) foreigners who contribute to the promotion of Turkey and Turkish culture, and who work for Turkish national interests on the international level. The Turquoise Card Regulation is based on the 11th and 25th articles of the Act No. 6735 (Act No. 6735, 2016). In the 11th article, the law states that Turquoise Card could be given to aforementioned classes of foreigners who made the relevant contributions in fields of economy, science, technology, culture, arts or sports or who are expected to make those contributions. The Turquoise Card system also foresees a point system. Although the Regulation states that the point system would take into account the aforementioned factors, the details of the point system (of how many points would be assigned to which factor, etc.) have not been made public yet.

4 Final Remarks and Policy Implications

We regard the process of immigration as a demand and supply mechanism. The government presents some opportunities and a place in the society to prospective immigrants with skills that would contribute to its economy and prestige. With the implementation of various point systems, attracting high-skilled labor became a systematic process. It now means a stable and regulated demand from countries towards high-skilled labor. On the other hand, prospective high-skilled immigrants evaluate and then choose to be residents of the country. But, skilled prospective immigrants have a wide range of options, countries offering residency to them. Recent study shows that migration flow is effected by factors like common culture and common language, among others. In order to design a more realistic point system for skilled immigration, Turkey must take into account this side of the story, namely the present choice structure of potential high-skilled immigrants. A point system taking into account the 'choice' side of the immigration process would not only be a more realistic policy, but also this approach would be more useful for future and further integration of the skilled labor to Turkey. According to a study by Kaçar (2016) that consists of interviews made with Turks who have gone abroad and settled and who have returned to Turkey, there are social and political reasons rather than material conditions behind the brain drain. Furthermore, the lack of the sense of belonging to that country, although being most prominent in singles rather than couples, is also effective on the subjects' decisions to return. In this vein, the relationship by affinity present between Turkey and its neighboring countries could also be an advantage in attracting foreign skilled labor. Moreover, political instabilities that erupt in its vicinity also increases Turkey's potential to become a center of attraction. The data presented by Directorate General of Migration Management (2017) supports this phenomenon: 384,000 people from Greece during 1922–1938; 800,000 people from the Balkans during 1923–1945; 800 people from Germany between 1933and 1945; 51,542 people from Iraq in 1988; 345,000 people from Bulgaria in 1989; 467,489 people from Iraq during the First Gulf War in 1991; 20,000 people from Bosnia during 1992–1998; 17,746 people from Kosovo in 1999; 10,500 people from Macedonia in 2001 approximately.

Tab. 9.1: Foreign Academics Working in Turkey During the 2016–2017 Academic Year—First 10 Nationalities. Source: Republic of Turkey Ministry of Interior Directorate General of Migration Management (2016).

USA	367	Syria	331	Azerbaijan	255	United Kingdom	211	Iran	192
TRNC*	104	Germany	85	Egypt	75	Greece	73	Canada	72

Note: *Turkish Rep. of Northern Cyprus

Tab. 9.2: Foreign Academics Working in Turkey as of 2019 Academic Year—First 10 Nationalities. Source: Council of Higher Education-YÖK (2019).

USA	353	Syria	319	Azerbaijan	291	United Kingdom	186	Iran	315
TRNC*	105	Germany	176	Egypt	91	Greece	79	Russia	89

These massive migrations occur during years of political instability, wars and so on. Another one of the determinants of our migration policy is our geography. Throughout our history, we have witnessed the immigration of high-skilled labor towards Turkey both in the form of forced and voluntary migration from other countries, with the effect of our physical and cultural atmosphere and geography. Thus, it is important for Turkey to maintain its relative stability. Another data shows the distribution of work permits according to nationalities. If we look at the distribution of work permits given according to nationality between 2013 and 2016, Georgia, Germany, China, Azerbaijan, Ukraine, Uzbekistan, Russian Federation, Syria, USA, Kyrgyz Republic, Turkmenistan, India, Moldova and Iran stand out as the most frequent nationalities (Rep. of Turkey Ministry of Interior Directorate General of Migration Management [2013–2016] Annual Migration Reports, March 15, 2019). These countries could be thought as countries from which Turkey could attract skilled labor from most easily. In all four years, Georgia is at the top of the list of work permit numbers. But it is not possible to say the same thing in terms of skilled labor. When we look at the data provided by the Council of Higher Education-YÖK (2019), for instance, the number of academics of Georgian nationality working in Turkish universities is only 34.

As the work permit data comes until 2016, we provided numbers of academics for both 2016–2017 and for 2019 to enable a healthy comparison illustrated in Tab. 9.1 and Tab. 9.2 respectively. According to these data, academics from the

United States of America are in the first place and academics from Syria come second according to data, and Iranian academics are in the third place. If foreign academics working in Turkish universities are to be used as proxies for skilled labor, it would be evident that the profile of skilled labor (academics) is non-consistent with the general profile of labor that is given permit to work in Turkey. From this point, we can argue that policies designed to attract skilled labor from our potential profile shown must be designed. The new Turquoise Card and its point system also should take this fact into account. Furthermore, there is the issue of language when it comes to attracting skilled labor to Turkey.

References

Australian Government. (2019). Department of Home Affairs website, https://immi.homeaffairs.gov.au (15.03.2019).

Chiswick, B. R. (2011). *Immigration: High Skilled vs. Low Skilled Labor?*. (IZA Policy Paper), No. 28. Bonn: Institute for the Study of Labor (IZA).

Government of Canada. (2019). Comprehensive Ranking System (CRS), Retrieved from https://www.canada.ca/en/immigration-refugees-citizenship/services/immigrate-canada/express-entry/eligibility/criteria-comprehensive-ranking-system/grid.html#a1. (15.03.2019).

Kaçar, G. (2016). *Türkiye'de Beyin Göçü ve Tersine Beyin Göçü Olgularının Değerlendirilmesi* (Yayınlanmamış Yüksek Lisans Tezi). Eskişehir Anadolu Üniversitesi Sosyal Bilimler Enstitüsü, Eskişehir.

Kırışçı, K. (2007). Turkey: A Country of Transition from Emigration to Immigration. *Mediterranean Politics*, 12(1), 91–97.

Leblang, D., Fitzgerald, J. & Teets, J.(2009). Defying the Law of Gravity: The Political Economy of International Migration. Available at SSRN, http://dx.doi.org/10.2139/ssrn.1421326 (09.06.2017).

Martinez-Herrera. (2008). Attractive Immigration Policies: "A worldwide fight for the skilled workers". *IUSLabor*, 1/2008. Retrieved from https://core.ac.uk/download/pdf/39042472.pdf (19.03.2019).

New Zealand Immigration (2019). Self-assessment Guide for Residence in New Zealand, INZ 1003. Retrieved from https://www.immigration.govt.nz/documents/forms-and-guides/inz1003.pdf (15.03.2019).

Perruchoud, R. & Redpath-Cross, J. Göç Terimleri Sözlüğü, İkinci Baskı, Uluslararası Göç Örgütü (IOM), No:31, Retrieved from http://publications.iom.int/system/files/pdf/iml31_turkish_2ndedition.pdf (14.08.2019).

T. C. İçişleri Bakanlığı Göç İşleri Genel Müdürlüğü. (Republic of Turkey Ministry of Interior Directorate General of Migration Management). (2016).

2016 Türkiye Göç Raporu. Retrieved from http://www.goc.gov.tr/files/2016_yiik_goc_raporu_haziran.pdf (02.04.2019).

T. C. İçişleri Bakanlığı Göç İşleri Genel Müdürlüğü (Republic of Turkey Ministry of Interior Directorate General of Migration Management). (2015). 2015 Türkiye Göç Raporu. Retrieved from http://www.goc.gov.tr/files/files/2015_yillik_goc_raporu.pdf (02.04.2019).

T. C. İçişleri Bakanlığı Göç İşleri Genel Müdürlüğü (Republic of Turkey Ministry of Interior Directorate General of Migration Management). (2014). 2014 Türkiye Göç Raporu. Retrieved fromhttp://www.goc.gov.tr/files/files/2014_yillik_goc_raporu.pdf (02.04.2019).

T. C. İçişleri Bakanlığı Göç İşleri Genel Müdürlüğü (Republic of Turkey Ministry of Interior Directorate General of Migration Management). (2013). 2013 Türkiye Göç Raporu. Retrieved fromhttp://www.goc.gov.tr/files/files/2013_yillik_goc_raporu.pdf (02.04.2019).

T. C. İçişleri Bakanlığı Göç İşleri Genel Müdürlüğü (Republic of Turkey Ministry of Interior Directorate General of Migration Management). (2017). Göç Tarihi. Retrieved fromhttp://www.goc.gov.tr/icerik/goc-tarihi_363_380 (02.10.2017).

Sirkeci, İ. & Martin, P. L. (2014). Sources of Irregularity and Managing Migration: The Case of Turkey. *Border Crossing*, 4(1–2), 1–16.

Tani, M. (2014). Using a Point System for Selecting Immigrants. *IZA World of Labor*, 24, 1–10.

Turkuaz Kart Yönetmeliği. (2017, March 17). Resmi Gazete, Sayı:30007.

Act No. 6735 on International Labour Force (2016, July 28), Resmi Gazete, 2016-08-13, No. 29800.

United Nations. (2017a). Department of Economic and Social Affairs, Population Division, I International Migration Policies: Data Booklet (ST/ESA/ SER.A/395).

United Nations. (2017b). Department of Economic and Social Affairs. International Migration Report, (ST/ESA/ SER.A/403).

Yale-Loehr, S. & Hoashi-Erhardt, C. (2001). A Comparative Look at Immigration and Human Capital Assessment. Cornell Law Faculty Publications, Paper 1097.

YÖK (Council of Higher Education). (2017). Retrieved from https://istatistik.yok.gov.tr (05.10.2017).

YÖK (Council of Higher Education). (2019). Retrieved fromhttps://istatistik.yok.gov.tr (15.03.2019).

H. Işıl Alkan

10. The Gender Impact of Last Global Crisis on Labour Markets

Abstract The 2008 global crisis, which was described as the biggest economic crisis following the Great Depression, caused a substantial economic slowdown and affected production and employment significantly across the world. Millions of workers have lost their jobs and many labourers have experienced cuts in hours work. Moreover, thousands of workers have been forced to leave their homeland in order to work as migrant labour. The crisis has affected the labour markets of both developed and developing countries; however, it has created more negative consequences for those who are employed vulnerably. As vulnerable employment is more prevalent among developing countries and women, the related process has significantly affected women as well as men who lost their jobs. The aim of this chapter is to examine the impact of global crisis on labour markets through a gender perspective both in developed and developing regions. The chapter also targets to develop policy implications regarding the issue.

Keywords: Global crisis, labour market, female employment, vulnerability

1 Introduction

The last global crisis, known as the second major economic crisis after the Great Depression of 1929, erupted in United States of America in 2008 and spread rapidly to Europe and then to the developing geographies with the accelerant impact of globalization. Each country experienced the crisis in varying proportions depending on its financial structure and intensity of commercial relations. In this context, it can be stated that developed markets with more complex financial structures and with more derivative products were impressed more negatively by the crisis. However, this does not mean that developing countries were not affected by the crisis as the main buyers of export products produced in developing countries were developed countries. Hence, restricted demand of developed world directly impacted on the production and employment structure of developing countries.

The aim of this chapter is to examine the impacts of last global crisis on labour markets through a gender perspective both in developed and developing regions. Therefore, in the second section, general impacts of economic downturns on female labour are examined and in the third section impacts of last global crisis on labour markets are explained. The fourth section of the chapter targets to unearth gender impacts of last global crisis on labour markets of developed and developing regions. The chapter is complemented by the conclusion section in which the findings of the study are summarized.

2 General Impacts of Economic Downturns or Crises on Female Labour

From the beginning, the status of women in labour markets is inferior compared to men across the world. Key indicators, such as the employment and unemployment rate, confirm the disadvantageous position of women in the labour market (ILOSTAT, 2019). Another characteristic of female labour is its high vulnerability to crises. The crises that occur frequently in market economies disperse rapidly to other countries and regions by courtesy of the globalization period and affect production and employment (Toksöz, 2009: 4). East Asia crisis is an example, the crisis that began in Thailand in July 1997 spread rapidly to all South East Asia, thus labour, especially the female labour has become more vulnerable as a result of the crisis. Male-breadwinner, female-caregiver bias was frequently seen during 1997 Asian crisis and particularly in Korean enterprises, married women and women in general were previously dismissed from men. According to World Bank, export-processing plants decreased regular workers who were predominantly women and rehired them as piece workers with lower wages. In many countries, informal employment accelerated in line with the decline in wage employment. In Indonesia, wage employment reduced while self-employment and unpaid family working increased between 1997 and 1998. The number of unpaid family workers who were mainly women increased in Korea. Similarly, many workers lost their jobs in cities and migrated to rural areas and started work as unpaid family workers in Indonesia and Thailand. Since women are more inclined to work informally in times of hardships, women living in poor families tend to work informal with low wages or unpaid (ILO, 2009: 5–6; Pilwha, 1998; Aslanbeigui, 2000; Betcherman and Islam, 2001; Lim, 2000; Islam et al., 2001; Horton et al., 2001; McGee et al., 1999; Focus on Global South and Save the Children 2001; Gragnolatti 2001; Jones et al., 2000). In brief, female labour has become more vulnerable and invisible with Asian crises.

Moreover, financial and economic downturns expand the informal labour market which is not gender blind in many regions. Women are more affected due to the changing working arrangements, especially due to the loss of benefits in line with these changes. For instance, the loss of maternity benefits significantly restricted the wage employment ability of women. On the other hand, rise in female labour force participation was observed in some countries in the post-crisis period, women became involved in the labour market to compensate for family income. The findings in Russia, Latin America and East Asia support this argument. The main reason for the increase in female labour force participation in the related countries was the economic crisis. However, the upgrade in employment did not refer to an increase in vulnerable employment as the stated women mainly engaged in temporary or part-time work (such as domestic workers, laundresses, piece workers or part-time factory workers). Besides, in certain countries, working in poor conditions in entertainment industry was widespread. It is reported that the number of sex workers increased in Russia and Lao PDR during the crisis (Tutnjevic, 2002: 6–7).

Another negative impact of economic crisis on women is the reduction of social state expenditures in countries with the crisis. Public expenditures which are reduced owing to the strict macro-economic policies have the most negative impact on the poorest in societies. Since women have the greatest share in the poorest, women become more vulnerable during the application of strict macro-economic policies. In times of economic crisis, the number and quality of meals are reduced, children are withdrawn from their schools and health-related expenditures are postponed in the poorest segments of society. The mentioned facts lead to continuation of the poverty cycle and ensure the permanence of poverty (UN, 2009: 4).

3 General Impacts of Last Global Crisis on Labour Markets

The last global crisis that occurred in 2008, but has been rooted in the past, has been described as the second largest crisis in the world after the Great Depression (IMF, 2019). The origin of the last global crisis is the largest real estate and credit bubble in history. In the beginning, mortgage loans, which were mostly given to high-quality customers, started to turn towards lower-quality customers and the volume of subprime mortgage loans increased significantly. In the related process, low-income customers preferred more floating-rate loans, however, with the increasing interest rates of FED and decreasing of real estate prices, these customers could not repay the loan they received. Additionally, in the same process, not only the real estate prices have been swollen, but also the transactions made on the bases of real estate prices have increased and the derivative products have ascended. Briefly, the 2008 crisis is a financial crisis created by the combination of mortgage loans given to marginal individuals and institutions, and the derivative products that are not limited (Eğilmez, 2018: 295–297).

The crisis that began in USA has spread rapidly to the world through domino effect by courtesy of globalization and neoliberal policies and it has significantly reduced demand, growth and investments in all over the world starting from the developed regions. Many financial institutions have gone bankrupt in various countries and global markets have narrowed significantly during the crisis. Bank loans were restricted for a long period of time and banks' lending standards increased significantly. These credit conditions have reduced investments, particularly residential investments, in many countries, and thus unemployment rates have increased all over the world (Erdönmez, 2009: 88). Due to the advanced structure of financial derivatives markets in developed countries, the effects of the crisis were felt more intensively in developed countries (Akın and Ece, 2009: 155–156). But the crisis also affected the developing world. According to Eğilmez, the crisis has affected developing countries for many reasons. The rapid rise in commodity prices is one of these reasons. In addition, hot money, which is directed towards developing countries, has started to withdraw due to the increase in risk. With the decline in growth in developed countries, the demand reduced, hence the exports of developing countries decreased and their economies shrank (2018: 297–298). World Trade Report reveals that both GDP and exports decreased significantly in all regions of the world in 2009 (WTO, 2011: 22). Owing to the global crisis, the unemployment rates have increased significantly all over the world and could not return to the pre-crisis level for several years. On the other hand, it should be emphasized that the impacts of the crisis on labour markets vary from region to region and even from country to country in the world (Tab. 10.1).

The world has witnessed a deceleration in employment growth by 2008. Employment growth decreased from 1.9 percent to −1.2 percent between the beginning of 2008 and the second quarter of 2009 in G20 countries. By the beginning of 2009, employment losses were in historical levels in various countries. The job losses in the EU-15 have reached 6.2 million between the third quarter of 2008 and first quarter of 2010. However, the impacts of the last global crisis on labour markets differ significantly worldwide. When developed G20 countries are analyzed, the countries less affected by the crisis in terms of loss in output and employment are Australia, Belgium, Canada, France and Republic of Korea. Countries having the highest output losses were Finland, Japan and Luxembourg. While Spain is the country with the highest loss in employment, Ireland is the most disadvantaged country in terms of both criteria. When developing G20 countries are analyzed, it is observed that Latvia, Estonia and Lithuania are the most disadvantaged both in terms of output and employment loss. Besides, Turkey and Russia are countries that have suffered more in output loss but less in employment loss compared to others (Eichhorst, 2010: 3–9).

Tab. 10.1: Unemployment Rates Between 2008–2011 by Regions (%). Source: ILOSTAT, 2019.

	2008	2009	2010	2011
World	5	5,6	5,5	5,4
Africa	6,4	6,7	6,8	7
Latin America and the Caribbean	6,5	7,6	7	6,5
Northern America	5,8	9,2	9,5	8,8
Arab States	7,5	7,3	7,3	7,2
Asia and Pacific	3,8	3,9	3,7	3,7
Northern, Southern and Western Europe	7,3	9,3	9,8	10
Eastern Europe	6,3	8,2	8	7,5
Central and Western Asia	8,2	9,4	8,6	7,7

4 Gender Impacts of Last Global Crisis on the Labour Markets

Since women's labour force participation and employment patterns differ in the context of regions and countries around the world, the effects of the global crisis on female labour have also varied in developed and developing countries. Therefore, the gender impacts of the global crisis on labour markets of developed countries and developing countries are examined separately. In this regard, it should be underlined that North America and European countries, which will be examined under the title of developed countries, are the countries where women's labour force participation is high. The employment rates of women in these countries are above the world average and gender inequalities in these countries are considerably less compared to developing geographies (ILOSTAT, 2019 and WEF, 2018).

4.1 Impacts on developed countries

The last global crisis, which began in the United States and spread to European countries and then spread all over the world, deeply influenced the developed countries. However, depending upon the intensity of financial relations, the share of derivative products in the financial system and financial fragility, each country has experienced the crisis differently. The decline in economic growth and the increase in unemployment rates are more evident in the countries that feel the financial crisis deeply. In the United States where the crisis started, the unemployment rate in 2008 was 5.8 %, however, it increased to 9.2 % in 2009 and to 9.5 % in 2010. The country did not recover easily in the post-crisis period, thus

Tab. 10.2: Unemployment Rates by Sex after the Global Crisis in Developed Regions (%). Source: ILOSTAT, 2019.

Years	Northern America		Northern, Western and Southern Europe	
	Male	Female	Male	Female
2008	6,2	5,4	6,9	7,9
2009	10,2	7,9	9,3	9,2
2010	10,3	8,5	9,9	9,8
2011	9,2	8,3	9,9	10
2015	5,6	5,3	9,9	10,1
2018	4,3	4	7,4	8

unemployment rates were only close to their former levels in 2015. The same situation was experienced in North, South and Western Europe: the average unemployment rate of the region was 7.3 % in 2008, however, increased to 9.3 % in 2009 and to 9.8 % in 2010. The recovery process in Europe took longer, such that, the unemployment rate was 11.2 % in 2013. Even though the related ratio was in a descending trend over time, it could not return to its previous level even in 2018 in North, South and Western Europe (ILOSTAT, 2019). When these ratios are analyzed in terms of gender, it is observed that unemployment rates of men are in a faster increase trend compared to women in developed countries after the global crisis (Tab. 10.2).

In line with the last global crisis, the basic indicators of labour markets in USA have undergone a major change and the country has witnessed the highest unemployment rates since the World War II (CNN, 2009). Since the crisis that started in USA hit the sectors where men were heavily concentrated, the men were more affected by the crisis than women. The unemployment rates of women and men, which were close to each other before the global crisis in USA, significantly changed after the crisis, and the rapidly increasing unemployment of men put women in the status of breadwinner. Many women became the primary breadwinners in many families, thus the traditional roles changed. However, this kind of change put a big strain on the back of families because the difference between the earnings of women and men was quite high. In a typical family where both spouses work, the income of the women corresponds to one-third of total income in the country, so it has become quite difficult to live with substantially reduced income. Another troubling situation is that many families could access to health insurance through the employers of their husbands (CAP, 2019). According to Seguino, unemployment rates do not tell the whole story in the country. Since the budget of many states in the USA has given a deficit, budget restrictions have been made and important limitations have been made in health, education and social services. As the occupations related to the targeted interruptions are professions that predominantly belong to women, joblessness of women has become natural (Seguino, 2011: 18–19).

The devastating effects of the crisis were felt in Europe as well. Périvier researched employment trends in eight European countries (Denmark, Germany, Spain, France, Italy, Sweden, United Kingdom and Greece) during the economic crisis and asserted that sex segregation is a common feature of European labour market and sectors where men are over-represented have been hit more than those where women are over-represented in terms of job destruction. She revealed in her study that the employment has narrowed more in manufacturing, construction, wholesale and retail trade, repair of motor vehicles and motorcycle sectors where men are concentrated. She mentioned that, in order to cope with the public deficits created by the crisis, austerity policies have been put into effect in the second half of 2010 and the public expenditures and social expenditures have been significantly reduced in many countries of Europe. Moreover, these policies have been implemented with a gender blind manner; the unequal position of women and men in the labour market has been ignored. Therefore, austerity policies have created more negative effects for women because women are mostly employed in social services sectors and the constraints in the related field cause the disappearance of job opportunities. Secondly, the working conditions of employees have been deteriorated. Austerity policies obstruct to harmonize workfamily life as women are the main users of social services that are limited by these policies. Furthermore, the findings of the study reveal that the relevant policies are more disadvantaged for women in Spain, the United Kingdom and Denmark (2014: 42–82).

According to Rodriguez, the social spending cuts in Spain, which started in May 2010, transformed the crisis into "female crisis" and destroyed various jobs held by women. In the years 2010, 2011 and 2012, destruction in female employment occurred, and in the stated years 61000, 186,800 and 308,300 jobs were lost respectively. The main reason for this situation is that women work predominantly in the sectors where social expenditures are limited. On the other hand, in the post-crisis period, there was an explicit decrease in the number of women working with social security in Spain. Especially, the social security of the domestic employees has been interrupted significantly in the post-crisis period (2013: 5–11). For another study, similar to the previous recessions in the United Kingdom, the working hours of women were reduced, employers tended to employ women in more flexible manners and women's full-time work

Tab. 10.3: Employment Rates in Developed Regions by Sex (%). Source: ILOSTAT, 2019.

Years	Northern America-Male	Northern America-Female	Northern-Western-Southern Europe-Male	Northern-Western-Southern Europe-Female
2008	67,5	55,6	61,4	46,5
2009	63,7	53,9	59,4	46
2010	62,9	53,1	58,7	45,7
2011	63,2	52,7	58,4	45,7
2015	64,8	53,4	57,9	46,3
2018	65,4	54,3	59,2	47,7

was transformed into part-time work considerably in the post-crisis period (Grimshaw and Rafferty, 2011: 525).

Employment rates of males and females decreased more in Northern America compared to Europe between 2008 and 2011. Male employment decline between the stated years in Northern America is notable (Tab. 10.3).

Even though statistics in developed countries mostly indicate that men are affected more negatively by the global crisis, it should not be ignored that female labour in these regions has been more adversely affected by the crisis than reflected in the statistics.

4.2 Impacts on developing countries

The last global crisis, which is felt more heavily in developed countries, has also affected the developing world. However, each of the developing geographies felt the crisis different from the other because of the differences in the economic structure. When examined in more detail, the countries that are mainly engaged in export-oriented production experienced the crisis even deeper. As observed in Tab. 10.4, women in developing countries are also affected by the increase in unemployment as much as men. Even in Latin American countries, women's unemployment rates increased more in the post-crisis period. Again, the rise in unemployment rates in South Asia is the same for men and women.

Asia is one of the significant developing geographies highly affected by the crisis. Exports have a significant share in the phenomenal growth of Asia; however, developed countries have significantly reduced spending in the post-crisis period. In the related process, the demand of United States and Europe for labor-intensive exports including toys, games, footwear, clothing, substantially decreased, thus, export-oriented production in Singapore, Malaysia, Cambodia,

Tab. 10.4: Unemployment Rates in Developing Regions after Global Crisis (%). Source: ILOSTAT, 2019.

Years	Sub-Saharan Africa		Latin America and the Caribbean		Arab States		Eastern Asia		Southern Asia	
	Male	Female	Male	Female	Male	Female	Male	Female	Male	Female
2008	5,3	6	5,3	8,3	6,5	13,2	4,9	3,9	2,3	3,5
2009	5,8	6,3	6,3	9,4	6,2	13,8	5,2	4,1	2,7	3,9
2010	5,8	6,5	5,9	8,8	6,1	14,1	5	3,9	2,5	3,7
2011	5,7	6,3	5,4	8,1	5,9	14,6	4,9	3,9	2,5	3,9
2015	5,6	6,2	5,7	8	5,6	15,5	4,9	3,9	2,8	4,9
2018	5,6	6,2	7	9,6	5,8	15,6	4,6	3,7	2,7	4,5

Thailand, Korea, PRC and Vietnam has decreased significantly. Thus, production in many factories in Asia has been reduced. Foreign direct investments (FDI) are the other important factors that enable the growth of Asian economies. In Singapore, Cambodia and Fiji, FDI contributed significantly to growth. Again, FDI contributed significantly to the formation of capital in Vietnam, Thailand, Malaysia and the Philippines. However, with the global crisis, FDI declined significantly, decreasing by 30 % in 2009. The significant reduction in the volume of production and investment flow naturally affected labour markets in Asian world. The first negative consequence was the increase in job losses, millions of workers in export industries lost their jobs. Subcontractors and temporary workers were the first to lose their jobs in this process because they were easier to conceal, less costly to dismiss. Gender-disaggregated analyses reveal that women tend to be harder hit by rising unemployment than men. In the developing countries of Asia, especially in Thailand, Philippines and Vietnam, the crisis has hit sectors that are particularly labour-intensive and where women workers work intensively. Moreover, the crisis has transformed formal employment patterns into informal and vulnerable employment patterns. High poverty and insufficient social security are the features of the latter patterns (Huynh et al., 2010: 3–7).

Similarly, to the Asia, the Latin American geography, which has a high export potential, has been significantly affected by the global crisis. Recession starting in the fourth quarter of 2008 has deeply affected Latin American economy by deterioration in the terms of trade, shrinking remittances from emigrants, and the massive withdrawal of private capital from financial markets (Guillen, 2011: 196). With this recession, production in Latin America has decreased significantly, while unemployment rates have increased significantly. As can be seen in Tab. 10.3, in the post-crisis period women's unemployment rates increased more than men. Since women in the region are more intensively employed in the export-oriented sectors,[1] they have faced higher unemployment with the decline in global demand and descending prices of commodities. Seguino asserts that the figures for job losses are not always fully reported in the region. In fact, the loss of jobs for women working in the export-oriented industries in Latin America and for women working in tourism in the Caribbean is more than that of men when explained in full terms (2011: 19). Moreover, those affected negatively by the crisis in Latin America are not only those working in export-oriented industries. Employees in the agricultural sector in the region have also been exposed to the negative effects of the crisis. It is determined in a study that the majority of workers in agribusinesses of ICA Valley are women in Peru and most of them are temporary workers. Together with the crisis, these women were forced to work longer without extra pay and the hours of rest were shortened. Furthermore, it is asserted that women workers who are about to complete five years in the workplace and are about to qualify for permanent employment are also dismissed and replaced with new ones (Arguello, 2011: 81–83).

Since the Arab States[2] were not integrated to the international economy as were the other regions, the production potential of these countries and the basic indicators of the workforce have not changed much. In other words, the impacts of the crisis on the region have been limited, however, Dubai which has been substantially exposed to global market trends was the most affected by the crisis (Behrendt et al., 2009: 1–4). The decline in oil prices in the relevant period affected exporting countries such as Saudi Arabia, Algeria and Yemen, hence workers' remittances abroad decreased. The decline in remittances put Egypt, Morocco and Jordon in difficult social and financial position (Orozco and Lesaca, 2009: 2).

In developing countries, the ways in which women experience crisis differ from each other. For example, in Sub-Saharan Africa, women are mostly employed in

---

1   In the relevant sector, employers choose women workers because they have "nimble fingers", are obedient, are prone to monotonous and repetitive jobs, and can be employed with lower wages than men. In addition, in this sector, working hours are long, social security payments and social assistance are few. Therefore, women's labour, which can be employed as low-paid and flexible, is a type of labour that makes global competition possible (Toksöz, 2011. 147).

2   Arab states are geographies where women's labour force participation is marginal. Solid patriarchal structures and traditional norms in these countries considerably exclude women from the labour market (Toksöz, 2011).

subsistence agriculture, so they were affected differently from women in Asia and Latin America from the crisis. Reductions in public expenditures, including health, education and other services, and decline in remittances are the factors that negatively affect women in the region (Seguino, 2011: 19).

Migrant labour has been adversely affected by the last global crisis as well as resident employees. As known, there is considerable demand for the labour of migrant workers in export industries, tourism, entertainment sector and agriculture worldwide. Moreover, "care economy" is vastly dependent upon migrant work.[3] Women constitute at least 50 % of migrant workers from Africa and Latin America and at least 80 % of migrant workers from South and Southeast Asia. However, immigrant workers have had less chance of getting a formal job in developed geographies such as Netherlands, Germany and the United Kingdom with the global crisis. Therefore, remittance flows decreased, household incomes reduced, livelihood conditions became difficult and migrants were increasingly condemned to work in precarious jobs (AWID, 2010: 19; Alberdi, 2009; AWID, 2009; WB, 2009).

On the other hand, the increase in poverty and the reduction of employment opportunities in the developing world increase illegal immigration activities, and illegal immigration leads especially women to become victims of trafficking. The same victimization can also occur for women who migrate legally, it is determined that women who have been legally sent to Japan to be employed in the entertainment sector have often been forced into prostitution (Toksöz, 2011: 191–192). Nominately, migration causes greater vulnerability to human trafficking. Women and young girls migrating to work outside their country can easily fall into the hands of human traffickers, hence, forced labour and commercial sexual exploitation can take place (USAID, 2019). Although the increase in the number of women employed in such patterns in the post-crisis period is known by everyone, it is not reflected in the official statistics, in other words this type of work is undocumented. Moreover, apart from human trafficking, women workers are more vulnerable to harassment and violence in the places where they predominate (domestic work and factories) (Gibb, 2009: 6).

Employment rates have not changed much in the developing world while unemployment rates notably changed (Tabs. 10.5 and 10.4). Therefore, based on

---

3 According to Toksöz, the increase in female labour force participation in developed countries and the "care services gap" in the rich countries increase the demand for the labour of migrant women working in the care service. For this reason, millions of women in the world migrate from poor to rich countries to provide care services (Toksöz, 2011: 195).

Tab. 10.5: Employment Rates in Developing Regions (%). Source: ILOSTAT, 2019.

Years	Sub-Saharan Africa-Female	Latin America-Male	Latin America-Female	Arab States-male	Arab States-Female	Eastern Asia-Male	Eastern Asia-Female	Southern Asia-Male	Southern Asia-Female
2008	59,5	74,6	46,5	70,3	14,7	74	60,4	79,4	27
2009	59	73,7	46,5	70,7	14,4	73,5	59,8	78,7	26,2
2010	58,6	73,7	46,5	71,2	14,6	73,3	59,4	78,4	25,5
2011	58,7	73,7	46,5	71,6	14,7	73,2	59,3	78	24,6
2015	58,8	73,1	47,1	72,9	15,2	72,5	58,7	77	24,2
2018	59,1	71,7	46,7	72,8	15,5	71,9	57,9	76,8	24,8

the findings of the study mentioned above, it is possible to state that the ratios of employment in the developing world have not significantly changed in the post-crisis period, however, employment patterns have changed especially in certain regions.

Finally, growing financial difficulties increase school dropout rates among children in the developing world. Country case studies conducted in Mongolia and Pakistan in 2009 by UNESCO indicated that vulnerable households could not afford the school costs of their children, and as a consequence, school dropouts and child labour increased in these countries (Chang, 2010: 18–19). As to gender, girls are more disadvantageous in this regard. Girls are more likely to dropout because their parents can push them into the workforce to supplement scarce household incomes (USAID, 2019). In brief, similar to other crises, the global crisis also limited the educational opportunities of girls living in poor families.

5 Conclusion

The last global crisis, which began as a financial crisis in the United States in September 2008 and spread to the whole world in a short period, affected all countries in varying proportions by the domino effect. Developed countries were more severely affected by the crisis, and the production and trade volumes of the stated countries narrowed significantly. The decline in the economic activities of the developed countries adversely impressed the exports of developing countries and the direct capital investments decreased significantly in the related process. The recession in economies has rapidly influenced labour markets and job losses and unemployment rates have increased dramatically. When we look at the impacts on labour markets from a gender perspective, statistics reveal that men are more adversely affected by the global crisis in developed countries. However, it should be noted that women are disadvantaged compared to men in terms of basic indicators of the labour market in the pre-crisis period and the effects of the crisis on women's labour are not fully reflected in the statistics. For example, with the crisis, social service expenditures in the developed countries were severely limited and this policy had a negative impact on women's labour as women's working hours were increased, wages were reduced and flexible employment became widespread.

Impacts on developing geographies are varying; in some regions where export-oriented production is intense, women's unemployment rates exceeded men's in the post-crisis period, and in some regions it remained equal. The formal employment of women in the developing countries has been transformed into informal employment and precarious employment has become a widespread form of work. The crisis also affected the migrant women's labour negatively as the demand for immigrant labour in developed countries reduced. On the other hand, migrant women who were able to find employment opportunities were forced to work under more adverse working conditions. The decline in remittances put many families in difficult social and financial conditions in the developing world. Moreover, school dropouts and child labour increased in the post-crisis period.

Eventually, the effects of the last global crisis on women's labour are more than the statistics indicate. This is because the last global crisis has created "invisible burden" on women's labour in both developed and developing countries. If the impacts of the last global crisis on women's labour are to be calculated in a realistic manner, the losses and invisible burdens (that are not reflected or are reflected on a limited scale in the figures/statistics) should be taken into account comprehensively.

References

Akin, F. & Ece, N. (2009). Küresel Finansal Kriz ve Bankacılık Sektörü İstihdamı Üzerindeki Etkileri. *Marmara Üniversitesi İİBF Dergisi*, 17(2), 153–168.

Alberdi, I. (2009). The World Economic and Financial Crisis: What Will It Mean for Gender Equality?. Speech given at the *5th Annual meeting of Women Speakers of Parliament*, Vienna, Austria, 13 July 2009.

Argulello, R. (2011). Securing the Fruits of Their Labours: The Effect of the Crisis on Women Farm Workers in Peru's Ica Valley. In *Gender and the Economic Crisis*, edited by R. Pearson and C. Sweetman. United Kingdom: Practical Action Publishing & Oxfam.

Aslanbeigui, N. & Summerfield, G. (2000). The Asian Crisis, Gender and the International Financial Architecture. *Feminist Economics*, 6(3), 81–103.

AWID. (2010). The Impact of the Crisis on Women: Main Trends across Regions. Author: Natalie Raaber, *Brief 11*, Published by the Association for Women's Rights in Development (AWID).

AWID. (2009). The Impact of the Crisis on Women in Central and Eastern Europe, by Ewa Charkiewicz. Retrieved from https://www.awid.org/sites/default/files/atoms/files/icw_2010_easterneurope.pdf (02.04.2019).

Behrendt, C., Haq, T., & Kamel, N. (2009). The Impact of the Financial and Economic Crisis on Arab States: Considerations on Employment and Social Protection Policy Responses. *Policy Note*, ILO: Regional Office for Arab States.

Betcherman, G. & Islam, R. (eds.). (2001). *East Asian Labor Markets and the Economic Crisis. Impacts, Responses and Lessons*. Washington, DC: The World Bank and International Labour Office.

CAP. (2019). Women Breadwinners, Men Unemployed. Center for American Progress. Retrieved from https://www.americanprogress.org/issues/economy/news/2009/07/20/6314/women-breadwinners-men-unemployed/ (07/04/2019).

Chang, G. (2010). Monitoring the Effects of the Global Crisis on Eductaion Provision. *Current Issues in Comparative Education*, 12(2), 14–20.

CNN. (2009). Worst Year for Jobs Since '45. Retrieved from https://money.cnn.com/2009/01/09/news/economy/jobs_december/ (07.04.2019).

Eğilmez, M. (2018). *Makro Ekonomi, Türkiye'den Örneklerle*. İstanbul: Remzi Kitabevi.

Eichhorst, W., Escudero, V., Marx, P., & Tobin, S. (2010). The Impact of the Crisis on Employment and the Role of Labour Market Institutions. Discussion Paper Series. Geneva: ILO.

Erdönmez, P. A. (2009). "Küresel Kriz ve Ülkeler Tarafından Alınan Önlemler Kronolojisi", *Bankacılar Dergisi*, Sayı 68, 2009, s: 85–101.

Focus on Global South and Save the Children. (2001). The Asian Financial Crisis and Filipino Households: Impact on Women and Children. Retrieved from https://focusweb.org/sites/www.focusweb.org/files/pdf/women%20and%20children.pdf (29.03.2018).

Gibb, H. (2009). Impacts of the Economic Crisis: Women Migrant Workers in Asia. *IWG-GEM Conference, Gender and Global Economic Crisis*, July 2009, New York.

Gragnolatti, M. (2001). The Social Impact of Financial Crisis in East Asia: Evidence from the Philippines, Indonesia and Thailand, East Asia

Environment and Social Development Unit. EASES Discussion Paper Series. World Bank.

Guillen, R. A. (2011). The Effects of the Global Economic Crisis in Latin America. *Brazilian Journal of Political Economy*, 31(2), 187–202.

Grimshaw, D. & Rafferty, A. (2011). Social Impact of the Crisis in the United Kingdom: Focus on Gender and Age Inequalities. In *Work Inequalities in the Crisis, Evidence from Europe*, edited by D. Vaughan-Whitehead. United Kingdom: Edward Elgar, Switzerland: ILO, 525–569.

Horton, S. & Mazumdar, D. (2001). Vulnerable Groups and the Labor Market: The Aftermath of the Asian Financial Crisis. In *East Asian Labor Markets and the Economic Crisis*, edited by G. Betcherman et al., Washington, DC: The World Bank, 379–422.

Huynh, P., Kapsos, S., Kim, K. B., & Sziraczki, G. (2010). Impacts of Current Global Economic Crisis on Asia's Labor Market. ADBI Working Paper Series, No. 243, Asian Development Bank Institute.

IMF. (2019). Recession, When Bad Times Prevail. Retrieved from https://www.imf.org/external/pubs/ft/fandd/basics/recess.htm (13.05.2019).

ILO. (2009). Asia in the Global Economic Crisis: Impacts and Responses from a Gender Perspective. *Technical Note*, Asian Decent Work Decade 20062015, ILO Bangkok, ILO Geneva.

ILOSTAT. (2019). International Labour Office. Retrieved from https://www.ilo.org/ilostat/faces/oracle/webcenter/portalapp/pagehierarchy/Page3.jspx?MBI_ID=2&_afrLoop=1652890328110875&_afrWindowMode=0&_afrWindowId=1dsqmgpom6_1#!%40%40%3F_afrWindowId%3D1dsqmgpom6_1%26_afrLoop%3D1652890328110875%26MBI_ID%3D2%26_afrWindowMode%3D0%26_adf.ctrl-state%3D1dsqmgpom6_57 (06.04.2019).

Islam, R., Bhattacharya, G., Dhanani, S., Iacono, M., Mehran, F., Mukhopadhyay, S., & Phan, T. (2001). The Economic Crisis: Labor Market Challenges and Policies in Indonesia. In *East Asian Labor Markets and the Economic Crisis*, edited by G. Betcherman et al., Washington, DC: The World Bank, 43–96.

Jones, G. W., Hull, T. H., & Ahlburg, D. (2000). The Social and Demographic Impact of the Southeast Asian Crisis of 199799. *Journal of Population Research*, 17(1), 39–62.

Lim, J. Y. (2000). The Effects of the East Asian Crisis on the Employment of Women and Men: The Philippine Case. *World Development*, 28(7), 1285–1306.

McGee, T. G., Angeles, L., Bautista, C., Daud, S., Hainsworth, G., Scott, S., Setiawan, B., Suksiriserekul, S., & Anh, V. T. (1999). The Poor at Risk: Surviving the Economic Crisis in Southeast Asia. *Final Report of the Project*

Social Safety Net Programs in Selected Southeast Asian Countries, 19972000. Canadian International Development Agency.

Périvier, H. (2014). Men and Women During the Economic Crisis Employment Trends in Eight European Countries, *Revue de l'OFCE - Débats et politiques*, 133(2), 41-84.

Pilwha, C. (1998). The Impact of the Economic Crisis on Women Workers in the Republic of Korea: Social and Gender Dimensions. Working paper prepared under the AIT/ILO Research Project on the Gender Impact of the Economic Crisis in Southeast and East Asia. Geneva: ILO.

Rodriguez, M. L. (2013). Effects of the Economic Crisis on Gender Equality, the Spanish Case. *Labour Law Research Network Inaugural Conference*, Barcelona.

Seguino, S. (2011). The Global Economic Crisis, Its Gender and Ethnic Implications and Policy Responses. In *Gender and the Economic Crisis*, edited by R. Pearson and C. Sweetman. United Kingdom: Practical Action Publishing & Oxfam, 15-36.

Orozco, O. & Leseca, J. (2009). Impact of the Global Economic Crisis in Arab Countries: A First Assessment. *Preparatory Document*, Casa Arabe and the Club of Madrid.

Toksöz, G. (2011). *Kalkınmada Kadın Emeği*. İstanbul: Varlık Yayınları.

Toksöz, G., 2009. Kriz Koşullarında Toplumsal Cinsiyet Perspektifinden İşgücü Piyasaları. Retrieved from http://www.keig.org/wp-content/uploads/2016/03/GulayToksoz-toplcinsisgucu-Rapor2009.pdf (01.04.2019).

Tutnjevic, T. (2002). Gender and Financial Economic Downturn. Working Paper 9, Recovery and Reconstruction Department. Geneva: ILO.

UN. (2009). Emerging Issue: The Gender Perspectives of the Financial Crisis, Written Statement, Submitted by Shamika Sirimanne. *United Nations Economic and Social Commission for Asia and the Pacific, Commission on the Status of Women Fifty-Third Session*, New York, 2 -13 March 2009.

USAID. (2019). The Economic Crisis: The Impact on Women. *Fact Sheet*, United States Agency. Washington, DC.

WORLD BANK. (2009). *Migration and Development Brief 11*. Washington, DC: World Bank.

WEF. (2018). *The Global Gender Gap Report 2018*. World Economic Forum.

WTO. (2011). *World Trade Report 2011*. Retrieved from https://www.wto.org/English/res_e/booksp_e/anrep_e/world_trade_report11_e.pdf (06.04.2019).

Mehmet Kenan Terzioğlu

11. Poverty, Income Inequality, Unemployment and Human Capital in the Democratization Process

Abstract Democracy, governance by the people, is the subject of different definitions and practices as a result of the economic, social and political events experienced during the historical process. The level of development and experiences of societies expand the scope of democracy. In developed nations, differences can be observed in how democracy is practiced due to structural diversities. On the other hand, developing economies can ensure the functioning of the democratization process by taking measures to improve their social and economic structures. As democracy emerged as a result of economic developments, the interaction of democracy with macroeconomic magnitudes is essential in defining the process of democratization. The aim of this chapter is to reveal the link between economic development and democratization process by considering macroeconomic variables such as poverty, income inequality, unemployment and human capital. The poverty-risk ratio is taken as the poverty variable, the Gini equivalent disposable income coefficient is taken as the income inequality variable, the unemployment rate between the ages of 15–74 and the share of unemployed for 12 months or longer is taken as the unemployment variable, and the total share of health and education expenditures of countries in comparison with their total expenditures is taken as the human capital variable.

Keywords: Democratization, Human capital, Unemployment, Poverty

1 Introduction

The perception of democracy is mentioned in situations such as the establishment and participation of organization, freedom of expression, participation in public life, free election and right to vote and easy access to different sources of information (Dahl, 1973). In many countries where the level of human capital development is high, the form of government emerges as democracy. There is a positive relationship between the increase in welfare levels of societies and their willingness to sustain democracy (Lipset, 1959). Countries with moderate economic development are more likely to transition to democracy. Therefore, the social and economic development situation of countries is an essential factor indicating the degree of democratization (Jackman and Bollen, 1989).

Income distribution refers to the division of the total income of goods and services and the welfare increases created in a country by the individuals and social classes living in that country. If this indicator is unequal, the population that gets less share of the income distribution becomes weighted in the population, and the phenomenon of 'injustice' emerges in economic and social terms. Although this situation causes income inequality, it adversely affects the development of countries. The disparity of income levels, which arises from the lack of balanced fairly distribution of economic resources among people, is considered to be one of the significant causes of poverty. Poverty can be considered under two categories: those who lack the potential to use labor power and those who have the potential to use labor power. Absolute poverty, the deprivation of income and consumption that can meet the basic needs of the individual or household, makes an assessment on the minimum subsistence needs and limits poverty to material deprivation. In addition to basic physical needs, social needs and the average standard of living of the society are expressed as relative poverty (Haralambos and Holborn, 2008). While there is an increase in the number of people employed in the countries where economic growth occurs, in some cases, employment cannot be provided and the informal economy, which expresses all economic activities contributing to the officially calculated gross national product, emerges for both developed and developing countries in the reduction of unemployment (Thomas, 1999).

The effect of human capital, which is the source of growth and development and expresses concepts such as education level, health status and place in social relations, varies according to the development levels of countries. Human capital accumulation, demonstrating the ability to compete internationally, is an important indicator in explaining macroeconomic performance. Development that brings equal opportunity and improvement of living (education, health, social security, technological investments, income, etc.) provides a balanced and sustainable growth process. Human capital is also considered to be one of the factors of production, such as physical capital (Lucas, 1998). Since human capital is effective in increasing knowledge and assimilating the acquired technology, human capital becomes an effective factor in the economic growth process, if production scales and potential are used effectively. Economic growth and development make an increase in welfare and quality of life.

Within this study, it is aimed to reveal the effects of factors such as income distribution inequality, poverty, human capital, industrialization, technological development, market development and economic growth on democratization process and sustainability. After giving general information about democratization process and selected macroeconomic variables in the second part, econometric method used in modeling is presented in the third part. The relationship between democratization process and macroeconomic variables is examined in the fourth part. The last section is devoted to the findings.

2 Economic Literature

Poverty is an important phenomenon that separates societies both economically and spatially with the change and diversification of production factors and economic mechanism. With globalization, which has a social, economic and political dimension, the countries that have integrated their economies into international markets approach the growth targets faster and reduce poverty. On the other hand, unsuccessful experiences in economic policies that are being tried to be implemented in developing countries can contribute to impoverishment by increasing the depth between the social strata. The fight against poverty, which requires the creation and implementation of new and sustainable social policies, comes into prominence in developed and developing countries' policies as it brings social, political, and economic changes. According to the development level of countries, social supports are generally funded by taxes and since they are developed by the countries as a mechanism for alleviating, delaying and eliminating the phenomenon of poverty, they are provided according to certain content and genre. However, cost increases and scope changes affect the sustainability of social supports (Terzioğlu et al., 2018). Neutel and Heshmati (2006) show that globalization reduces income distribution inequality and poverty. Maertensa et al. (2011) emphasize that globalization is effective in reducing poverty and increasing income levels.

Per capita income and income inequality positively and negatively affect the speed and possibility of transition to democratization (Bourguignon et al., 2000). Since there is an inverse U relationship between income distribution and economic growth, as income increases, inequality increases first, and when it reaches a certain level of development, growth rate increases and income inequality decreases (Kuznets, 1955). Especially in developing countries, the convenience of entry and exit in the informal system and the failure to create necessary employment in the formal sector provide the development of the informal economy. Since most of the informal income is spent in the formal sector, the economy is active and contributes to the indirect tax increase (Terzioğlu et al., 2018). Schneider and Enste (2000) state that the act of spending a large portion of the informal income in the economy has a positive value indicator. Kızılot and Çomaklı (2004) pointed out that while the growth of the formal economy was adversely affected by the increase in the informal economy in developed countries, this relationship is positive in developing countries. Chen (2012) found that unemployment and underemployment in developed countries decreased with economic growth, while in developing countries, they increase with economic growth. In this context, the labor force that emerged in the industrial and service sectors along with economic development leads to the growth of the middle class in cities and strengthens democracy (Muller, 1995). The concept of urbanization that emerged with the development of trade is an important variable in democratic development. Also, with the development of capital markets, the spread of capital to the base may lead to an increase in the middle class. The growth of the middle class with modernization and the adoption of electoral policies in order to gain benefits put the middle class at the heart of the democratization movement (Huntington, 1993). On the other hand, income inequality that emerges with development leads to polarization between classes and has a negative impact on the sustainability of democracy (Muller, 1995). In this case, not only the quality but also the sustainability of democracy is essential (Beetham et al., 2005). Since democracy is the rule of the majority, the majority of the poor classes make democratization difficult. While the majority of the poor or upper class makes the democratization movement difficult, the middle class emerges as a democratic power (Rueschemeyer et al., 1992). In developing democracies, different segments of society can get a share from development through sustainable development.

Human capital, which provides benefits in the long run by raising the level of health, education and activity, has a positive effect on national economies by increasing productivity. Denison (1962) states that education contributes to the increase of national income by improving the skills and productivity capacity of the workforce. As knowledge and skills are developed as a result of investments in human capital, positive returns are obtained at economic and social levels with the increase in productivity. Societies with higher education levels can promote economic growth and development in the long run by developing technological innovations. The level of education increases the marginal productivity of physical capital and labor, and resultantly, increases the national income of the country. Education and health are two main elements in the human capital stock. Educated and healthy societies improve living standards in the short, medium and long run and contribute positively to economic growth (Terzioğlu et al., 2018). Mayer (2001) shows that health is more important for human capital in the long term than education. Gyimah et al. (2004) show that health shocks cause an increase in per capita income. Duman (2008) emphasizes that education expenditures have a reducing effect on income inequality. Education, making able to build cause-effect relationship and to develop the freedom of thought, is effective in the development of democratization and in increasing the demand for democracy (Tavares et al., 2001). There is a relationship between support on democratic values and quality of education (Smith, 1948). The perspectives that have been developed with the development of the level and quality of education decrease the adoption of radical doctrines (Lipset, 1959). The entrepreneurial factor, which is one of the factors of production, stands out in the market economies (Pech and Cameron, 2006). The difference between the development levels of countries which occur as a result of the progress in industrialization, entrepreneurship and technological advances affects the growth and sustainable development of the countries. Developing societies aim at sustainable development, even if their priority and order of importance changes. Education and technological progress are essential for economic and social development. In countries with high educational quality, productivity increases in production and competitive advantage is ensured with technological developments and industrialization. When advanced technology and specialization are combined with an advanced management style, significant differences occur in the development levels of countries (Terzioğlu et al., 2018).In the countries that cannot produce technology, cannot transfer technology correctly, and cannot take place in the production processes, growth process has slowed down. Entrepreneurship, which reorganizes the economic structure as a result of new product production and process development, new export-markets and new raw-material resources and new combinations, makes the operation of socio-economic development effective by creating a basis for development (Döm, 2008). Since the development of technology and industrialization occur together with social change and technological development brings along the increase in production level and industrialization, there is a positive relationship and mutual causality between the wealth that emerges as a result of industrialization and democracy (Huntington, 1993).

Lipset (1959) states that as the income level increases, the democracies of the countries strengthen. Coleman (1960) emphasizes that socio-economic development positively affects democracy. Neubauer (1967) stresses that the possibility of socio-economically developed societies to be democratic is not clear. Cutright (1969) shows that the level of political development is positively affected by the level of communication, economic development, education, and urbanization. Kim (1971) emphasizes that there is no strong and positive relationship between socio-economic development and measures of democracy. Arat (1988) states that the relationship between economic development and democracy is not important for developed countries and that there is a positive relationship between economic development and democracy at lower stages of economic development in middle-income countries. Huber et al. (1993) mention that poor economic performance is a critical problem for democracy. Helliwel (1994) reveals that democracy is positively affected by income. Burkhart et al. (1994) conclude that economic development increases democratic performance. Muller (1995) states that income inequality affects democracy and counteracts the positive impact of economic development on democracy. Londongen et al. (1996) emphasize that income has a positive effect on democratization. Przeworski et al. (1997) states that income increases the possibility of maintaining democracies. Barro (1999) shows that democracy is positively affected by gross national product and education; while it is negatively affected by urbanization and natural resources. Glasure et al. (1999) stress that economic development leads to lower levels of democracy in developing countries. Colaresi et al. (2003) state that economic development supports democratization. Boix et al. (2003) express that economic development positively affects democratization. Gould (2003) emphasizes that economic growth provides democratization. Milanovic (2005) shows that income is essential for the transition to higher levels of democracy. Acemoglu et al. (2005) point out that when historical factors are not taken into consideration, there is a positive relationship between democracy and income, but when historical factors are taken into account, there is no long-term effect. Robinson (2006) shows the relationship between income and democracy. Epstein et al. (2006) state that economic development has an impact on the emergence of democracies and that income, growth, population density, and percentage of the urban population affect democratization.

3 Limited Dependent Panel Data Model

The probability of occurrence of the event at time t and the unit i^{th} is expressed, with the explanatory variables, as $P_{it} = E(Y_{it}|X_{it}) = F(\beta'X_{it})$ In the $Y_{it} = \beta'X_{it} + \mu_i + u_{it}$ model where the unit effect (μ_i) is assumed to be constant, $P(Y_{it} = 0)$ is calculated as

$$P(u_{it} \geq -\beta'X_{it} - \mu_i) = \int_{-\beta'X_{it} - \mu_i}^{\infty} f(u_{it}) \, du_{it} = 1 - F(-\beta'X_{it} - \mu_i).$$

The probability density function of u_{it} is $P(u_{it} \geq -\beta'X_{it} - \mu_i) = F(\beta'X_{it} - \mu_i)$ if the $f(u_{it})$ is symmetric as in normal and logistic density functions. A fixed effect or random effect model is formed according to the structure of the relation between the explanatory variables and the unit effect μ_i that represent the heterogeneity of unobservable units. If the $f(\mu_i/x_{it})$ conditional distribution is independent of x_{it}, then the random effects model is formed and there is no constraint on the distribution of heterogeneity. When the distribution is unconstrained, the fixed effects model is formed because μ_i and x_{it} can be correlated (Greene, 2003).

Under the assumption of fixed effects, the estimation of the limited dependent variable panel data model is performed by the maximum likelihood method. Even if estimators are consistent while the observation size (T) goes to infinity, due to the fact that the number of μ_i parameters increases with the number of cross-sectional units but does not provide additional information for μ_i and the observation size, generally, is short, there is a limited number of Y_{it} observations containing μ_i information and it raises incidental parameter problem (Neyman and Scott, 1948). All estimates of μ_i are biased if the observation size is finite. In linear regression models, when the size of the observation is finite, β and μ_i are asymptotically non-correlated and β's consistent estimators can be obtained. In nonlinear models, although the number of cross-sectional units goes to infinity, the maximum likelihood estimator of β is inconsistent since the estimation of β and μ_i is not independent. In other words, if the observation size is infinite in non-linear models, then the maximum likelihood estimator is consistent; if it's less, inconsistent estimators are obtained. In small samples, the conditional likelihood function is used to estimate parameters under the assumption of fixed effects (Andersen, 1970). If there is a minimum sufficient statistics (τ_i) for the unit effect and this statistic is independent of the structural parameters, the conditional probability density function is $f(Y_i|\beta,\tau_i) = \frac{(Y_i|\beta,\mu_i)}{(\tau_i|\beta,\tau_i)}$ for $g(\tau_i|\beta,\mu_i)$ Since the dependence of β to μ_i has been eliminated, $(\tau_1,\ldots,\tau_N)$, a consistent estimation of β is obtained by maximizing the conditional density function of $(Y_1,\ldots,Y_N)$. In the logit model, when the joint probability of Y_i is

$$P(Y_i) = \frac{\exp\{\mu_i(\sum_{t=1}^{T} Y_{it}) + \beta'(\sum_{i=1}^{T} X_{it} Y_{it})\}}{\prod_{i=1}^{T}[1+\exp(\beta' X_{it}+\mu_i)]},$$ taking the partial derivative of log-likelihood function according to μ_i and performing necessary operations, $\sum_{t=1}^{T} Y_{it}$ which is the minimum sufficient statistic of μ_i can be obtained as $\sum_{t=1}^{T} Y_{it} = \sum_{t=1}^{T} \left[\frac{e^{\beta' X_{it}+\mu_i}}{1+e^{\beta' X_{it}+\mu_i}} \right]$ for i=1,...,N. While $\tilde{B}_i = \{(d_{i1},\ldots,d_{iT})|\ d_{ij}$ is equal to 0 or 1 and $\sum_{t=1}^{T} d_{it} = \sum_{t=1}^{T} Y_{it}\}$,

$$P\left(\sum_{t=1}^{T} Y_{it}\right) = \frac{T!}{\left(\sum Y_{it}\right)(T - Y_{it}!)} * \frac{\exp\left[\mu_i\left(\sum_{t=1}^{T} Y_{it}\right)\right]}{\prod_{i=1}^{T}[1 + \exp(\beta' X_{it} + \mu_i)]} * \left[\sum_{d \in B_i} \exp\left(\beta' \sum_{t=1}^{T} X_{it} d_{it}\right)\right]$$

and the conditional probability of $\left(\sum_{t=1}^{T} Y_{it}\right)$ for Y_i can be shown as

$$P\left(Y | \sum_{t=1}^{T} Y_{it}\right) = \frac{\exp\left\{\beta'\left(\sum_{t=1}^{T} X_{it} Y_{it}\right)\right\}}{\sum_{d \in B_i} \exp\left(\beta' \sum_{t=1}^{T} X_{it} d_{it}\right)} * \frac{\left(\sum Y_{it}\right)! \left(T - \sum Y_{it}\right)!}{T!} \quad (1)$$

The conditional likelihood function is independent of the exogenous parameters. When g(.) is the density and θ shows the model-specific parameters, in a nonlinear model, the log-likelihood function for T observation and for N unit can be expressed as

$$LogL = \sum_{i=1}^{N}\left[\sum_{t=1}^{T_i} \log g(Y_{it}, \beta' X_{it} + \mu_i, \theta)\right], i = 1, \ldots, N, t = 1, \ldots, T \quad (2)$$

In random effects model, it is assumed that μ_i is not correlated with X_i in random effects models. Log-likelihood function for a random sample obtained from univariate G distribution

$$Log\, L = \sum_{i=1}^{N} \log \int \prod_{t=1}^{T} F(\beta' X_{it} + \mu_i)^{Y_{it}} [1 - F(\beta' X_{it} + \mu_i)]^{1 - Y_{it}} dG(\beta, \mu_i) \quad (3)$$

Consistent estimators for β and μ_i are obtained by maximizing Equation (3) while N goes to infinity. If there is a correlation between μ_i and X_{it}, the estimators are inconsistent. In the random effects model, there is a restriction of μ_i and X_{it} being uncorrelated. The orthogonality conditions for μ_i and X_{it} are examined using the independence test (Avery et al., 1983). It is necessary to know the distribution of μ_i with respect to X_{it} should be known in the estimated model, in the absence of externality. There are different opinions for the relations between μ_i and X_{it} (Mundlak,1978; Chamberlain,1980). If the univariate distribution function for η is shown as G*, using the relation that is put forward by Chamberlain (1980) in Equation (3), the log-likelihood function

$$Log\ L = \sum_{i=1}^{N} log \int \prod_{t=1}^{T} F(\beta'X_{it} + m'X_i + \eta)^{Y_{it}} [1 - F(\beta'X_{it} + m'X_i + \eta)]^{1-Y_{it}} dG^*(\eta)$$

(4)

is obtained (Tatoğlu, 2012).

4 Democratization and Macro-Economic Variables

While democracy is difficult to measure, there are many measures for democracy based on international comparisons. In this study, the effects of poverty (PVT), income inequality (GINI), unemployment (UNEMP) and human capital (HC) on democratization process using the democracy index published by Economist Intelligence Unit (EIU) are investigated using limited dependent panel data model. The democracy index is being created for 165 independent countries and two regions, covering the vast majority of the world's population and countries. It is based on five categories, which are electoral process and pluralism, civil liberties, government functioning, political participation, and political culture. According to the scores obtained from the indicators in these categories, four types of countries (full democracy, flawed democracy, hybrid regime, and authoritarian regime) are classified. Tab. 11.1 shows the classification for the year 2018 according to the type of regime in the world.

The data on poverty, income inequality, unemployment, and human capital are obtained from Eurostat and OECD database. In the study, Western European countries are divided into two as full and flawed democracy. Tab. 11.2 shows the inter-regional regime structure and democracy index for 2017 and 2018 for Western and Eastern European countries. Tab. 11.3 shows the Western European Countries' regime structures in 2018. All countries with full democracy and flawed democracy in Western European countries were included in the analysis by considering the 2018 period. The dependent variable used in this study is

Tab. 11.1: Democracy Index 2018 by Regime Type. Source: The Economist Intelligence Unit.

	Number of Countries	Country Share	World Population Share
Full Democracy	20	12.0	4.5
Flawed Democracy	55	32.9	43.2
Hybrid Regime	39	23.4	16.7
Authoritarian Regime	53	31.7	35.6

Tab. 11.2: Democracy Among Regions. Source: The Economist Intelligence Unit.

Year	Number of Countries	Democracy Index	Full Democracy	Flawed Democracy	Hybrid Regime	Authoritarian Regime
West Europe						
2018	21	8.35	14	6	1	0
2017	21	8.38	14	6	1	0
East Europe						
2018	28	5.42	0	12	9	7
2017	28	5.40	0	12	9	7

qualitative variable, which shows if the country have flawed democracy or not by taking 1 and 0 values. A value of 1 indicates that the country has a flawed democracy, and a value of 0 indicates that the country has full democracy. The independent variables are poverty, income inequality, unemployment, and human capital.

The democracy index is based on the score of 60 indicators divided into five categories with a scale of 0 to 10 in each category. The total index is obtained by taking the simple average of five category indices. Index values are used to classify countries into one of four regime types. If the index value is greater than 8, it indicates a full democracy; greater than 6 and less than or equal to 8 indicates a flawed democracy; greater than 4 and less than or equal to 6 indicates a hybrid regime; and if it is less than or equal to 4, it indicates an authoritarian regime. Full democracy expresses the structure in which the political freedoms and civil liberties are respected, the political culture favorable to the development of democracy, the functioning of the government is satisfactory, the media is independent and diverse, there is an effective control and balance system, the judiciary is independent and applied. Flawed democracy refers to a structure that has free/fair elections, respecting fundamental civil liberties. However, the problems of governance include weaknesses such as underdeveloped political culture, low levels of political participation, and violations of media freedom. The hybrid regime refers to a structure in which there are irregularities that prevent elections from being free/fair, there are weaknesses in political culture, the functioning of the government and political participation, there is a corruption tendency, the rule of law is weak, the civil society is weak, there are pressures on the media, and the judiciary is not entirely independent. The authoritarian regime refers to the structures with dictatorship, although they are officially democratic institutions.

Tab. 11.3: Western-Eastern Europe Regime Structure, 2018. Source: The Economist Intelligence Unit.

Western Europe	Full Democracy	Norway	Ireland	Luxembourg	Malta
		Iceland	Finland	Germany	Spain
		Sweden	Switzerland	England	
		Denmark	Netherlands	Austria	
	Flawed Democracy	Portugal	Belgium	Cyprus	
		France	Italy	Greece	
	Hybrid Regime	Turkey			
	Authoritarian Regime	-			
Eastern Europe	Full Democracy	-			
	Flawed Democracy	Estonia	Lithuania	Poland	Serbia
		Czech Rep.	Slovakia	Hungary	Romania
		Slovenia	Bulgaria	Croatia	
	Hybrid Regime	Albania	Montenegro	Kyrgyzstan	
		Macedonia	Ukraine	Bosnia and Her.	
		Moldova	Georgia	Armenia	
	Authoritarian Regime	Belarus	Russia	Uzbekistan	Turkmenistan
		Kazakhstan	Azerbaijan	Tajikistan	

Before the analysis is performed, it is necessary to examine whether the series are stationary over time to ensure that the results are not misleading. After it is found no cross-section dependence, the first generation tests; Im, Pesaran and Shin test, Fisher Philips-Perron test, Hadri test (which gives consistent results when T is small) are examined in Tab. 11.4, and according to the results of unit root tests, it is observed that variables are stationary at level. The Hausman test is used to make a choice between the fixed effects and random effects model and it is determined that the random effects model is appropriate and its estimators are consistent.

In Tab. 11.5, it is seen that all of the coefficients of the independent variables in the models are statistically significant in explaining the dependent variable. Factors affecting democratization are poverty, human capital, income inequality, and unemployment. Since the model is not linear, the odds ratios are obtained. It is observed from the Wald test result that the model is statistically significant. In the variance component, the logarithm of the

Tab. 11.4: Unit Root Test Results. Source: Author's own calculations

		PVT I(0)	HC I(0)	GINI I(0)	UNEMP I(0)
Hadri		7.11**	14.42**	5,25**	10,52**
Im, Pesaran and Shin		−2.48**	4.14	−2,33**	−1,37**
Fisher-Philips-Perron	Inverse $\mathcal{X}^2$	132.41**	124,49**	106,13**	163,75**
	Inverse normal	−2.41**	−2,43**	−2,53**	−1,96**
	Inverse logit	−5.07**	−4,89**	−4,15**	−5,05**
	Modified inv. Inverse $\mathcal{X}^2$	10.33**	9,44**	7,39**	13,83**
Pesaran's CADF		2.61**	2,61**	2,61**	2,61**

Notes: *,** and *** represent 10 %, 5 % and 1 % significance levels respectively.

variance of the unit effect $ln\sigma_\mu^2$, the standard error of the unit effect σ_μ, and the unit-effect variance in the total variance is ρ. When ρ closes to zero, the unit effect variance becomes insignificant, and the panel estimation has no differences from the pooled estimation. After testing the equality of ρ to zero by likelihood ratio (LR) statistics, random effects logit model is preferred over pooled logit model. Random effects logit model is estimated by the maximum likelihood estimation method. According to Durbin-Watson autocorrelation testing, Levene-Brown-Forsythe test and Friedman test, respectively, it can be said that there is no autocorrelation, heteroscedasticity and correlation between units. In random effects logit model, the variable which is statistically significant and has a odds ratio value greater than 1, refers to a variable that has a significant effect on dependent variable. Significant variables that have a large odds ratio and have an effect on democratization are defined as income inequality (odds ratio is 3.5119) and unemployment (odds ratio is 1.6331). Poverty, income inequality and unemployment significantly and positively affect the possibility to be in the flawed democracy, while the increase in human capital significantly and negatively affects the possibility to be in the flawed democracy. In other words, while each unit increase in poverty, income inequality and unemployment increases the possibility of having flawed democracy; the increase in human capital decreases the possibility of having flawed democracy and increases the possibility of transition to full democracy.

Poverty, Inequality, Unemployment and Human Capital

Tab. 11.5: Random Effects Model Estimation Results. Source: Author's own calculations

Logit Model		
DCY	Coefficient	Std. Deviation
PVT	.6597784**	1.121995
HC	-.7280079**	.594944
GINI	3.511975**	3.947776
UNEMP	1.633153**	.4414307
Const.	.00633*	.000543
$ln\sigma_\mu^2$	5.892381	.6824123
σ_μ	19.03331	6.494281
ρ	.9910004	.0060862
LR test of ρ =0:79.66	Wald $\mathcal{X}^2$ =13.68	Log Likelihood = −11.443856
Probit Model		
DCY	Coefficient	Std. Deviation
PVT	.0909288*	.1891689
HC	-.1100149 *	.0853388
GINI	.2718882**	.1656375
UNEMP	.086413**	.0282832
Const.	.2236634*	1.0360160
$ln\sigma_\mu^2$	1.206584	.5390496
σ_μ	1.828127	.4927256
ρ	.769694	.0955547
LR test of ρ =0:80.02	Wald $\mathcal{X}^2$ =29.43	Log Likelyhood = -11.461161
Complementary Log-Log Model		
DCY	Coefficient	Std. Deviation
PVT	1.73708**	.8169357
HC	-.8553667**	.1874206
GINI	1.304357**	.4882236
UNEMP	1.164308*	.0984978
Const.	.0000137**	.0003674
$ln\sigma_\mu^2$	4.377929	1.122911
σ_μ	8.925964	5.011532
ρ	.9797715	.0222553
LR test of ρ =0: 80.44	Wald $\mathcal{X}^2$ =15.97	Log Likelihood = −11.975544

Notes: *,**,*** represent 10 %, 5 % and 1 % significance levels respectively.

5 Conclusion

The continuous change/development of the production systems and the increasing population reveal the phenomenon of unemployment in developed and developing countries as a social problem. Market imbalances which arise with unemployment cause social problems, inadequate demand, excess supply and idle labor force. There is a positive correlation between poverty and income inequality and a negative correlation between economic growth and poverty. The poverty process, which started with the increase in income distribution inequality, slows down economic growth and increases poverty and income distribution inequalities, even more. The social supports that arise as a result of the increase in poverty levels due to income inequality and unemployment problem are not enough to prevent poverty, and in some cases, can even act as a dynamic that increases the outflows from the labor market. Human development, which includes meeting the basic needs of the people comfortably, maintaining their lives in a healthy environment and ensuring their development through education, is used as a measure of the level of development. The impoverishment resulting from the unfair distribution of income affects human capital structure by limiting access to services such as education, health and social security. When the findings obtained from the study are examined, the results are consistent with the theory. It is concluded that the increase in poverty, income inequality and unemployment have a negative impact on the democratization level while the increase in human capital has a positive impact on the democratization process.

Since the increase in income is provided by the production of high-technological products and the increase in exports, development in human capital is an important factor in poverty and income injustice. Therefore, it is necessary to give importance to qualified education in order to reduce poverty and to minimize the injustice of income distribution within the scope of the determined development goals. Moreover, the decrease in income distribution inequality positively affects the development of countries and healthy living conditions. Maintaining the basic needs comfortably in a healthy environment and benefiting from adequate educational opportunities positively affect employment and development, also, they are important in the process of democratization. The production relations that arise in the transition from agricultural to industrial society and from industrial to information society cause changes in social structure. Research and development activities increase human capital and contribute to economic growth due to the fact that the increase in human capital supports high-tech exports and this interaction takes place mutually. Democratization efforts are becoming increasingly widespread due to capital mobility, increasing foreign trade, technological advances and industrialization. The country/society structure and quality have an impact on the sustainability and process of democracy. Since the sustainability of democracy depends on the continuity of economic development, it should be ensured that the awareness of democratization is established and spread by supporting the policies of countries' economic development process.

References

Acemoglu, D., Johnson, S., Robinson, J. A., & Yared, P. (2008). Income and democracy. *American Economic Review*, 98(3), 808–42.

Andersen, E. B. (1970). Asymptotic properties of conditional maximum-likelihood estimators. *Journal of the Royal Statistical Society: Series B (Methodological)*, 32(2), 283–301.

Avery, R. B., Hansen, L. P., & Hotz, V. J. (1983). Multiperiod probit models and orthogonality condition estimation. *International Economic Review*, 24(1), 21–35.

Arat, Z. F. (1988). Democracy and economic development: Modernization theory revisited. *Comparative Politics*, 21(1), 21–36.

Barro, R. J. (1999). Determinants of democracy. *Journal of Political economy*, 107(S6), S158S183.

Beetham, D., & Boyle, K. (2005). *Demokrasinin temelleri*. Translated by A. Z. Kopuzlu, Ankara: Adres Yayınları.

Boix, C., & Stokes, S. C. (2003). Endogenous democratization. *World Politics*, 55(4), 517–549.

Bourguignon, F., & Verdier, T. (2000). Oligarchy, democracy, inequality and growth. *Journal of development Economics*, 62(2), 285–313.

Burkhart, R. E., & Lewis-Beck, M. S. (1994). Comparative democracy: The economic development thesis. *American Political Science Review*, 88(4), 903–910.

Chamberlain, G. (1980). Analysis of covariance with qualitative data. *Review of Economic Studies*, 17, 225–238.

Chen, M. A. (2012). *The informal economy: Definitions, theories and policies* 1(26), 9014190144. WIEGO Working Paper.

Colaresi, M., & Thompson, W. R. (2003). The economic development-democratization relationship: Does the outside world matter? *Comparative Political Studies*, 36(4), 381–403.

Coleman, J. S. (1960). Conclusion: The political systems of the developing areas. In Almond, G. A., Coleman, J. S., eds., *The Politics of Developing Areas*. Princeton: Princeton University Press, 532–581.

Cutright, P. (1969). National political development: Measurement and analysis. In *Comparative Government*, 29–41. London: Palgrave.

Dahl, R. A. (1973). *Polyarchy: Participation and Opposition*. New haven and London: Yale University Press.

Denison, E. F. (1962). Education, economic growth, and gaps in information. *Journal of Political Economy*, 70(5, Part 2), 124–128.

Döm, S. (2008). *Girişimcilik ve küçük işletme yöneticiliği*. Ankara: Detay Yayıncılık.

Duman, A. (2008). Education and income inequality in Turkey: does schooling matter? *Financial Theory and Practice*, 32(3), 369–385.

Epstein, D. L., Bates, R., Goldstone, J., Kristensen, I., & O'Halloran, S. (2006). Democratic transitions. *American Journal of Political Science*, 50(3), 551–569.

Glasure, Y. U., Lee, A. R., & Norris, J. (1999). Level of economic development and political democracy revisited. *International Advances in Economic Research*, 5(4), 466–477.

Gould, J. A. (2003). Out of the blue? Democracy and privatization in postcommunist Europe. *Comparative European Politics*, 1(3), 277–311.

Greene, W. H. (2003). *Econometric Analysis*. Upper Saddle River, NJ: Prentice Hall.

Gyimah-Brempong, K., & Wilson, M. (2004). Health human capital and economic growth in Sub-Saharan African and OECD countries. *The Quarterly Review of Economics and Finance*, 44(2), 296–320.

Haralambos, M., & Holborn, M. (2008). *Sociology: Themes and Perspectives*. UK: HarperCollins.

Helliwell, J. F. (1994). Empirical linkages between democracy and economic growth. *British Journal of Political Science*, 24(2), 225–248.

Huber, E., Rueschemeyer, D., & Stephens, J. D. (1993). The impact of economic development on democracy. *Journal of Economic Perspectives*, 7(3), 71–86.

Huntington, S. P. (1993). *The Third Wave: Democratization in the Late Twentieth Century* (Vol. 4). Oklahoma: University of Oklahoma Press.

Jackman, R. W., & Bollen, K. A. (1989). Democracy, stability and dichotomies. *American Sociological Review*, 54(4), 612–621.

Kızılot, Ş., & Çomaklı, Ş. E. (2004). Vergi Kayıp ve Kaçakları ve Kayıt Dışı Ekonomi İlişkisi ve Boyutlarının Mevzuat Açısından Değerlendirilmesi. *Türkiye Maliye Sempozyumu*, 19, 1014.

Kim, C. L. (1971). Socio-economic development and political democracy in Japanese prefectures. *American Political Science Review*, 65(1), 184–186.

Kuznets, S. (1955). Economic growth and income inequality. *The American economic review*, 45(1), 1–28.

Lipset, S. M. (1959). Some social requisites of democracy: Economic development and political legitimacy. *American Political Science Review*, 53(1), 69–105.

Londregan, J. B., & Poole, K. T. (1996). Does high income promote democracy? *World Politics*, 49(1), 1–30.

Lucas Jr., R. E. (1988). On the mechanics of economic development. *Journal of Monetary Economics*, 22(1), 3–42.

Maertens, M., Colen, L., & Swinnen, J. F. (2011). Globalisation and poverty in Senegal: a worst case scenario? *European Review of Agricultural Economics*, 38(1), 31–54.

Mayer, D. (2001). The long-term impact of health on economic growth in Latin America. *World Development*, 29(6), 1025–1033.

Milanovic, B. (2005). Relationship between Income and Emergence of Democracy Reexamined, 18202000: A non-parametric approach. Retrieved from https://papers.ssrn.com/sol3/papers.cfm?abstract_id=812164 (12.08.2019).

Muller, E. N. (1995). Economic determinants of democracy. *American Sociological Review*, 60(6), 966–982.

Mundlak, Y. (1978). On the pooling of time series and cross section data. *Econometrica: Journal of the Econometric Society*, 46(1), 69–85.

Neubauer, D. E. (1967). Some conditions of democracy. *American Political Science Review*, 61(4), 1002–1009.

Neutel, M., & Heshmati, A. (2006). Globalisation, Inequality and Poverty Relationships: A Cross Country Evidence. IZA Discussion Paper No. 2223, Retrieved from http://ftp.iza.org/dp2223.pdf (12.08.2019).

Neyman, J., & Scott, E. L. (1948). Consistent estimates based on partially consistent observations. *Econometrica*, 16(1), 1–32.

Pech, R. J., & Cameron, A. (2006). An entrepreneurial decision process model describing opportunity recognition. *European Journal of Innovation Management*, 9(1), 61–78.

Przeworski, A., & Limongi, F. (1997). Modernization: Theories and facts. *World Politics*, 49(2), 155–183.

Robinson, J. A. (2006). Economic development and democracy. *Annual Review of Political Science*, 9, 503–527.

Rueschemeyer, D., Stephens, E. H., & Stephens, J. D. (1992). Capitalist Development and Democracy. Chicago: University of Chicago Press.

Schneider, F., & Enste, D. H. (2000). Shadow economies: Size, causes, and consequences. *Journal of Economic Literature*, 38(1), 77–114.

Smith, G. H. (1948). Liberalism and level of information. *Journal of Educational Psychology*, 39(2), 65.

Tatoğlu, F. Y. (2012). *Panel veri ekonometrisi*. Istanbul: Beta Yayınları.

Thomas, J. (1999). Quantifying the black economy:'measurement without theory' yet again? *The Economic Journal*, 109(456), 381–389.

Tavares, J., & Wacziarg, R. (2001). How democracy affects growth. *European Economic Review*, 45(8), 1341–1378.

Terzioğlu, M. K., Kandemir, H., &Akgün, G. (2018). Yoksulluk ve Sosyal Transferler Kapsamında Göç Dinamikleri. IX. International Non-Governmental Organizations Congress Proceedings e-book, 283–293, Retrieved from https://cdn.comu.edu.tr/cms/ngo/files/4-stktammetinrev431aralikaaaa.pdf (12.08.2019).

Terzioğlu, M. K., Bulut, M., & Erkut, E. N. (2018). Göç: Girişimcilik ve Bilgi Teknolojilerinin Etkisi, *ICOAEF'18 IV. International Conference on Applied Economics and Finance extended with Social Sciences Full Paper Proceedings*, November 28-30, 2018, Kuşadası, Turkey, 1227–1240, Retrieved from http://www.icoaef.com/wp-content/uploads/2018/08/05.01.ICOAEF-final-TAM-MET%C4%B0N.pdf (12.08.2019).

Terzioğlu, M. K., Can, D., & Degigoğlu, İ. (2018).Göç, Turizm ve Kayıtdışılık, *ICOAEF'18 IV. International Conference on Applied Economics and Finance Extended with Social Sciences Full Paper Proceedings*, November 28-30, 2018, Kuşadası, Turkey, 877–888. Retrieved from http://www.icoaef.com/wp-content/uploads/2018/08/05.01.ICOAEF-final-TAM-MET%C4%B0N.pdf (12.08.2019).

Terzioğlu, M. K., Sert, S., & Aygün, M. (2018). Göç ve Beşeri Sermaye: Doğu Avrupa Ülkeleri, *ICOAEF'18 IV. International Conference on Applied Economics and Finance & Extended with Social Sciences Full Paper Proceedings,* November 28–30, 2018, Kuşadası, Turkey, 889–901. Retrieved from http://www.icoaef.com/wp-content/uploads/2018/08/05.01.ICOAEF-final-TAM-MET%C4%B0N.pdf (12.08.2019).

Ufuk Bingöl

12. Employment Policy Goals in Turkish Government Programs in Terms of Social Policies: Justice and Development Party Governments[1]

Abstract The subject of this chapter is to examine governments established by Justice and Development Party winning all the elections in Turkey since 2002 in terms of employment policy objectives with the help of socio-political view. Although the unemployment problem has been relatively controlled thanks to the policies implemented by the early governments of Justice and Development Party; on subsequent periods, unemployment rates becoming increasingly chronic because of economic, political reasons and with the emergence of the issues of poverty and social security constitutes the problem of the study. The aim of the study is to analyze the employment policy commitments of the Justice and Development Party government programs in order to increase the welfare in the fight against social exclusion and poverty and to create a future projection. In this context, seven governments established since 2002 and one provisional government (63rd government) were subjected to content analysis with embedded theory qualitative research design and NVIVO 11 Computer Assisted Qualitative Data Analysis Software in order to reveal the targets related to employment policies within the programs (Glaser & Strauss, 1967). The study is categorized and coded under the theme of "Employment" which is one of the "Big Five" also described as in the literatureas "Education", "Employment", "Housing", "Health", "Social Security" themes (Alcock, Daly and Griggs, 2008: 1). As a result of the analyzes, it was revealed that all governments mention it should be considered as a structural problem by repeating the objectives of sustainability of employment, increasing the harmony between education and employment, increasing female labor force participation and preventing informal employment. By being a part of political parties, it arises that the governments of Turkey in the coming period should formulate polices regarding employment-creating growth, informal employment issues, providing quality professional and vocational education, increasing and promoting women's labor force participation and emloyment.

---

1 This section is derived (from a section) and expanded from the Doctoral Dissertation Thesis which is titled "Qualitative data analysis of academic publications and government programs based on social policy basics for the period 2005–2015 in Turkey", by Dr. Ufuk BİNGÖL who was under the supervision of Prof. Yılmaz ÖZKAN.

Keywords: Employment, Government program, Employment policies, Unemployment, Poverty

1 Introduction

The most important elements of employment, which are the most significant tool that social policy instruments use in the fight against social exclusion and poverty in terms of the welfare of society can be listed as the development of human resources for sustainable and quality improvement; improving working conditions. Particularly in terms of disadvantageous and vulnerable individuals and providing equal opportunities, creating adequate social protection systems and establishing an honest dialogue with social partners. In this chapter, the "employment-oriented objectives" programs of the Justice and Development Party (Adalet ve Kalkınma Partisi/AKP) (which came to power elected by the citizens of the last 17 years in Turkey) governments were discussed in terms of a socio-political point of view.

The programs of all governments involved in the research subject were examined through computer-aided qualitative data analysis software (CAQDAS) and by primary text mining method. The government programs were analyzed by Miles-Huberman coding method and grounded theory approach after importation into the software. According to this coding method, 'descriptive coding' and 'inference coding' (regarding employment policies) were made in the scope of the government programs under investigation, in terms of the basic elements of social policy (Patton: 2014). In grounded theory, the researcher reveals the theory embedded in the data when collecting or interpreting data and can access new concepts and theories throughout the research. Also, analysis with data collection is carried out simultaneously. This process is called "continuous comparative analysis" (Glaser and Strauss, 1967). According to this approach, rather than adapting the data to the previously defined categories, the theory was produced from the theories revealed by the participants. Instead of revealing the social policy theories hidden in the content of the AKP period government programs and including a direct argument in the hypothesis of this study. It is an embedded theory study (based on qualitative data analysis), which is made in response to the research question about "What kind of socio-political employment commitments are offered?" in government programs (Auerbach and Silverstein, 2003: 7).

In the context of the reliability of the study, two different coding studies (QSR: 2018) were performed with the support of CAQDAS by obtaining an expert opinion. Within the scope of the study, the first and second level themes and coding were determined according to Miles and Huberman (1994), and the coding congruence percentage was achieved with an average of 84 % agreement on the themes mentioned in Graph 12.2 In the study, data visualization was performed in the context of data reduction and clustering analyzes and thematic analyses were performed. Completion and verification studies were conducted between the analysis to check the reliability. To see consistency, each analysis was compared with the other one. Following the introductory section, the relationship between Social Policy and Employment titles and politics was examined in general terms. In the third, fourth and fifth sections, the thematic analysis findings were examined in terms of socio-political elements in which government programs, defined as the "road map" by the government, about the employment policies. The study ended with comments and conclusions.

2 Relations Between Social Policy, Employment and Politics

It is known that there is no exact decision on Social Policy as a term and it is continuing to discuss, but its scope expands day by day. According to Lishman (2007:13), who referred to the study of Alcock, Payne ve Sullivan (2004), it is similar to an elephant. In this respect, he pointed out that it has a very large structure. However, it is difficult to understand what it looks like and how to define. It is a well-known discipline from this aspect.

The most important topics in terms of social policy are increasing employment in a sustainable and quality way, improving working conditions especially in terms of disadvantageous and vulnerable individuals, establishing appropriate social protection systems, developing human resources for a sustainable employment structure, and to provide equal employment opportunities to both women and men, while fighting against social exclusion and poverty. Assuming that the majority of the society provides for its livelihood with the labor it offers to labor markets, it is clear that it needs to be protected by means of social policy. To ensure the sustainability of employment, it is inevitable that social policy will primarily use educational instruments. Thanks to education, the quality and performance of the work of the individual who will be part of the future labor force will increase, and this will also contribute positively to production and growth. Therefore, the interrelated relationships between the education and employment sub-components of social policy are permanent and essential.

Political parties and politicians use the public research, analysis and policy documents of some institutions and think tanks to guide their policies to ensure success and stability in the country's management (McGANN, 2007: 18). They offer many commitments in their declarations, particularly during the election periods, and in their programs (if they win elections). These commitments are based on the main components in the social policy field, which is described in the literature as "BIG FIVE" (Alcock, Daly, and Griggs, 2008: 1). These issues can be shown as "Employment," "Education," "Social Security," "Housing," and "Health" (Bingöl and Özkan, 2018).

The AKP, which won all the elections in the process since its establishment, has included a separate title on social policies in its program, and detailed its objectives within these headings (in particular the disadvantaged groups) and committed social policies for the welfare and happiness of all citizens (The AKP, 2002). The government or those who should provide support policies for their citizens, in terms of job placement, legal arrangements, policies to increase labor demand and legal arrangements and so on (Luo and Fan, 2017). The AKP, especially in the period of 2002–2011 brought a lot of regulations (with the effect of the liberal approach adopted in economic and social life), especially for women and disadvantaged groups (including those who make their living by labor), and has undertaken many international obligations (Altuntaş and Demirkanoğlu: 2017). The AKP has taken measures in this regard due to facts that 2001 crisis created chronicity of rising unemployment leading to a "growth without employment," and because that past decision makers usually have perceived this situation as "just a growth problem." These measures have been long-term goals to reduce unemployment and has determined policies for the development of new production capacities (Karagöl, 2013). Unemployment rates did not fall significantly and remained above the OECD averages despite the incentive and similar active labor market policies, the share of the agricultural sector in employment, the current account deficit due to import-dependent growth, the increase in the active population and the economic crises (OECD: 2017). Although the national income increased seven times between 2002 and 2018, the number of unemployed raised by five times between the same years (TSI, 2018). Despite the increase in national income, 10 percent unemployment ratio between 2002 and 2016 raises the question of whether income distribution is fair or not. In this context, it is essential to examine the socio-political aspect of the issue in order to increase the employment by decreasing unemployment and to understand the fact that the AKP still maintains its power despite the partial achievement of the desired targets. From past to present, it is considered that the commitments of the AKP governments regarding employment and unemployment related to macroeconomic and social policies and their actions towards the solution of the problem can arise with the analysis of the programs of these government.

When the scope of literature review of studies in Turkey, contrary to the government's program, the academic interest in the election manifestos have been seen more (Aymankuy et al., 2016; Aytac, 2017; Bingöl and Özkan, 2018; Ertük and Şeşen, 2017; Lamba, 2015; Özkaynar, 2015; Tiyek, 2015; Yildirim et al., 2016), however, it has been found that government programs have been subject to various academic studies before, including the interests of social policy (Arslan, 2017; Elmas, 2018; Karkın and Öztepe, 2017; Solak and Sürmeli, 2015). In international studies, it is seen that election manifestos are examined under various titles (Ashworth, 2000; Volkens, Bara and Budge, 2009; Dolezal et al., 2016; Netswera, 2008; Wesley, 2014; Brouard, S. et al., 2018;). In Turkey, there are no qualitative analysis studies observed except Lamba's (2015). On the other hand, the lack of descriptive analysis in both national and international research has increased the interest in the subject. To this end, in this chapter, socio-political analysis of programs made by six separate governments [from 3 November 2002 parliament elections until the beginning of 2019 – Excluding the 63, the Government program (Election Government) and 66 Government (Presidential System of Government)]; programs which the Prime Ministers of Turkey submitted for a vote of confidence in Parliament and read personally, have taken a vote of confidence and passed on to the minutes of the parliament (T.G.N.A., 2018) (WIKIPEDIA, 2018).

3 Social Policy and Growth-Employment-Unemployment Cases in Government Programs

With the help of CAQDAS, in Graph 12.1, the government programs under study were categorized according to the themes of Education, Employment, Housing, Health, Social Security, which are the main components of social policy science [also known as the BIG FIVE (Alcock, Daly and Griggs, 2008: 1)], and pattern coding study was conducted. Also, concerning the value of the reference coding, the distribution of these themes by governments is shown (Onwuegbuzie, Frels and Hwang, 2016).

In Graph 12.1, all six national programs within the scope of the study come to the fore with commitments regarding the theme of "employment." Immediately after the employment, "education" commitments are included. Within the scope of the thematic analysis of government program commitments, the distribution of "word cloud" and sub-categories of employment theme according to the governments is presented in Graph 12.2, as a result of coding studies performed according to categories and sub-categories:

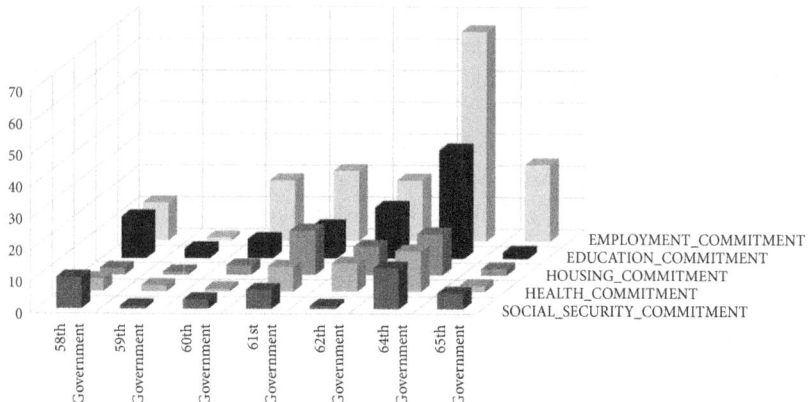

Graph 12.1: The distribution of social policy core components commitment in Government Programs. Source: This graphic is produced by the author with the support of CAQDAS, according to the Reference Coding value.

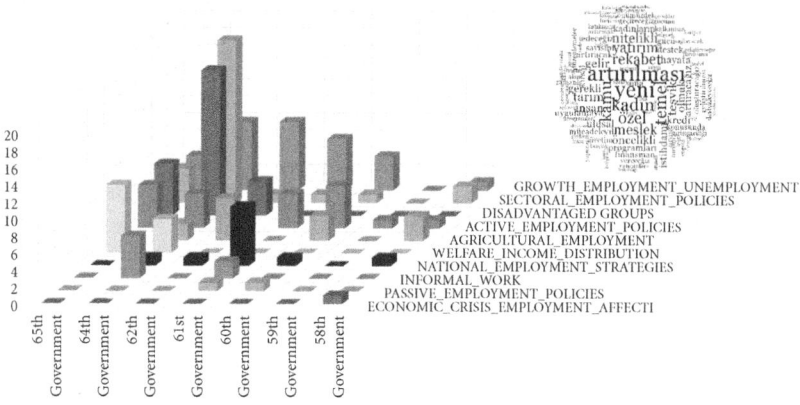

Graph 12.2: Employment commitments by Categorical Pattern coding study. Source: This graphic is produced by the author with the support of CAQDAS, according to the Reference Coding value.

As a result of the categorical pattern coding study, it is seen that governments aim to use a wide range of instruments for the unemployment problem. If we look at the distribution of the commitments according to government periods, it is determined that there is elaboration in the employment policies since the

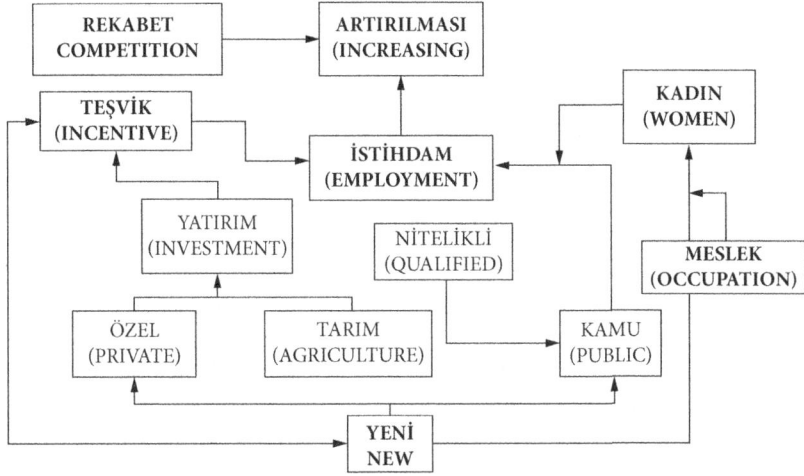

Fig. 12.1: In the Government Programs, the expression analysis of the most frequently used words related to 'employment'. Source: This figure is written by the author with the support of CAQDAS, resulting from Text Search Query filtered by Treemap.

Note: English equivalents are under the words in the programs

60th government period. The analysis of the statements performed in order to examine the areas where the most common words are used in a sentence in the word cloud in Graph 12.2 is shown in Fig. 12.1. There are common themes shown in all government program. Some of these are commitments focused on the working life of women (in the context of disadvantaged groups), increasing employment and promoting investments. In Graph 12.2, which is the result of CAQDAS, the specified categories will be examined, respectively.

From the 58th government program, it is possible to see that employment and unemployment concepts are primarily associated with investments and growth. If we look at the expression analysis in Fig. 12.1, the commitment could be stated to bring the corporate, private sector in Turkey "to a level that can compete with the whole world" by encouraging foreign investors to invest in Turkey. Thus by this, it is stated that the goal of "growth-creating employment" can be achieved. The 58th61st government programs promised to increase employment with international direct investments (with similar expressions). In the 62nd government program (despite of the global economic crisis of the period), it was emphasized that the growth continued, and that *'the economy that offers new employment opportunities to millions of people'* will be among the priorities of the government.

For the first time, however, the private sector or the real economy statement was elaborated in the 62nd government program. Priority sectors in terms of incentives were stated as "*Iron and steel, textiles, apparel, electricity and electronics, chemicals and especially in machinery and automotive sectors would continue to be the leading sectors of our export strategy. Aircraft engines and their parts production system will be strengthened the incentive system and domestic production capacity would be increased*". In addition, it is stated that the service and tourism sectors would be detailed and incentives would be provided to priority sub-sectors. In the 64th government program, it was emphasized that investment and employment incentives for the regions behind the development would be provided, not only in terms of the growth and employment relationship but also the improvement of human resources in the relevant regions and increasing the welfare of citizens. In the 65th government program, the expression "*We will make the real economy stronger and more competitive in a way to become a country that grows, creates employment and develops exports*" aimed to gather the statements of all previous programs under one. Accordingly, it is stated that the focus would be on the manufacturing industry and high value-added sectors. The 65th government program once again combined all and of all previous governments' statements declaring the employment-related objectives: "*The main priority of our government will be supported by investments in production, growth, employment and more qualified private sector.*"

4 Employment Policies

4.1 Agricultural employment and sectoral employment policies

In addition to industrial production and investment for employment policies and efforts to increase employment, governments also have sector-based commitments, particularly in agricultural employment. Under the scope of the 60th Government Program's "*Rural Development Strategy Document*" it is pledged to promote agricultural employment in order to increase and diversify job opportunities in rural areas. It is stated that the most important objective of the rural development strategy is to increase agricultural employment by providing water, electricity, roads, and similar infrastructure services. In addition, by providing consultancy services to farmers with expert personnel support, trainings are undertaken to increase production through agricultural technologies. Government programs in agricultural employment also aim at to balance urbanrural income and welfare. In this sense, as mentioned in all government programs (except for the 65th government program), income support, incentives, loans, and similar economic arrangements would be carried out for the inputs of farmers in agricultural activities. Considering the fact that youth unemployment is an important problem, the 64th government's commitment to grant support for young farmers towards agricultural employment is noteworthy.

Promoting entrepreneurship is another tool that government programs use as a basis for employment promotion strategies. With the incentives like supporting SMEs for technological development, incentives in organized industrial zones, grant or zero interest loan and so on, SMEs are expected to increase their competitiveness and make a positive contribution to employment. The 64th government promises to support entrepreneurs in the early incubation period by establishing a national fund for entrepreneurs. The provision of entrepreneurship training and consultancy services to all citizens is under the commitment of all governments. With the "Women Entrepreneurship Program" in the 65th government program, it has been promised to facilitate women's entry into business life, and it is stated that young people who want to establish a business will receive interest-free support up to a 100,000 Turkish liras. Increasing R&D centers and mobilization of "reverse brain drain," governments are expressing the goal of qualified employment through the development of science and technology. Energy, health, space and aviation, and others. Sectors are identified as priority sectors. To increase qualified employment in the fields *"Qualified Labor Engagement Program,"* it is aimed to turn our country into a qualified labor center.

Governments are committed to the creation of a National Employment Strategy (UIS) for all sectors, including increasing agricultural employment. According to UIS's definition in the 61st government program, it is planned to find permanent solutions to the unemployment problem until 2023. It is stated that the definition is composed of 22 concrete objectives and 102 separate policies aimed to this goal by accepting the problem of structural unemployment in our country and seeking solutions in this direction. According to government programs, it is aimed to contribute to employment positively by facilitating flexible working styles through UIS. Governments undertake to ensure that their flexible work will be integrated into the social security legislation in line with the EU acquis, thereby underlining that social protection will be ensured without compromise.

4.2 Commitments for active and passive labor market policies

In all of the governments included in the study, the active employment policy tool applied to increase employment is employment incentives. Accordingly, it was promised that legal arrangements would be made in the first governments to ease the burdens of employers on employment, reduce unregistered employment and rigidity in working life, and to encourage workers in forced employment. In the 60th government program, the statement "*The opening of private employment offices will be encouraged*" is remarkable. In the period of the 58th, 59th and 60th governments, it was aimed to increase the active labor force programs that are committed to be made through Turkish Employment Agency (İŞKUR) and to reach the labor force needed by the labor markets. In the 61st government program, it was stated that the "National Occupational Standard" would be taken into consideration, and a job and vocational counselor would be assigned to all unemployed citizens enrolled in İŞKUR. In the same program, for the first time, it is guaranteed that the registered employees, who are on the poverty line, will be supported with social assistance. The 64th government program appears to have experienced a paradigm shift concerning previous governments' commitments (although there has been no change in the context of the tool in employment policies). Accordingly, increasing the effectiveness of İŞKUR programs, simplification, and clarification of incentive practices, providing employer support for the insurance premiums of employed and socially supported citizens are important points marked for the following period. It is stated that an informal economy inventory will be created and an important data collection will be carried out in order to ensure equality in income distribution and employment and the audit capacity will be increased to prevent informality. In the 65th government program, focal points are expressed as the revision of the public personnel regime, effective training programs to ensure the quality production required by labor markets, and mechanisms in which entrepreneurship is encouraged to provide high added-value areas.

Starting from the 58th government program to the 65th government program, current policy targets were set to increase "passive employment policies," which aim to eliminate the effects of unemployment in all government programs temporarily, took place with shorter expressions. The 58th government program stated that the implementation of social welfare projects would be focused on people who were affected the most by the economic crisis which took place in 2001. Furthermore, it is seen that the first intention of inclusion in social security is put forward with the expression that "*all segments of the society will be covered by social security*." Subsequently, the 60th government program promises to facilitate the use of unemployment insurance and to use the unemployment fund for active employment policies. The 61st government program committed ensuring that those who are in a position to work from the members of the disadvantageous groups/families (which include people who are disabled, old age, and others) will be referred to the labor markets. The 64th government program included the creation of the "*Social Risk Map*" (based on the family structure) in which the title "employment" is present. Additionally, in order to be realized within the scope of equality of opportunity, the "*Social Welfare Law*" is promulgated to keep the welfare of the people employed while receiving social welfare, until the income level of their social benefits is at the determined threshold. Also, in the 64th and 65th government programs, the term "*strengthening the employment link with social assistance*" is expressed.

4.3 Commitments to disadvantageous groups in employment

It can be stated that the main objective of the government programs is the participation of women in the working life and the solution of youth unemployment. Starting from the 62nd government program, it is committed to ensuring equal opportunity in employment, education, and social security to women. In the 64th government period, incentives were given to women to facilitate their entry into employment by child care facilities and nursery services, the right to part-time-but-full-paid work after childbirth and the right to work part-time until their newborn start school. In addition, the adoption of seasonal and "*home-based*" working women in the social security system, granting premium support to the employment of young people, women and unemployed people receiving vocational training and the programs supporting the entrepreneurship of young people and women were guaranteed by the government.

The commitment of "*National Youth Employment Strategy Document*" in the 64th government program is remarkable in the context of youth unemployment. In this context, according to commitment, it will be under the responsibility of providing career consultancy services to the transition to higher education to strengthen the harmonization of education and employment. All governments within the scope of the study have committed to entrepreneurship incentives at various rates. In addition, it also expresses one year salary payments from the state to the newly employed young people, development of flexible types of work to increase the young employment, not cutting the grants or loans for young people in higher education who are expected to be employed as part-time workers and strengthen the employment with social welfare.

About the employment of persons with disabilities, an effective supervision system was promised with the sentence, "*We will continue to take measures to fill the disability quota in public and private sectors.*" In the period of the successive governments, it is pledged the continuity of the incentive policies in education and employment towards the disabled people in the working life and their participation in social life. In the context of child labor prevention, the 64th government program includes the expression *"protection with social support programs"* for the children.

5 Education—Employment Commitment

The most important point for the elimination of structural unemployment problems, the modeling and future implementation of future labor needs projection is to train a labor force that meets the needs of the labor market. All governments within the scope of the study offered various commitments to achieve this issue. The main focus of governments is 'vocational education' and 'equality of opportunity' in education. According to governments, the commitment statement graphic and word cloud for vocational education and training of disadvantaged groups are presented in Graph 12.3.

According to the word cloud in the right side of the Graph 12.3, opportunity (fırsat), occupation (meslek), our women (kadınlarımızın) are the most commonly used words in government programs. Accordingly, it is pledged to facilitate the access of the whole community to education, especially girls. Particularly with open education, distance education and lifelong learning tools, women's participation in labor markets and social life will be increased (in almost all government programs). In addition, it is stated that the equalization of access to education will be ensured by increasing the enrollment rate. The government's vocational training commitments is based on the qualification issue in vocational-technical education. In all the programs other than the

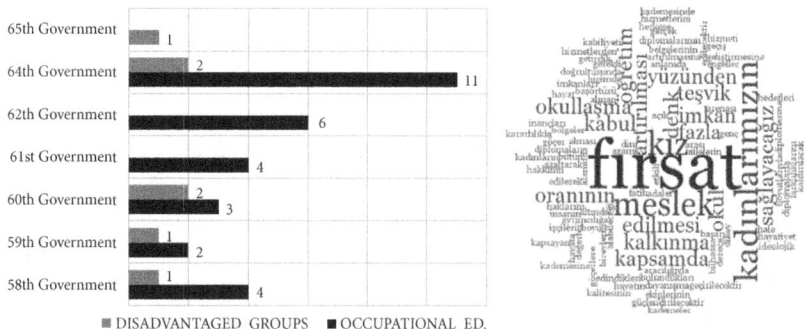

Graph 12.3: Training-Employment commitments according to Categorical Pattern coding study. Source: This graphic is produced by the author with the support of CAQDAS, according to the Reference Coding value.

65th government program within the scope of the study, it is stated directly that the schools which have a vocational and technical education and especially secondary education institutions cannot produce qualified labor force.

In a commitment about vocational training by the 58th Government Program, this is expressed as follows "*The existing secondary education system, which cannot provide sufficient qualifications for graduates, will be re-examined and vocational education programs will be expanded.*" In the 60th government program, the curriculum of vocational schools is committed to be set together with the actors of the labor markets. In the 61st government program, it is stated that the main needs of the labor markets will be determined by establishing Vocational Training Centers, Vocational Qualification Institution and National Occupational Standard. In the 62nd government program, it is stated that in addition to the previous ones, the business establishment and development centers will be created, and the 'entrepreneurship' will be encouraged in the transition from education to employment. In the 64th government program, the promotion of employment of vocational high school and university graduates, increasing private sector cooperation in vocational schools, and "Priority Transformation Program for the Development of Basic and Professional Skills" are important elements. In the 65th government program, it is stated that vocational education dimension will be added to the programs of community benefit.

Tab. 12.1: Cluster analysis of government programs. Source: The author created this table with CAQDAS and Pearson Correlation Coefficient Similarity Criterion.

WORD SIMILARITY CLUSTERING						
	58th Govt.	59th Govt.	60th Govt.	61st Govt.	62nd Govt.	64th Govt.
59th Govt.	0,969457	-				
60th Govt.	0,923623	0,928136	-			
61st Govt.	0,71245	0,76419	0,73217	-		
62nd Govt.	0,944119	0,947921	0,94083	0,81076	-	
64th Govt.	0,951611	0,94841	0,927021	0,75186	0,96342	-
65th Govt.	**0,001323**	**0,007741**	**0,000875**	**0,10627**	**0,003413**	**0,005679**
CODING SIMILARITY CLUSTERING						
59th Govt.	0,014770	-				
60th Govt.	0,073372	0,016436	-			
61st Govt.	**0,155643**	0,048272	0,013744	-		
62nd Govt.	**0,138597**	0,059776	0,039129	0,013873	-	
64th Govt.	0,037643	0,007012	0,027991	0,081636	**0,108541**	-
65th Govt.	**0,185958**	**0,205237**	**0,224213**	**0,172208**	0,095707	**0,369971**

6 Conclusion

It is seen that the subject of "employment" is the focus in the socio-political contents of the the AKP governments that have been elected consecutively in Turkey since/after the 2001 economic crisis. Growth-employment-unemployment, Provision of training-employment compliance, sectoral and active-passive employment policies and policy commitments for the employment of disadvantaged groups have been promised by the the AKP governments in their respective periods. An examination of the similarity of expression of government programs with each other is important for the conclusion of the study. Based on the analysis of coding similarity in this context, the cluster tables of government programs are shown in Tab. 12.1.

According to the analysis of the word "similarity" in the upper part of Tab. 12.1, it is seen that the government programs express their commitments on employment with similar words. It was found that the 65th government program had a low level of similarity with previous programs. However, in terms of qualitative coding similarity, it is seen that the 65th government shows the highest coding similarity in terms of all other government programs. Accordingly, it is considered that it would be possible to determine more precise and concrete objectives with less expression in government programs.

In the light of clustering analysis, it is seen that all governments within the scope of our study have made commitments in order to increase employment by increasing employment, reducing unemployment and its effects, strengthening education employment compliance, increasing women's labor force participation, and solving unregistered employment problem in the same ways. Therefore, this commitment similarity shows that governments are aware of the need to address the unemployment problem as a *structural problem*. In order to solve the problem, concept targets that can balance the labor supply and demand for the needs of labor markets are put forward in all government programs. The commitments in line with these concept goals are becoming the vision of governments in employment together with simpler expressions. However, due to economic, bureaucratic, political, national, international and social reasons, the targets have not been achieved yet. In this sense, the policies adopted by the "Turkish Semi-Presidential System" model—adopted as a result of the 2018 general elections—should be followed.

In the future, the most important titles in the employment policies of governments likely to serve in Turkey (as in all the AKP government) would be seen as employment-creating growth, the informal economy, women's employment, and vocational training. In this context, especially the access of young people to qualifying vocational education and increasing women's employment are the long-term focuses. Also, the participation of young people and women in working life is also likely to contribute to the social security context and affect other functions of the social state. Considering that the effects of low-level and often unregistered women on social security systems are negative in terms of cost, it will be possible to increase employment rates by increasing the education levels of women in the qualified education-employment-social security cycle. In this way, it is necessary to raise awareness on the provision of actuarial balance by paying premiums as a result of women being a dependent member of the social security system. In addition, it is possible to make a positive contribution to the solution of production-based growth and unemployment problem by ensuring equality of women's access to vocational education opportunities compatible with labor markets. It is absolute that young people and women are the primary focus group of all employment policies in the following periods.

References

AK Party Programme. (2002). V. Md. Sosyal Politikalar ve Sosyal Politika Anlayışımız. Retrieved from https://www.akparti.org.tr/parti/parti-programi/ (14.01.2018).

Altuntaş, N. & Demirkanoğlu, Y. (2017). Adalet ve Kalkınma Partisi'nin Kadına İlişkin Söylem ve Politikalarına Bakış: Muhafazakâr Demokratlıktan Muhafazakârlığa Doğru Evrilişin İzdüşümleri. *Akademik Yaklaşımlar Dergisi*, 8(1), 65–96.

Alcock, C., Sullivan, M., & Payne, S. (2004). *Introducing Social Policy*, revised edition. Harlow: Pearson Prentice Hall.

Alcock, C., Daly, G., & Griggs, E. (2008). *Introducing Social Policy*, second edition. Milton: Taylor & Francis.

Arslan, A. Ş. (2017). Kadın Odaklı Politikaların Hükümet Programlarına Yansımaları Üzerine Bir Analiz. *Türkiye Barolar Birliği Dergisi*, 2017(2017), 412–440.

Ashworth, R. E. (2000). Party Manifestos and Local Accountability: A Content Analysis of Local Election Pledges in Wales. *Local Government Studies*, 26(3), 11–30.

Auerbach, C. F. & Silverstein, L. B. (2003). *Qualitative Studies in Psychology*. Qualitative Data: An Introduction to Coding and Analysis. New York: New York University Press.

Aymankuy, Y., Demirbulat, G. Ö., & Aymankuy, A. (2016). Türkiye'de Siyasi Partilerin Seçim Beyannamelerinde Turizmin Yeri – Haziran 2015 Genel

Seçimleri Örneği. *Mehmet Akif Ersoy Üniversitesi Sosyal Bilimler Enstitüsü Dergisi*, 8(16), 292-302.

Aytaç, S. E. (2017). Türkiye'de Siyasi Partilerin Seçim Beyannamelerindeki Politika Öncelikleri, 2002-2015. *SİYASAL: Journal of Political Sciences*, 26(2), 7-26.

Brouard, S., Grossman, E., Guinaudeau, I., Persico, S., & Froio, C. (2018). Do Party Manifestos Matter in Policy-Making? Capacities, Incentives and Outcomes of Electoral Programmes in France. *Political Studies*, 66(4), 903-921.

Dolezal, M., Ennser-Jedenastik, L., Müller, W., & Winkler, A. (2016). Analyzing Manifestos in Their Electoral Context a New Approach Applied to Austria, 2002-2008. *Political Science Research and Methods*, 4(3), 641-650.

Ertürk, K. Ö. & Şeşen, E. (2017). Bir Halkla İlişkiler Aracı Olarak Seçim Beyannameleri: 2000'li Yıllar Üzerine bir İnceleme. *Erciyes İletişim Dergisi*, 5(1), 60-80.

Elmas, A. (2018). 1923-2016 Yılları Arası Hükümet Programlarında Dezavantajlı Gruplar. *Social Studies Journal*, 4(15), 945-953.

Glaser, D. G. & Strauss, A. L. (1967). *The Discovery of Grounded Theory: Strategies for Qualitative Research*. New Brunswick: Aldine Transactions.

Karagöl, E. T. (2013). *AK Parti Dönemi Türkiye Ekonomisi*. Ankara: Seta Yayınları.

Karkın, N. & Öztepe, M. C. (2017). Değerler Üzerinden Türk Kamu Yönetiminde Değişim Algısı: 20022015 Dönemi Hükümet Programları. *Çankırı Karatekin Üniversitesi İİBF Dergisi*, 7(1), 113-142.

Lamba, M. (2015). Türkiye'de Yeni Kamu Yönetimi Anlayışının Yansımaları: Hükümet Programları Üzerinden Nitel Bir İnceleme. *Süleyman Demirel Üniversitesi İktisadi ve İdari Bilimler Fakültesi Dergisi*, 20(1), 127-141.

Lishman, J. (2007). *Handbook for Practice Learning in Social Work and Social Care*. London: Jessica Kingsley Publishers.

Luo, X. & Fan, Y. T. (2017). Rethinking the Government Role of Promoting Employment Policies for Persons with Disabilities: A Case Study of Anhui Province in China. *Open Journal of Social Sciences*, 5, 209-219. Retrieved from https://doi.org/10.4236/jss.2017.58017 (14.03.2019).

Mcgann, J. G. (2007). *Think Tanks and Policy Advice in the United States Academics, Advisors and Advocates*. Washington, DC: Routledge Press.

Miles, M. B. & Huberman, M. (1994). *Qualitative Data Analysis an Expanded Sourcebook*, second edition. California, USA: SAGE Publications.

Netswera, F. (2016). The Rhetoric of Political Election Manifestos: An Analysis of the African National Congress Local Government Elections Manifestos Between 1995 and 2011. *Loyola Journal of Social Sciences* xxx(1), 247-266.

OECD. (2018). *Education at a Glance 2018: OECD Indicators*. Paris: OECD Publishing. Retrieved from http://dx.doi.org/10.1787/eag-2018-en (14.01.2019).

OECD. (2017). *Harmonised Unemployment Rate OECD – Turkey*. Retrieved from https://data.oecd.org/unemp/harmonised-unemployment-rate-hur.htm#indicator-chart (10.1.2019).

Onwuegbuzie, A. J., Frels, R. K., & Hwang, E. (2016). Mapping Saldaña's Coding Methods onto the Literature Review Process. *Journal of Educational Issues*, 2(1), 130–150.

Özkan, Y. & Bingöl, U. (2018). Siyasi Parti Seçim Beyannamelerinde Sosyal Politika: 2018 Cumhurbaşkanlığı ve Milletvekili Seçimleri Nitel Analizi. In *Çalışma Ekonomisi ve Endüstri İlişkileri Seçme Yazılar—II*, edited by E. Erdoğan, 33–79. Sakarya-Turkey: Sakarya Yayınevi.

Özkaynar, K. (2015). Siyasi Partilerin 2011 ve 2015 yılı Seçim Beyannamelerine Bakış: Bir Doküman İncelemesi ve İçerik Analizi Çalışması. *Akademik Sosyal Araştırmalar Dergisi*, 3(17), 439–452.

QSR. (2018). NVivo Coding Query & How Is the Kappa Coefficient Calculated? Retrieved from https://help-nv.qsrinternational.com/12/win/v12.1.55-d3ea61/Content/queries/coding-comparison-query.htm (11.12.2018).

Patton, M. Q. (2014). *Qualitative Research and Evaluation Methods*, fourth edition. London: Sage Publications, Inc.

Turkish Grand National Assembly Archive. (2018). Governments Programs of Republic of Turkey. Retrieved from https://www.tbmm.gov.tr/kutuphane/e_kaynaklar_kutuphane_hukumetler.html (03.11.2018).

Tiyek, R. (2015). Sosyal Politika Kapsamında Seçim Bildirgelerinin Değerlendirilmesi. *HAK-İş Emek ve Toplum Dergisi*, 4(9), 36–63.

Solak, G. S. & Sürmeli, İ. (2015). Hükümet Programları ve Kalkınma Planları Ekseninde Çevre Politikası Analizi. *Yasama Dergisi*, 10(30), 22–43.

TSI (Turkish Statistical İnstitute). (2018). İşsizlik Oranları 20002018, Retrieved from https://biruni.tuik.gov.tr/medas/?kn=72&locale=tr (30.01.2019).

WIKIPEDIA. (2018). Türkiye hükûmetleri listesi. Retrieved from https://tr.wikipedia.org/wiki/T%C3%BCrkiye_h%C3%BCk%C3%BBmetleri_listesi (17.10.2018).

Volkens, A., Bara, J. & Budge, I. (2009). Data Quality in Content Analysis. The Case of the Comparative Manifestos Project. *Historical Social Research*, 34(1), 234–251.

Wesley, J. J. (2014). The Qualitative Analysis of Political Documents. In *From Text to Political Positions: Text Analysis Across Disciplines* [Discourse Approaches to Politics, Society and Culture 55] 201, edited by B. Kaal, I.

Isa Maks, and A. Van Elfrinkhof, Amsterdam: John Benjamins Publishing Company, 135–160.

Yıldırım, M., Demirkol, Ö., Yıldırım, L. & Solmaz, C. (2016). Tourism in the Electoral Programs of Turkish Political Parties: An Analysis of the 2015 General Elections in Turkey. *Uluslararası Yönetim, İktisat ve İşletme Dergisi*, 12(30), 81–96.

SECTION 3: Inequality and Poverty

Ayşe Aylin Bayar, Bengi Yanık-İlhan, and Nebile Korucu-Gümüşoğlu

13. The Effects of Elders' Earnings on Turkish Income Inequality[1]

Abstract A structural transformation occurs during the last decades and due to the changing conditions of the demographic structures of developing countries, the living standards have improved. Therefore, this has led to an aging population in the Turkish economy. Based on the changing conditions, not only the older workers' participation in the labor market but also distributional consequences of the increased integration to the world have been increased. Although the issues on income inequality attract the researchers' attention, the role of the older workers on income inequality is not being questioned in these studies. According to the literature, this study seems to be the first that investigates the impacts of elders' earnings on income inequality for Turkey. A counterfactual distribution of income for the specific years, obtained from Turkish Income and Living Conditions Survey, is generated and then the examination of the inequality measures and the decomposition of income inequality have been done. One of the important findings is the fact that the impact of elders' earnings is seen to be equalizing. In addition to that, examining the contribution of elders' earnings to overall inequality, it has a minor contribution on the inequality.

Keywords: Earnings, Inequality, Decomposition, Elder, Turkey

1 Introduction

For the developing countries, not only income inequalities but also the aging population is a rising problem in all over the world. Aging is one of the outcomes of longevity and declining fertility (Wang et al., 2017). Talking with the numbers, in 2017, although the people older than the age of 60 were around 962 million, the expectation of this age group is to be 2.1 billion in 2050 and 3.1 billion in 2100.[2] In addition to this, there is an increasing trend in population aging and decreasing trend in fertility rate.[3]

---

1 This study was presented at the 39th Annual MEEA Meeting in Conjunction with the Allied Social Sciences Association (ASSA) held in Atlanta, Georgia, US, January 4–7, 2019 by the support of Istanbul Kultur University.
2 https://www.un.org/development/desa/publications/world-population-prospects-the-2017-revision.html.
3 Calculated by using UN 2017 Population Projection.

Other projections about age group over older than 60 are made by different researchers. For example, the projection of Bloom et al. (2011) on this age group was that the number of them will be 1 billion by 2020 and almost 2 billion by 2050, while the ratio of the people aged 80 or over is expected to increase to 4 % from 1 % till 2050.

United Nations (UN) updates its future estimates every two years. According to the revision of UN (2017), the population of the world which is 7.6 billion is expected to reach 9.8 billion in 2050 and 11.2 billion in 2100. The upward trend is expecting for the world population with a downward trend in fertility rate.[4] According to the UN Population Facts Report (2017), developing countries have the demographic transition fact that increasing level of life expectancy at birth and decreasing level of lifetime fertility. According to 2025-2030 projection, below replacement level fertility (fewer than 2.1 births per woman) rate is expected to be 67 %[5]. Recently, fertility has a decreasing trending almost all over the world and the same decreasing trend is seen even in Africa. However, Europe has an increasing rate from 1.4 births per woman in 2000-2005 to 1.6 births in 2010-2015.[6] In this case, the fertility rate decreases from 2.4 % to 2.1 % during 2005-2015 for Turkey[7]. Therefore, there is a negative trend in the youth population and a positive trend in the elders' population below for Turkey.

There is a growing literature about the aging population and its effect as it is projected that the entire world is aging rapidly than before. Some of the studies analyze the effect of an aging population on different macro variables such as economic growth (Bloom et al., 2011) and poverty (Cameron, 2000; Goudswaard et al., 2012; Altman, 2015). After the year of 2002, Turkey has a more equalized distribution comparing the 1990-2000s.[8] However, this improvement throughout years was not enough for Turkey to catch up most of the OECD

---

4   For the UN rates, seehttps://www.compassion.com/multimedia/world-population-prospects.pdf, last access date: 20.12.2018.
5   The data used for this analysis are from United Nations, *World Population Prospects: The 2017 Revision*, New York: United Nations.
6   https://www.un.org/development/desa/publications/world-population-prospects-the-2017-revision.html, last access date: 20.12.2018.
7   United Nations Population Division, New York, World Population Prospects: The 2017 Revision, last accessed December 2018.
8   For Turkey, as there is a data limitation before 2002, researches have limited studies about that issue. Gürsel et al. (2000), which investigates these years, indicate that there occurs slightly increased in overall inequality from 1987 to 1994.

countries because overall inequality in Turkey is still higher than those OECD countries.[9] From this point of view, topics related to income inequality become one of the popular subjects for Turkish researchers, as well. Furthermore, an aging society is on the scene for Turkey, older workers' decisions about the labor market take the researchers' attention.

After retirement, older workers are at the edge of taking some decisions about the labor market. One of the choices is continuing to work, the other one is staying out of the labor market or re-entering the labor market. The results of these decisions have impacts on not only the labor market but also on income inequality. Re-entering the labor market or continuing to work in countries where retirement pensions are very low has more significant consequences via income inequality. Therefore, investigating the impacts of elderly population earnings on income inequality for Turkey is very important.

With the lack of literature on this issue in Turkey, the aging population is a common subject among researchers all over the world. Bloom et al. (2011) examined the economic impacts of aging, especially on economic growth. There will be an age structure effect on economic growth as analyzed for East Asia by Bloom and Williamson (1998). They stated the second effect as accounting effects such as aging would change the labor supply and saving behavior. Moreover, as increasing life expectancy, the retirement age also changes.

It is found out that comparing the older individuals who are out of labor force and the ones attached to the labor market, the ones who are out of labor force are less happy (Winkelmann and Winkelmann, 1998). In addition to that, the fact that older workers would prefer working shorter hours to longer hours and they would also prefer to shorter hours to move to full retirement is examined. From this perspective, having more elderly individuals in the labor market has an impact not only on economic growth but also on income inequality.

To deal with this inequality problem, social transfers can be an important way as suggested in Brown and Prus (2004). They analyzed the social transfer across countries as higher social transfers lead to lower old-age income inequality. They showed government transfers play an important role to reduce the old-age income inequality.

---

9   According to OECD (2016), Turkey possesses one of the worst income distribution records among the OECD countries. At the ranking, Turkish income inequality is only better than Chile and Mexico.

As China is becoming an aging society due to the implementation of the one-child in the 1970s (Dong et al., 2018), income inequality has been a serious problem across different age groups. Therefore, this problem for China has been studied by the studies such as Chen et al. (2017), Dong et al. (2018) and Zhong (2011). They emphasize that the impact of age is greater in urban areas compared to rural ones on inequality.

In addition to China, several other Asian countries have an aging problem. For instance, South Korea by Hong and Kim (2008), Japan by Ohtake and Saito (1998) and Indonesia by Cameron (2000). Hong and Kim (2008) analyzed the income inequality for South Korea. As South Korea has experienced rapid economic growth during the past 20 years, this caused changes in income inequality. As Gini coefficient is closer to 1.00 that means there is higher income inequality. They represented a higher Gini coefficient as years pass for people aged 61–65 compared to younger age groups. In their study, Ohtake and Saito (1998) follow the approach of Deaton and Paxson (1994) which found that both consumption and income inequality increase with population aging for the US, the UK, and Taiwan.

Moreover, there is a growing aging population and income inequality studies for Western countries such as the US (Osberg, 1991; Danzier and Gottschalk, 1995; Xiao et al., 1999; Rubin et al., 2000), the EU (Goudswaard et al., 2012), the UK (Jenkins, 1995).

There are a number of different ways that this chapter contributes to the existing literature. First, according to our knowledge, it is the first study on the effects of elders' earnings on the inequality for Turkey as a developing country that has the highest youth ratio in the EU. However, by utilizing Income and Living Conditions Surveys, Bayar and Yanık-İlhan (2014) examined the effect of wives' earnings on overall inequality in which they applied a similar methodology with this chapter. First, they found out that the highest within inequality belongs to wives' earnings for those years. Second, they claimed that although husbands' and wives' earnings show a correlation and wives' earnings are nonequalizing. Third, they discriminate the impact of wives' earnings and then they conclude that a small contribution on income inequality is seen via wives' earnings. Making a comparison with Bayar and Yanık-İlhan's (2014) paper and this paper results, one can compare the contribution of wives' earnings and elderly earnings for Turkey. Another motivation of this study is to look from this view to the study of Bayar and Yanık-İlhan (2014).

As to reveal the importance of the elders' position in the labor market and contribution to the income inequality, for empirical analysis Income and Living Condition Survey (SILC) data stem is used. The data belongs to the years 2006,

2011 and 2016. The comparison between the existing distributions for the investigated years and the hypothetical distribution that elders have no earnings is the main aim of this chapter.

There are four sections in this chapter. The methodology section is presented in Section 2 while data and empirical results are investigated. Section 4 is reserved for the conclusion.

2 Methodology

There exist some important points for the examination of the inequality measures and the decomposition of income inequality. In the present chapter, the required steps for the examinations are followed. At first, choosing not only the appropriate unit of analysis but also the right household's equivalent scale is determined.

As a unit of analysis, the whole household is not taken. Instead of this, the equivalent disposable household is taken.[10] Certainly, there are some assumptions for these households. One assumption is the equal sharing of individuals within a household. Therefore, the equivalent scale is calculated for the model.[11]

The empirical studies in the literature apply some common measures for revealing income inequality of the countries. Gini coefficient is one of the well-known measures. Generalized Entropy (I_a) class measures are the other measures that are used widely (Litchfield, 1999).[12]

Decomposing by Income Source

In his chapter, Shorrocks (1982) focuses on a complete decomposition of inequality of total income. The contributions of each earnings on inequality are revealed. Shorrocks (1982) explores a way of decomposing inequality in which the contribution of each income source of k to overall inequality s_k is expressed as follows[13]:

$$s_k = \text{cov}(Y_k, Y)/\sigma^2(Y) \tag{1}$$

where Y_k represents the income level of the individual in category k, while cov (Y_k, Y) is the covariance of income source Y_k, Y shows total income level and $\sigma^2(Y)$ is the variance of total income.

---

10 For the detailed definition of household disposable income, please see TurkStat (2011).
11 The details of the methodology please see Gürsel et al. (2000).
12 More details about the inequality measures please see Litchfield (1999).
13 The sum of inequality contribution of each income components gives total inequality.

Tab. 13.1: General Characteristics of the Samples. Source: Authors 'calculations based on SILC 2006, 2011 and 2016.

	2006	2011	2016
The Number Individuals in the Sample	42,795	56,437	77,324
The Number Household in the Sample	10,920	15,024	22,441
Mean of Households	3.91	3.76	3.45
Mean of Annual Income per hh	13,884	23,025	40,185
Mean of Equivalent Annual Income per hh	7,634	12,884	20,666
Head-Count Ratio	18.5	15.2	12,4
Gini Coefficient	0.42	0.39	0.38

3 Data and Empirical Results

For the empirical analysis, the Survey of Income and Living Conditions for the years 2006, 2011 and 2016 is utilized. Elders are defined as individuals who are older than 54 years old. As to be consistent with the main aim, overall households divided into two subgroups: one is the elders' earnings and all other earnings. In that respect, the different inequality measures of subgroup earnings and the decomposition analysis are investigated. In order to examine the significance of the difference between the contribution of different earnings on overall inequality, Shorrocks decomposition technique is applied.

In addition to these, in order to test whether there is an impact of elders' earnings on the overall inequality or not, a hypothetical distribution is created where elders' earnings are equal to zero. By comparing the income inequality in the counterfactual distribution and in the actual distribution, the impact of elders' earnings can be explored.

The general descriptive summary of the households for each year is given in Table 13.1. As seen from the table, household sizes are not changing too much during the investigated years. For the year 2006, it is nearly 10,000 while it is 15,000 in 2011 and 22,000 in 2016. On the other hand, the sample size varies from nearly 42,000 to 56,000 and then to 77,000. In addition to this, the mean households' size is around at the value of four that has the same value for all investigated years. Same as the mean equivalent annual incomes per household (7,600 TL in 2006, 13,000 TL in 2011 and 20,700 TL in 2016), the mean annual income per household is increasing (while it is 13,000 TL in 2006, it is around 23,000 TL in 2011 and 40,185 in 2016).

As the percentages are nearly the same for 2006, 2011 and 2016, according to the descriptive statistics, around 51 % of the population is females while this percentage increases to 53 % for the elderly. This is valid for all years. Besides, it is revealed that the elderly population (who are older than 54 years old) percentage in the whole population is around 15 %[14].

When we compare the elderly and the whole population through their education level, it is exposed that the whole population and elderly are mainly graduated from primary school. Primary graduates' percentages are 40 %, 35 % and 42 % for the years 2006, 2011 and 2016, respectively. These values are nearly the same as the elderly. However, for the case of the second highest percentages, illiterates take the place among the elderly while secondary school graduates take the place among the whole population. In addition to that, it can be said that younger individuals are more educated than older ones since higher education levels percentages' are higher for younger than the ones for older.

Examining the employment types for the elderly and the whole population, full-time workers have the highest values for investigated years (for the elderly it is 20 %, 18 %, and 19 %, respectively). For the case of retired individuals, around 7 % of the whole population is retired, on the other hand, for the elderly, this number increased to 23 %, 28 %, and 34 %.

In Turkey, paid workers are more likely to reach health care since they have social security coverage. As the ratio of paid workers among the elderly is lower compared to the one among the whole population, for the case of elderly social security coverage is very low, as well. The ratio of paid workers has a value of 42 %, 50 % and 60 % for the whole population while these values are 10 %, 14 % and 26 % for the elderly. Among the elderly, self-employment is more common. The ratio of self-employed individuals among the elderly is more than twice as much as the one among the whole population. For example, among elderly, it is 51 %, 53 %, and 44 % while it is 20 %, 24 % and 17 % among the whole population for the years 2006, 2011 and 2016, respectively. These results indicate that the elderly are more likely to be self-employed. Therefore, being self-employed may be the reason for the fact that the social security of the elderly is lower than the whole population.

The fact that older individuals' willingness to investing their human capital endowment is lower. Having constraints to not only travel for long distances, or migration for the search of employment but also to learn new technology may be counted as the reasons for them not to invest in their human capital. In this situation, self-employment comes to the forefront in the employment of

---

14 As to gain space in the main text, the detailed descriptive statistics are not given in the chapter. Upon request, these findings will be shared by the authors.

older workers since the self-employment decision is more likely to depend on individuals' choice compare to be a paid worker. An older individual will take the decision to participate according to the arrival rate of suitable offers and an older individual probably will give more value to leisure. Therefore, having higher self-employment ratios among the elderly is logical.

When different regions are explored more closely, Aegean and Mediterranean regions have the highest ratio of elderly (14.15 % and 8.96 %; 14.83 % and 10.36 %, 14.55 % and 11.51 % for the years 2006, 2011 and 2016, respectively).[15] These facts expose that for the case of living the elder individuals prefer seaside regions to others.

Counterfactual distribution

The impact of elders' earnings on overall inequality can be captured from the findings of Tab. 3. In the table, the findings of the counterfactual distribution of households are given. The first panel of the table exhibits the inequality within the household with elders' earnings, and the second panel shows the results of the hypothetical distribution in which the elders' earnings equal to zero.

At the first panel of Tab. 13.2, it can be seen that the households with elders' earnings inequalities are lower than the other panel. If the improvements within the inequality of households with elders' earnings for the different years are compared, Gini coefficient is 0.42 in 2006 and it decreases to 0.39 and 0.39 for the years 2011 and 2016. A slight improvement over time in all inequality measures is examined.

Besides, comparing the inequality of households with and without elders' earnings, clear evidence is obtained: The overall inequality is higher when the elders' earnings removed from the household's earnings. This means, even the number of elder individuals is low; they improve the income distribution of overall households. Although the elderly have a higher inequality within the group, their contribution to the income distribution is positive.

Tab. 13.3 indicates the within inequality measures of different income sources. The two common inequality measures are given in the table. These results represent supportive evidence for decomposition analysis. The elders' earnings inequality measures are higher than all other earnings. The Gini coefficient and Half Coefficient of Variation Squared are 0.49, 0.47 and 0.47 and 0.73,

---

15 According to SR1 level, for Turkey, Anatolian part is divided into five different regions as follows: Central Anatolia, West Anatolia, Middle East Anatolia, Southeast Anatolia and Northeast Anatolia.

Tab. 13.2: Inequality Measures. Source: Authors' calculations based on SILC 2006, 2011 and 2016.

	2006	2011	2016
Inequality within the household with Elders' Earnings			
Gini Coefficient	0.42	0.39	0.39
Theil Index	0.32	0.29	0.29
Coefficient of Variation	0.98	0.97	0.98
Mean Log Deviation	0.31	0.26	0.26
Half Coef. of Variation Squared	0.48	0.47	0.48
Inequality within the household without elders' earnings			
Gini Coefficient	0.45	0.43	0.44
Theil Index	0.36	0.35	0.36
Coefficient of Variation	1.05	1.07	1.08
Mean Log Deviation	0.38	0.35	0.37
Half Coef. of Variation Squared	0.55	0.58	0.58

Tab. 13.3: Within Inequality Measures of Different Earnings. Source: Authors' calculations based on SILC 2006, 2011 and 2016.

	2006	2011	2016
	Half Coefficient of Variation Squared		
Elders' Earnings	0.73	0.75	0.88
All Other Earnings	0.55	0.58	0.58
Total	0.48	0.48	0.48
	Gini coefficient		
Elders' Earnings	0.49	0.47	0.470
All Other Earnings	0.45	0.43	0.44
Total	0.42	0.39	0.39

0.75 and 0.88 for the investigated years for the elders' earnings. These numbers are higher than the other group. Therefore, these results indicate that the elders' earning group have more non-equal distribution within the group compared to the other group.

As mentioned above, inequality within the elderly is higher than the other group. This is the opposite of the findings for 15 EU since Goudswaard et al. (2012) pointed out that, for the majority of the countries, inequality among older people is lower than the ones younger than 65. Having a more unequal distribution within the group is probably because the elderly mainly work as self-employed or unpaid family workers. Self-employed individuals are probably the retired ones and they re-enter the labor market since they considered their retirement pensions are not enough. In addition, as it is stated in Yanık-İlhan and Bayar (2013), the jobs in non-agricultural sectors require skilled workers. Therefore, it is more difficult for older individuals to find a job in non-agricultural sectors. Among older workers, the highest share belongs to the service sector compared to manufacturing and construction. Yanık-İlhan and Bayar (2013) stated that among older individuals, the employment rate in the service sector is around 75 %, on the other hand, it is only 20 % in the manufacturing sector.

One can say that those elder individuals are unpaid workers in the agricultural sector and they are not retired. In other respects, self-employed ones are usually in non-agricultural sectors and they are the ones who re-entered the labor market after retirement. From this perspective, putting these two different older individuals' group will lead to a heterogeneity via income levels and their human capital and as a result income distribution within the elderly is more unequal. Although their distribution within the group is unequal, their contribution to overall inequality is positive. This is probably due to the Kuznets' Inverted-U hypothesis that is as per capita national income of a country increases, inequality in income distribution rises and after reaching a certain degree, income inequality falls. More clearly, calculating the per capita income by setting the elders' earnings to zero leads to a lower value than calculating the per capita income by considering the elders' earnings. This leads to per capita income of a country increases. Therefore, calculating income inequality by considering the elderly earnings may be the reason for per capita to reach that peak value. Actually, there is no clear evidence on this cause and effect relationship, however, the results lead us to make a conclusion like that, this conclusion has to be investigated in further studies.

The Gini coefficient measure and coefficient of variation of the households without elders' earnings are 0.45 and 1.05; 0.43 and 1.07 and 1.08, for the years 2006, 2011 and 2016 respectively. All other inequalities represent the same deterioration at the distribution. This finding reveals consistency with theoretical expectations. Even the number of elders with any kind of earning is very low in Turkey; the income dispersion of the elders' is wide.

Decomposition of Income Inequality

The Shorrocks decomposition technique is employed in order to examine how different earnings affect total inequality for the years 2006, 2011 and 2016.

Tab. 13.4: Impacts of Different Earnings to Overall Inequality. Source: Authors' calculations based on SILC 2006, 2011 and 2016.

	2006	2011	2016
Factor shares in total income (1) (%)			
Elders' Earnings	14.03	15.05	17.61
All Others' Earnings	85.97	84.94	82.38
The proportional contribution of earnings to inequality (2) (%)			
Elders' Earnings	15.62	13.78	19.03
All Other Earnings	84.38	86.22	80.96
Relative inequality indicator [(2)/(1)]			
Elders' Earnings	1.11	0.92	1.08
All Other Earnings	0.98	1.01	0.98

Therefore, the contribution of different earnings on overall inequality is revealed. In accordance with the main aim, as to explore the contribution of the elders' earnings on overall inequality, the household's income is divided into two subgroups. One of the subgroups belongs to elders and the other one belongs to all other individual's earnings within the household and all the other type of earnings including interest and transfer payments.

The decomposition analysis is presented in Tab. 13.4. In each column of the table, different years' results are given. The first panel of the table shows the shares of the different earnings in overall equivalent disposable income while the second panel represents the proportionate contribution of earnings to overall inequality. The third panel, named relative inequality indicator, is obtained by dividing the second-panel results to first panel ones.

While the first column is examined, it is seen that elders' earnings have a smaller share of total income with respect to all other earnings. The share of the elders' earnings is around 14 % in the year 2006 and increased to 15 % in the year 2011 and nearly 18 % in the year 2016. All other earnings share in total income is around 86 % in the year 2006 and 85 % in the year 2011 and 82 % in 2016. These findings explore that elders' earnings are responsible for a small part of the total income for all investigated years.

According to the findings of the Shorrocks decomposition analysis, the contribution of elders' earnings on overall inequality is lower than all other earnings contribution. While the proportionate contribution of elders' earnings on overall inequality is around 16 %, 14 % and 19 % for the years 2006, 2011 and 2016, respectively and the contribution of all other earnings on overall inequality is around 84 %, 86 % and 81 % for the same years. The results reveal that elders' earnings are not as responsible as all other earnings for the overall inequality.

The relative inequality indicator is calculated by dividing panel (2) with panel (1). The third panel of the table presents those relative inequality indicators. In those panels, given numbers demonstrate the relative importance of each income sources. In this respect, if the value of relative inequality is above unity, then it can be stated that the contribution to inequality is significant. However, if it is below unity, then it can be stated that the contribution is unimportant.

The relative inequality indicator of elders' earnings is 1.11, 0.92 and 1.08, for the investigated years that is higher than unity. From this point of view, comparing with all other earnings and elders' earnings, even all other earnings generate greater inequality, the elders' earnings contribution to overall inequality compared to their share in total income is higher. However, the relative importance of elders' earnings is decreasing over time. That means the elders' earnings lose their importance for creating overall inequality. Bayar and Yanık-İlhan (2014), for the case of wives' earnings, found that the relative inequality indicator is higher than unity (1.56) for 2006. In addition to that, the smallest share of the total income belongs to their earnings. On the other hand, it is lower than one for the year 2011. For the case of married couples, their share is 7.30 % and 10.03 % of total income. This result indicates that although wives' earnings have a low share of total income, they create more inequality. On the other hand, according to the findings of this chapter, although elderly earnings have a higher share of total income compared to wives' earnings, elderly earnings do not generate more inequality.

4 Conclusion

Investigating the effects of elders' earnings on income inequality for Turkey, it is clear from the findings that, which are consistent with the expectations, when elders' earnings are removed from households' earnings, it is examined that within inequality is higher compared to the one when the calculations are done with elders' earnings. Besides, elders' earnings equalize the household's income. This can be interpreted as the fact that their earnings lead to an improvement in the income distribution of households. However, Bayar and Yanık-İlhan (2014) found the opposite impact of wives' earnings on inequality, which is: wives' earnings are non-equalizing.

Besides, the decomposition analysis shows that elders' earnings contribute less on overall inequality than the others while their share in total income is low. In that respect, the contribution of elders' earnings on overall inequality is beyond their share in total income.

In addition, the contribution of elders' earnings is decreased over time, which is the opposite of the existing empirical studies which find larger impacts of elders' earnings on income inequality for developing countries compared to developed countries (Zhong, 2011). Besides the increasing of income inequality over the years, the aging population has played a role like an accelerator. As a policy implication of this chapter, governments should upgrade their policies to decrease income inequality by taking into account the impact of the aging population. However, more investigations on this issue should be done to understand which employment type and sector for the elderly is more equalizing.

Moreover, there should be some policy changes in the aging population in Turkey. Each age groups (60–80) and 80+ have different needs. Older individuals tend to work and save less while they need more health care services. Moreover, the 80+ group of people need more full-time care and financial support (Bloom et al., 2011). To conclude, according to most of the literature, the positive impact of aging on income inequality is seen that may be caused by a lower share of labor income (Wang et al., 2017).

References

Altman, D. (2015). Seniors and Income Inequality: How Things Get Worse with Age. *Wall Street Journal*, Retrieved from https://blogs.wsj.com/washwire/2015/06/11/seniors-and-income-inequality-how-things-get-worse-with-age/https://blogs.wsj.com/washwire/2015/06/11/seniors-and-income-inequality-how-things-get-worse-with-age/ (10.12.2018).

Bayar, A. A. & Yanık-İlhan, B. (2014). Do Wives' Earnings Have an Impact on Income Inequality?: Evidence from Turkey. *Topics in Middle Eastern and African Economies*, 16(2), 105–123.

Yanık-İlhan, B. & Bayar, A. A. (2013). The Labour Market Attachment of an Aged Population: An Empirical Analysis from Turkey. *METU Studies in Development*, 4(1), 49–82.

Bloom, D. E., Canning, D., & Fink, G. (2011). Implications of Population Aging for Economic Growth, NBER Working Papers Series, No: 16705.

Bloom, D. E. & Williamson, J. G. (1998). Demographic Transitions and Economic Miracles in Emerging Asia. *The World Bank Economic Review*, 12(3), 419–455.

Brown, R. L. & Prus, S. G. (2004). Social Transfers and Income Inequality in Old Age: A Multi-National Perspective. *North American Actuarial Journal*, 8, 4.

Cameron, L. A. (2000). Poverty and Inequality in Java: Examining the Impact of the Changing Age, Educational and Industrial Structure. *Journal of Development Economics*, 62, 149–180.

Chen, X., Huang, B. & Shaoshuai, L. (2017). Population Ageing and Inequality: Evidence from China. *The World Economy*, 41, 1976–2000.

Danzier, S. & Gottschalk, P. (1995). *America Unequal*. Cambridge: Harvard University Press.

Deaton, A. & Paxson, C. (1994). Intertemporal Choice and Inequality. *Journal of Political Economy*, 102(3), 437–467.

Dong, Z., Tang, C., & Wei, X. (2018). Does Population Aging Intensify Income Inequality? Evidence from China. *Journal of Asia Pacific Economy*, 23(1), 66–70.

Goudswaard, K., Vliet, O. V., Been, J., & Caminada, K. (2012). *Cesifo DICE Report*, 4, 21–26.

Gürsel, S., Levent, H., Selim, R., & Sarıca, Ö. (2000). Türkiye'de Bireysel Gelir Dağılımı ve Yoksulluk: Avrupa Birliği ile Karşılaştırma, TÜSİAD Working Paper, No: TÜSİAD-T/2000-12/295. Ankara.

Hong, B. E. & Kim, H. Y. (2008). EASP 5th Conference Welfare Reform in East Asia, Taiwan.

Jenkins, S. P. (1995). Accounting for Inequality Trends: Decomposition Analyses for the UK, 197186. *Economica*, New Series, 62(245), 29–63.

Litchfield, J. A. (1999). *Inequality: Methods and Tools*. The World Bank, Washington, DC. Retrieved from https://siteresources.worldbank.org/INTPGI/Resources/Inequality/litchfie.pdf (12.08.2019).

OECD. (2016). *Society at a Glance 2016*, OECD Books.

Ohtake, F. & Saito, M. (1998). Population Aging and Consumption Inequality in Japan. *Review of Income and Wealth*, 44, 3.

Osberg, Lars. (1991). *Economic Inequality and Poverty*. Armonk, New York: ME Sharpe.

Rubin, R. M., White-Means, S., & Daniel, L. M. (2000). Income Distribution of Older Americans. *Monthly Labor Review*, November, 19–30. Retrieved from https://www.bls.gov/opub/mlr/2000/11/art2full.pdf (12.08.2019).

Shorrocks, A. F. (1982). Inequality Decomposition by Factor Components. *Econometrica*, 50, 193–212.

TurkStat. (2011). Income and Living Conditions Survey. http://www.tuik.gov.tr/PreIstatistikMeta.do?istab_id=1378

United Nations. (2017). World Population Prospects 2017, Department of Economic and Social Affairs, Population Division, Retrieved from https://population.un.org/wpp/Publications/Files/WPP2017_DataBooklet.pdf (12.08.2019).

Wang, C., Wan, G., Luo, Z., & Zhang, X. (2017). Aging and Inequality: The Perspective of Labor Income Share. ADBI Working Paper Series, No. 764, July. Retrieved from https://www.adb.org/sites/default/files/publication/336016/adbi-wp764.pdf (02.12.2018).

Winkelmann, L. & Winkelmann, R. (1998). Why Are the Unemployed So Unhappy? Evidence from Panel Data. *Economica*, 65(257), 1–15.

Xiao, J. J., Malroutu, Y. L., & Yuh, Y. (1999). Sources of Income Inequality Among the Elderly. *Financial Counseling and Planning*, 10(2), 49–59.

Yanık-İlhan, B. & Bayar, A. A. (2013). The Labour Market Attachment of an Aged Population: An Empirical Analysis from Turkey. *METU Studies in Development*, 4(1), 49–82.

Zhong, H. (2011). The Impact of Population Aging on Income Inequality in Developing Countries: Evidence from Rural China. *China Economic Review*, 22, 98–107.

Mehmet Akif Destek

14. Liberalization, Globalization and Income Inequality in Emerging Economies

Abstract This chapter aims to investigate the relationship between liberalization, globalization and income inequality for the period 20002014 in 16 emerging economies. For this purpose, we utilize the panel dynamic GMM estimation methodology. In doing so, to examine the effects of liberalization more thoroughly, we consider five dimensions of liberalization. Moreover, three indicators of globalization are used to observe the effects of globalization on income inequality in detail. In case of liberalization, the results show that the reduction of the role of government in the economy leads to increase in income inequality. Similarly, high protection of property rights seems to benefit the high-income groups more. In addition, the deregulation policies on both credit and labor market leads to decrease in income inequality. In case of globalization, we found that both economic and political globalization reduces the income inequality.

Keywords: Liberalization, Globalization, Income inequality, GMM, Emerging economies

1 Introduction

Over the past decades, many studies have been conducted on the effects of economic freedom on economic growth and income distribution. A consensus has emerged that liberalization and globalization policies have positive effects on economic growth (see, e.g., De Haan and Sturm, 2000; Sturm and De Haan, 2001; Vega-Gordillo and Alvarez-Arce, 2003; Berggren and Jordahl, 2005; Dreher, 2006; Justesen, 2008), the inclusiveness of the increasing income level as a result of economic freedom has become a matter of debate.

This study aims to examine the second issue that the effects of liberalization and globalization on income inequality in emerging economies for the period 20002014. In doing so, we consider different liberalization and globalization dimensions to better observe the effect of economic freedom on income inequality. The selected country group is chosen because most of the emerging economies have begun in both the trade and financial liberalization reforms in the early 1990s (Das and Mohapatra, 2003). In addition, these liberalization reforms have also been shown as the main reason for the high growth rates that emerging countries have caught in the 2000s (Bekaert et al., 2001).

The contributions of this study are threefold. First, to the best of our knowledge, this is the first study to investigate the effects of liberalization and globalization on income inequality in emerging economies. Second, considering different indicators of both liberalization and globalization will give us a chance to more detailed policy implications. Third, since the estimation of a bivariate model may lead to inconsistent findings, empirical analyses of this study are based on the multivariate framework.

2 Theoretical Explanations and Literature Review

It is known that the terms of liberalization and globalization are multidimensional phenomena. Therefore, following Bergh and Nilsson (2010), we utilized with the Economic Freedom of the World Index (EFI) of Gwartney et al. (2013) which includes five dimensions of economic freedom and also used KOF Globalization Index which covers three globalization indicators. Based on this reason, we categorized the literature part depend on different indicators of both liberalization and globalization.

2.1 Liberalization and income inequality

2.1.1 Size of government

The first indicator of EFI measures the role of government in the economy, which is called EFI1. This indicator contains the government consumption as a share of total consumption, total transfers and subsidies as a share of GDP, government enterprises and top marginal tax rate. In case of this index, the low level of government consumption, transfers and subsidies means higher-index value. Similarly, smaller government enterprises and low-level marginal tax rates lead to high-index value. It is theoretically expected that public transfers will have an equal effect in countries where the size of the public sector is high. On the other hand, income inequality may increase when the targeted groups are wrongly selected for public transfers.

In the literature on government and inequality nexus, it seems contradictory results have been found. For instance, Lee (2005) examined the nexus for the period 19701994 in 64 countries and concluded that increasing size of the government increases income inequality. However, Roine et al. (2009), Niehues (2010) and Doerrenberg and Peichl (2014) found that income inequality decreases with the increasing public sector size.

2.1.2 Legal system and property rights

The second indicator of EFI measures the importance given to the rule of law which is called as EFI2. This dimension includes judicial independence, impartial courts, protection of property rights, military interference in rule of law and politics, integrity of legal system, legal enforcement of contracts, regulatory costs of the sale of real property, reliability of police and business costs of crime. Theoretically, similar to the EFI1 dimension, the effect of this dimension on income inequality is ambiguous. It is expected that high level protection of property rights benefits the high-income groups with more property. However, Sonin (2003) and Bergh and Nilsson (2010) argues that low level protection of property rights may be relatively beneficial to already wealthy groups.

2.1.3 Sound money

The third dimension of the EFI measures the predictability of money supply and inflation rate, which is called as EFI3. This indicator consists of money growth, standard deviation of inflation, inflation and freedom to own foreign currency bank accounts. When the literature is examined, it can be seen that the inconsistent arguments are valid about the effect of inflation on income inequality, too. Romer and Romer (1998), Easterly and Fisher (2001) and Albanesi (2007) argue that high rate of price inflation is correlated with inequality and poverty. On the other hand, Galli and Hoeven (2001) argue that the relationship between inflation and income inequality varies depending on the initial rates of inflation. According to this study, income inequality may decrease with the increasing inflation rates when the initial inflation rate is low, and vice versa.

2.1.4 Freedom to trade internationally

The fourth indicator of the EFI measures the freedom to trade which is called as EFI4. This dimension covers the tariffs, regulatory trade barriers, black market exchange rates and the controls of the movement of capital and people. According to the standard foreign trade theory (Stolper-Samuelson theorem), as a result of trade liberalization, the price of scarce factors decreases and the price of the abundant factor increases. In this direction, income inequality is expected to decrease due to trade liberalization and foreign trade. However, Stiglitz (1998) argued that trade liberalization could further increase income inequality by increasing the yield of education and skill-related factors. For this reason, there is generally no accepted view of the effects of commercial liberalization on inequality.

2.1.5 Regulation

The fifth indicator of the EFI measures the level of the deregulation policies of the countries which is called as EFI5. This dimension includes credit market regulations, labor market regulations and business regulations. Similar to the other liberalization dimensions, it is a controversial issue that affects the deregulation policies on inequality. Greenwood and Jovanovic (1990), Galor and Zeira (1993) and Banerjee and Newman (1993) argue that increasing credit accessibility reduces income inequality by creating job opportunities for low-income segments. However, income inequality may increase even more in case of the intervention of political elites to the deregulation policy (Bergh and Nilsson, 2010: 490).

2.2 Globalization and Income Inequality

2.2.1 Economic globalization

The first indicator of the KOF Globalization Index measures the economic integration level of the countries which is called as KOF1 indicator. Even though KOF1 indicator is similar to the EFI4 index, it differs from the EFI4 index in terms of including direct investments and portfolio investments. Bluestone and Harrison (1982) argue that the effects of foreign direct investments on inequality differ according to the level of development of countries. In case of developed countries, capital outflow increases the marginal product of the capital and reduces the marginal product of labor and thus reduces the price of labor. However, in case of less developed or developing countries, multinational corporations often demand high-skilled labor, which increases the difference between the wages of low-skilled labor and high-skilled labor (Alderson and Nielsen, 2002: 10).

2.2.2 Social globalization

The second dimension of the KOF Globalization Index measures the country's tendency to integrate with global social norms which is called as KOF2 index. This index covers some social indicators such as telephone traffic, foreign populations, internet users, the number of McDonald's restaurants, the number of IKEA, trade in newspapers and trade in books. Atkinson (1997) argues that social globalization may affect the income inequality by changing the behavior of unions. In empirical literature, Gaston and Rajaguru (2009) examined the nexus between social globalization and income inequality spanning the period of 1970–2001 in Australia and concluded that social globalization increases inequality. In addition, Destek (2018) probed the nexus between social globalization and income inequality for the period from 1991 to 2013 in 11 transition economies and found that social globalization increases inequality in Russia while it reduces inequality in Belarus and Poland.

2.2.3 Political globalization

The political component of the KOF Globalization Index which is called as KOF3 indicator measures the international political connections of the country. KOF3 indicator includes embassies in the country, membership in international organizations, participation in United Nations Security Council Missions and international treaties.

In the literature, the studies which use both the Economic Freedom Index and KOF Globalization Index to examine the effect of liberalization or globalization on income inequality are very limited. Berggren (1999) utilized with Economic Freedom Index (EFI) to examine the nexus between liberalization in 69 countries and found that liberalization reduces income inequality. Scully (2002) also used EFI and concluded that liberalization decreases inequality. Krieger and Meierrieks (2016) investigated the causal relationship between economic freedom (EFI) and inequality in panel of 100 countries and found that there is unidirectional causality from inequality to economic freedom. Sturm and De Haan (2016) searched the relationship between financial liberalization and inequality utilizing with EFI in 121 countries and concluded that financial liberalization increases the income inequality. Yay et al. (2016) used with both EFI indicators and KOF Globalization Index to examine the relationship between liberalization, globalization and inequality in 90 developed and developing countries and found that smaller government, deregulated markets, trade freedom and political globalization increases the income inequality. Bergh and Nilsson (2010) looked at the relationship between liberalization, globalization and inequality in 78 countries and this study confirms that the social globalization, freedom to trade and deregulation lead to increase in income inequality. Dreher and Gaston (2008) searched the relationship between globalization (KOF index) and income inequality in OECD and non-OECD countries and argued that globalization increases inequality in OECD countries. Perez-Moreno and Angulo-Guerrero (2016) examined the relationship between liberalization and income inequality using with EFI indicators in 28 EU member countries for the period of 2000–2010 and found that only the government size has positive significant effect on Gini coefficient.

3 Data, Model and Methodology

In order to examine the relationship between liberalization, globalization and income inequality, we construct two empirical model as follows;

$$lnINE_{it} = a_0 + a_1 lnLIB_{it} + a_2 lnX_{it} + \varepsilon_{it} \qquad (1)$$

$$lnINE_{it} = \beta_0 + \beta_1 lnGLO_{it} + \beta_2 lnX_{it} + \mu_{it} \qquad (2)$$

where INE is Gini coefficient and implies the income inequality, LIB includes six liberalization indicator (overall liberalization index, size of government, legal system and property rights, sound money, freedom to trade internationally and regulation) and indicates the liberalization level of the countries, GLO covers three globalization indicators (economic globalization, social globalization and political globalization) and implies the globalization level of the countries. In addition, we use X control variable which includes the gross domestic product and human capital. The annual data covers the period from 2000 to 2014 in 16 emerging countries: Argentina, Brazil, Chile, China, Colombia, Greece, Hungary, Indonesia, South Korea, Malaysia, Mexico, Pakistan, Philippines, Peru, Russia and Turkey. The data of INE is sourced from Standardized World Income Inequality Database (SWIID) 6.0 database of Solt (2016), LIB indicators are obtained from Fraser Institute database and the data of GLO is downloaded from KOF Globalization Index database. In addition, gross domestic product per capita is measured in 2010 constant US dollars and obtained from World Development Indicators of World Bank, human capital series are sourced from PWT 9.0 database.

This study uses the generalized method of moments (GMM) procedure to detect the impact of explanatory variables on income inequality. There are some reasons for choosing this approach. The first and most important reason is that this technique can be used in case of the number of cross-section is larger than the time period. In addition, the technique overcomes the possible endogeneity, heteroskedasticity and simultaneous reverse causality problems. Considering the high correlation risk between the explanatory variables in the empirical models established and the fact that the number of cross-section is larger than time period, this technique is the most appropriate technique to use. The GMM procedure was first developed by Arellano and Bond (1991), the system GMM estimation which both allows the level and first-differenced variables is constructed by Blundell and Bond (1998). In addition, this study employs two-step system GMM instead of one-step system GMM because using the two-step system GMM leads to more consistent findings than one-step test in case of the existence of possible heteroskedasticity problem.

4 Empirical Results and Discussion

In the first step of empirical analysis, we observe the correlation between liberalization and income inequality as seen in Tab. 14.1. According to the results, the

Tab. 14.1: Correlation Between the Variables for Liberalization-Inequality Nexus. Source: Author's own calculations

	INE	GDP	HC	EFI	EFI1	EFI2	EFI3	EFI4	EFI5
INE	1.000								
GDP	0.295	1.000							
HC	−0.369	0.779	1.000						
EFI	−0.195	0.530	0.724	1.000					
EFI1	0.155	−0.183	−0.198	0.180	1.000				
EFI2	0.266	0.648	0.681	0.622	−0.274	1.000			
EFI3	−0.194	0.422	0.583	0.729	−0.186	0.353	1.000		
EFI4	−0.044	0.515	0.585	0.692	−0.051	0.411	0.432	1.000	
EFI5	−0.110	0.122	0.434	0.627	0.010	0.327	0.333	0.275	1.000

explanatory variables are not highly correlated, so the endogeneity problem does not exist in the first model. In addition, real income, size of government, legal system and property rights are positively correlated with the income inequality while human capital, overall liberalization index, sound money, freedom to trade internationally and regulation are negatively correlated with inequality.

In the second step of empirical analysis, we examine the relationship between income inequality and different liberalization indicators for full sample (2000–2014) using panel dynamic GMM estimation and the results are given in Tab. 14.2. Before interpreting the results, we examine the validity of instruments and the existence of serial correlation problem. As seen in Tab. 14.2, Hansen J-test shows that instruments are valid and the test for second-order autocorrelation indicates that there is no serial correlation problem for all specifications. According to the results, the lagged value of income inequality positively, human capital index negatively affects the income inequality for all estimations. In addition, income inequality has increased with economic growth. In case of liberalization indicators, the effect of overall liberalization index on income inequality is positive but the coefficient is statistically insignificant. The coefficient of EFI1 indicators which decrease with bigger government size is positive and statistically significant. This means that the reduction of the role of government in the economy leads to increase in income inequality. Similarly, the coefficient of EFI2 is positive and statistically significant. Because of the bigger value of EFI2 requires high protection of property rights, this finding show that better protection of property right seems to benefit the high-income groups. The coefficient of EFI3 and EFI4 indicators are negative but statistically insignificant. Moreover,

Tab. 14.2: GMM Estimation Results on Liberalization-Inequality Nexus (2000–2014). Source: Author's own calculations

	(1)	(2)	(3)	(4)	(5)	(6)
INE(-1)	0.235***	0.168***	0.286***	0.232***	0.210***	0.319***
	[0.060]	[0.028]	[0.045]	[0.064]	[0.076]	[0.052]
GDP	0.019	0.052**	0.026*	0.020	0.046**	0.040**
	[0.017]	[0.020]	[0.013]	[0.016]	[0.018]	[0.017]
HC	−0.138***	−0.068	−0.133***	−0.134***	−0.169***	−0.098***
	[0.039]	[0.051]	[0.028]	[0.038]	[0.051]	[0.032]
EFI	0.018	-	-	-	-	-
	[0.052]					
EFI1	-	0.063*	-	-	-	-
		[0.032]				
EFI2	-	-	0.021***	-	-	-
			[0.007]			
EFI3	-	-	-	−0.003	-	-
				[0.021]		
EFI4	-	-	-	-	−0.032	-
					[0.102]	
EFI5	-	-	-	-	-	−0.072*
						[0.036]
Hansen J-test (p-value)	0.467	0.235	0.609	0.459	0.713	0.378
AB test for AR(1)	0.702	0.672	0.438	0.865	0.117	0.478
AB test for AR(2)	0.143	0.390	0.244	0.720	0.311	0.245

Note: The numbers in brackets are standard errors. *, ** and *** indicates the statistical significance at 10, 5 and 1 percent level, respectively.

we found that the coefficient of EFI5 indicator which implies the degree of deregulation policies of the country is negative and statistically significant. This result means that the deregulation policies on both credit and labor market leads to decrease in income inequality.

Moreover, to decompose the findings for two periods for before and after 2008 global financial crisis, we also employ the GMM estimation for these sample-periods. Tab. 14.3 shows the estimation results of the GMM estimation for 2000–2008 period. Based on the findings, it seems the sign of the coefficient of real income and human capital on inequality is same with the full sample findings. However, unlike the full sample, the impact of government size and protection of property rights is not statistically significant. In case of the period from 2009

Tab. 14.3: GMM Estimation Results on Liberalization-Inequality Nexus (2000–2008). Source: Author's own calculations

	(1)	(2)	(3)	(4)	(5)	(6)
INE(-1)	0.394*** [0.051]	0.167** [0.083]	0.333*** [0.089]	0.349*** [0.081]	0.138** [0.069]	0.150*** [0.032]
GDP	0.086*** [0.012]	0.042 [0.029]	0.039** [0.018]	0.024 [0.017]	0.049*** [0.014]	0.057*** [0.005]
HC	−0.196*** [0.049]	−0.472** [0.220]	−0.398** [0.170]	−0.267** [0.117]	−0.421*** [0.121]	−0.193** [0.085]
EFI	0.033 [0.050]	-	-	-	-	-
EFI1	-	0.022 [0.086]	-	-	-	-
EFI2	-	-	0.036 [0.024]	-	-	-
EFI3	-	-	-	−0.016 [0.021]	-	-
EFI4	-	-	-	-	0.002 [0.043]	-
EFI5	-	-	-	-	-	−0.065** [0.030]
Hansen J-test (p-value)	0.284	0.701	0.440	0.340	0.646	0.351
AB test for AR(1)	0.091	0.076	0.542	0.479	0.911	0.746
AB test for AR(2)	0.437	0.599	0.221	0.287	0.558	0.572

Note: The numbers in brackets are standard errors. *, ** and *** indicates the statistical significance at 10, 5 and 1 percent level, respectively.

to 2014, as shown in Tab. 14.4, the impact of human capital on inequality is also significantly negative. However, the impact of real income on inequality is negative and statistically insignificant. This finding means that increasing national income lost its regulatory effect on income distribution after the 2008 crisis.

In case of globalization, we first observe the correlation between globalization indices and income inequality. As presented in Tab. 14.5, the independent variables are not highly correlated therefore the endogeneity problem is not valid for the second model. Further, it can be seen that human capital, economic globalization, social globalization and political globalization are negatively correlated with income inequality while the real income is positively correlated with inequality.

Tab. 14.4: GMM Estimation Results on Liberalization-Inequality Nexus (2009–2014). Source: Author's own calculations

	(1)	(2)	(3)	(4)	(5)	(6)
INE(-1)	0.413***	0.349***	0.402***	0.405***	0.358***	0.360***
	[0.111]	[0.129]	[0.102]	[0.120]	[0.128]	[0.113]
GDP	−0.021	−0.033*	−0.010	−0.022	−0.018	−0.019
	[0.019]	[0.019]	[0.031]	[0.020]	[0.019]	[0.020]
HC	−0.162**	−0.172***	−0.160***	−0.196	−0.175***	−0.172***
	[0.063]	[0.063]	[0.057]	[0.086]**	[0.059]	[0.053]
EFI	0.080	-	-	-	-	-
	[0.090]					
EFI1	-	0.036	-	-	-	-
		[0.050]				
EFI2	-	-	0.029	-	-	-
			[0.046]			
EFI3	-	-	-	0.047	-	-
				[0.081]		
EFI4	-	-	-	-	0.034	-
					[0.076]	
EFI5	-	-	-	-	-	−0.121*
						[0.062]
Hansen J-test (p-value)	0.591	0.386	0.605	0.698	0.630	0.588
AB test for AR(1)	0.094	0.078	0.113	0.236	0.123	0.210
AB test for AR(2)	0.175	0.265	0.156	0.218	0.206	0.261

Note: The numbers in brackets are standard errors. *, ** and *** indicates the statistical significance at 10, 5 and 1 percent level, respectively.

We next investigate the relationship between income inequality and different globalization indicators using with system GMM estimator and the results are presented in Tab. 14.6. First, it can be seen that Hansen J-test results indicate the instruments are valid and Arellano-Bond test shows that there is no autocorrelation problem in the model. At a first glance, it seems that the lagged value of Gini coefficient and the real gross domestic product positively affects Gini coefficient while human capital negatively affects it. When the effects of globalization indicators on income inequality are evaluated, Tab. 14.6 shows that increasing economic globalization and political globalization reduces Gini coefficient.

Similar to the liberalization, we also observe the impact of globalization on inequality for the 2000–2008 and 2009–2014 sub-sample periods. According to

Tab. 14.5: Correlation Between the Variables for Globalization-Inequality Nexus. Source: Author's own calculations

	INE	GDP	HC	KOF1	KOF2	KOF3
INE	1.000					
GDP	0.295	1.000				
HC	−0.369	0.779	1.000			
KOF1	−0.181	0.631	0.740	1.000		
KOF2	−0.370	0.721	0.598	0.573	1.000	
KOF3	−0.427	0.276	0.189	0.118	0.264	1.000

Tab. 14.6: GMM Estimation Results on Globalization-Inequality Nexus (2000–2014). Source: Author's own calculations

	(1)	(2)	(3)
INE(-1)	0.435***	0.288***	0.773***
	[0.073]	[0.060]	[0.171]
GDP	0.028***	0.037*	0.038
	[0.085]	[0.019]	[0.048]
HC	−0.075**	−0.149***	−0.131**
	[0.033]	[0.041]	[0.066]
KOF1	−0.060**	-	-
	[0.025]		
KOF2	-	0.018	-
		[0.086]	
KOF3	-	-	−0.156*
			[0.088]
Hansen J-Test (p-value)	0.224	0.635	0.828
AB test for AR(1)	0.123	0.628	0.501
AB test for AR(2)	0.076	0.654	0.647

Note: The numbers in brackets are standard errors. *, ** and *** indicates the statistical significance at 10, 5 and 1 percent level, respectively.

the results from Tab. 14.7, the impact of increasing real income, human capital and globalization indices does not change for the period of 2000–2008. On the other hand, for the period from 2009 to 2014, only political globalization has significantly negative impact on income inequality.

To sum up, when the results are interpreted from the perspective of liberalization, we found that the deregulation policy is the most efficient indicator to

Tab. 14.7: GMM Estimation Results on Globalization-Inequality Nexus (2000–2008). Source: Author's own calculations

	(1)	(2)	(3)
INE(-1)	0.291***	0.360***	0.305***
	[0.088]	[0.057]	[0.063]
GDP	0.052***	0.105***	0.045***
	[0.014]	[0.030]	[0.016]
HC	−0.290*	−0.220**	−0.260**
	[0.161]	[0.106]	[0.121]
KOF1	−0.071***	-	-
	[0.026]		
KOF2	-	0.019	-
		[0.095]	
KOF3	-	-	−0.059*
			[0.034]
Hansen J-Test (p-value)	0.453	0.456	0.501
AB test for AR(1)	0.426	0.109	0.342
AB test for AR(2)	0.844	0.248	0.580

Note: The numbers in brackets are standard errors. *, ** and *** indicates the statistical significance at 10, 5 and 1 percent level, respectively.

reduce the income inequality in emerging countries. However, for a globalization perspective, political integration is more efficient than economic globalization to narrow the gap between the high-income and the low-income groups.

5 Conclusions and Policy Implications

The purpose of this chapter is to examine the effects of different dimensions of both liberalization and globalization on income inequality on 16 emerging economies for the period of 2000–2014. In doing so, the relationship between liberalization and income inequality is searched with six liberalization indicators (overall liberalization index, size of government, legal system and property rights, sound money, freedom to trade internationally and regulation). In addition, the nexus between globalization and income inequality is investigated with three dimensions of globalization (economic globalization, social globalization and political globalization). Moreover, the real gross domestic product per capita and human capital accumulation are used as control variables to the empirical

Liberalization, Globalization and Income Inequality 247

Tab. 14.8: GMM Estimation Results on Globalization-Inequality Nexus (2009–2014). Source: Author's own calculations

	(1)	(2)	(3)
INE(-1)	0.410***	0.371***	0.416***
	[0.072]	[0.125]	[0.110]
GDP	−0.001	−0.020	−0.015
	[0.018]	[0.019]	[0.019]
HC	−0.155***	−0.185**	−0.156**
	[0.049]	[0.084]	[0.065]
KOF1	−0.073	-	-
	[0.018]		
KOF2	-	−0.017	-
		[0.150]	
KOF3	-	-	−0.461*
			[0.253]
Hansen J-Test (p-value)	0.589	0.605	0.609
AB test for AR(1)	0.095	0.165	0.208
AB test for AR(2)	0.181	0.209	0.221

Note: The numbers in brackets are standard errors. *, ** and *** indicates the statistical significance at 10, 5 and 1 percent level, respectively.

model. In doing so, the dynamic panel GMM estimation method is used because of the limited sample period.

The results show that the real income per capita positively affects income inequality and the human capital index negatively affects income inequality in emerging economies. In case of liberalization, we found that low size of government and high protection of poverty rights increases the Gini coefficient therefore increases the income inequality. Based on this finding, it appears that public investments, public pioneering firms and publicly funded transfers are still an important tool to reduce inequality in emerging economies. On the other hand, deregulation policies reduce the income inequality. This finding means deregulation policies especially in credit market facilitates the credit access of low income-segment and leads to create new job opportunities for this segment. In case of globalization, increasing economic and political globalization reduces the income inequality while the coefficient of economic globalization is statistically insignificant.

Regarding the policy implications of the findings of this study, it can be suggested that the governments of emerging countries should be a pioneer in certain sectors until the private sector achieves its internal adequacy. In addition, the governments should pay attention to financial deregulation policies, especially to increase job opportunities, without ignoring the problems of repayment risk and adverse selection. Moreover, based on the finding that expresses the positive impact of political globalization on income distribution, policy agreements and harmonization laws should be maintained with a view to reducing income inequality

References

Albanesi, S. (2007). Inflation and inequality. *Journal of Monetary Economics*, 54(4), 1088–1114.

Alderson, A. S. & Nielsen, F. (2002). Globalization and the great U-turn: Income inequality trends in 16 OECD countries. *American Journal of Sociology*, 107(5): 1244–1299.

Arellano, M. & Bond, S. (1991). Some tests of specification for panel data: Monte Carlo evidence and an application to employment equations. *The review of economic studies*, 58(2), 277–297.

Atkinson, A. B. (1997). Bringing income distribution in from the cold. *The Economic Journal*, 107, 297–321.

Banerjee, A. V. & Newman, A. F. (1993). Occupational choice and the process of development. *Journal of Political Economy*, 101(2), 274–298.

Bekaert, G., Harvey, C. R., & Lundblad, C. (2001). Emerging equity markets and economic development. *Journal of Development Economics*, 66(2), 465–504.

Berggren, N. (1999). Economic freedom and equality: Friends or foes? *Public Choice*, 100(3), 203–223.

Berggren, N. & Jordahl, H. (2005). Does free trade really reduce growth? Further testing using the economic freedom index. *Public Choice*, 122(1), 99–114.

Bergh, A. & Nilsson, T. (2010). Do liberalization and globalization increase income inequality? *European Journal of Political Economy*, 26(4), 488–505.

Bluestone, B. & Harrison, B. (1982). *The Deindustrialization of America: Plant Closings, Community Abandonment, and the Dismantling of Basic Industry*. New York: Basic Books.

Blundell, R. & Bond, S. (1998). Initial conditions and moment restrictions in dynamic panel data models. *Journal of Econometrics*, 87(1), 115–143.

Das, M. & Mohapatra, S. (2003). Income inequality: the aftermath of stock market liberalization in emerging markets. *Journal of Empirical Finance*, 10(1), 217–248.

De Haan, J. & Sturm, J. E. (2000). On the relationship between economic freedom and economic growth. *European Journal of Political Economy*, 16(2), 215–241.

Destek, M. A. (2018). Dimensions of globalization and income inequality in transition economies: Taking into account cross-sectional dependence. *Eastern Journal of European Studies*, 9(2), 5.

Doerrenberg, P. & Peichl, A. (2014). The impact of redistributive policies on inequality in OECD countries. *Applied Economics*, 46(17), 2066–2086.

Dreher, A. & Gaston, N. (2008). Has globalization increased inequality? *Review of International Economics*, 16(3), 516–536.

Dreher, A. (2006). Does globalization affect growth? Empirical evidence from a new Index. *Applied Economics*, 38, 1091–1110

Easterly, W. & Fischer, S. (2001). Inflation and the poor. *Journal of Money, Credit and Banking*, 33(2), 160–178.

Galli, R. & Hoeven, R. (2001). *Is Inflation Bad for Income Inequality: The Importance of the Initial Rate of Inflation*. ILO. Employment Paper No. 29.

Galor, O. & Zeira, J. (1993). Income distribution and macroeconomics. *The Review of Economic Studies*, 60(1), 35–52.

Gaston, N. & Rajaguru, G. (2009). The long-run determinants of Australian income inequality. *Economic Record*, 85(270), 260–275.

Greenwood, J. & Jovanovic, B. (1990). Financial development, growth, and the distribution of income. *Journal of Political Economy*, 98(5, Part 1), 1076–1107.

Gwartney, J., Lawson, R., & Hall, J. 2013. Economic Freedom of the World: 2013 Annual Report. Fraser Institute.

Justesen, M. K. (2008). The effect of economic freedom on growth revisited: New evidence on causality from a panel of countries 1970–1999. *European Journal of Political Economy*, 24(3), 642–660.

Krieger, T. & Meierrieks, D. (2016). Political capitalism: The interaction between income inequality, economic freedom and democracy. *European Journal of Political Economy*, 45, 115–132.

Lee, C. S. (2005). Income inequality, democracy, and public sector size. *American Sociological Review*, 70(1), 158–181.

Niehues, J. (2010). Social spending generosity and income inequality: A dynamic panel approach. IZA DP No. 5178, Retrieved from http://ftp.iza.org/dp5178.pdf (13.08.2019).

Pérez-Moreno, S. & Angulo-Guerrero, M. J. (2016). Does economic freedom increase income inequality? Evidence from the EU countries. *Journal of Economic Policy Reform*, 19(4), 327–347.

Roine, J., Vlachos, J., & Waldenström, D. (2009). The long-run determinants of inequality: What can we learn from top income data? *Journal of Public Economics*, 93(7), 974–988.

Romer, C. D. & Romer, D. H. (1998). *Monetary Policy and the Well-Being of the Poor*. NBER. Working Paper Series, No. 6793.

Scully, G. W. (2002). Economic freedom, government policy and the trade-off between equity and economic growth. *Public Choice*, 113(1), 77–96.

Solt, F. (2016). The standardized world income inequality database. *Social Science Quarterly*, 97(5), 1267–1281.

Sonin, K. (2003). Why the rich may favor poor protection of property rights. *Journal of Comparative Economics*, 31(4), 715–731.

Stiglitz, J. E. (1998). More Instruments and Broader Goals: Moving Toward the Post-Washington Consensus. WIDER Annual Lecture 2. Helsinki: UNU-WIDER.

Sturm, J. E. & De Haan, J. (2001). How robust is the relationship between economic freedom and economic growth? *Applied Economics*, 33(7), 839–844.

Sturm, J. E. & De Haan, J. (2016). Finance and Income Inequality: A Review and New Evidence. CESifo Working Papers, 6079.

Vega-Gordillo, M. & Alvarez-Arce, J. L. (2003). Economic growth and freedom: a causality study. *Cato Journal*, 23, 199.

Yay, G., Taştan, H., & Oktayer, A. (2016). Globalization, economic freedom, and wage inequality: A panel data analysis. *Panoeconomicus*, 63(5), 581–601.

Kıymet Yavuzaslan

15. The Attitude of Income Inequality of Individuals: An Experimental Economics Approach

Abstract Beyond the individual income, the issue of how the income is shared in the whole society has an important place in individual decisions. An examination of the behavioral basis of individuals' attitudes to income inequality may lead to findings to prevent inequality of income distribution. Analyzing the effects of psychological processes, which are effective in people's behaviors and preferences but neglected in economics theories, can be possible with experimental economics studies. With the opportunities provided by experimental economics, humanistic cases in which traditional economics theories are insufficient can be revealed due to behavioral economics and neuroeconomics studies which are emphasized with theories for decision making under uncertainty and risk. Thus, it is possible to produce more reliable policies in the future by predicting economic events in advance. In this study, our aim is to investigate the behavioral and neurological basics of individual income distribution perception through experimental studies. As a result, the findings obtained from the experiments prove that individual characteristics of the society create the inequality in the income distribution. On the other hand, the inequality of income distribution also affects psychologically the individual decisions.

Keywords: Experimental economics, Behavioral economics, Neuroeconomics, Income distribution

1 Introduction

Considering the main assumptions on which classical economics theories are based, within the framework of homoeconomicus, consumers and firms always behave rationally. Consumers try to maximize their utilities, and firms try to maximize their profits. Microeconomic theories are formed without taking into consideration the many characteristics of consumers such as demographic structure, emotions, beliefs and culture and the characteristics of the firms, such as the shareholding structure of the firms, the status of their competitors, their future expectations and their targets. For example: while maximizing the utility is the main objective of a consumer, many consumers' characteristics are not considered (age, gender, marital status, etc.). However, particularly technological advances establish an environment in which economic decision units, who try to maximize their utility, behave influenced by social media, for instance, in the sense of emulation. In short, it is possible to see that economic analyses are inadequate in predicting the economic events that surpass the expectations due to the economic decision-making units that need to increase their utility to maximum levels (Simon, 1955; Vriend, 1996; Kahneman ve Tversky, 1979; Tsang, 2007; Riedl, 2010; Yavuzaslan, 2018).

It is thought that behavioral economics is only developed on the subjects that fall within the scope of microeconomics since it deals with the psychologies underlying the behavior of individuals, especially from economic decision-making units. As more comprehensive research can be done by using experimental economics methodology, it is possible to find examples showing that the responses to human psychology can be measured in macroeconomic policies. Inequality in the income distribution, one of the primary reasons of conflicts in today's societies, has become one of the most increasing and debated problems of the world in the 21st century (Piketty, 2014). Many studies in the field of behavioral economics enable people to discuss their decisions about sharing from different perspectives. In addition, it is important for the high-income individuals to share with low-income individuals, in order to maintain a peaceful society and future generations.

Assuming that the technological improvement is the most effective factor, that the traditionally developed theories are based on rational expectations and the hypothesis are far from reflecting the real world, it is crucial to predict the human behavior and improve healthier economic policies. Moreover, it is possible to say that experimental methods are quite useful to examine easily unobservable or immeasurable factors. The main research question in this article is how to respond to the perception of individual income distribution and its results in experimental economics. Because when psychological or physical reasons that affect the income distribution of countries are determined, it is possible to prevent inequality in the distribution of income. In this context, in the first section, the advantages of experimental economics methodology will be examined. In the second and third sections, in order to measure the attitudes of individuals towards the distribution of income, behavioral economics and neuroeconomics models that emerge with experimental economics methods will be discussed. In the last section, the results and importance of individuals' attitude towards income distribution will be evaluated.

2 Experimental Economics

Experimental economics, which is a highly effective new economic paradigm to collect and analyze data in order to examine the relations between economic agents in a controlled manner, has only a half-century history. For a long time, since the moment economics was accepted as a science until the present day, it has been always considered as difficult, or even impossible, to make an experiment in economics by the mainstream economists. Although many variables are effective in economic decision making, these variables are not physical objects as in technical sciences and laboratory control of these factors is difficult (Hey, 1994). However, the findings of behavioral economics and neuroeconomics from experimental economics show that this view is no longer valid. Under the leadership of the psychology discipline, it is possible to test the models and theories of the behavioral economics, which has an aspect to expand the basis of rational choice and equilibrium models rather than rejecting the approaches of mainstream economics. While experimental economics is defined as the creation and examination of the synthetic economic situations in which the subjects take the decision to answer one or more specific research questions, the theoretical framework is provided by the game theory. When the types of games constructed in the experiments about the attitude of individuals in income inequality are examined, the most commonly used types of games are ultimatum, dictator and public property games (Yavuzaslan, 2018).

The reputation of experimental economics, which has an interdisciplinary nature using concepts in the fields of psychology, biology, mathematics and sociology as well as economics, has grown considerably among the economists with the experimental economics studies of Vernon Smith who received the Nobel Prize in Economics in 2002. The projects that were made in the name of experimental economics from the 1960s as the period when Vernon Smith first worked to the present, have increased and become prominent. In this period, more than 100 experimental economics studies were published. Many of the first experimental studies on the game theory, individual choices and market mechanisms have been studied in the 1960s and nowadays they have been presented in new versions (Bett et al., 2016; Chowdhury and Jeon, 2014; Plott, 2006).

The most important development in experimental economics studies that made it more preferred is the unpredictable reasons in the economic sphere and the effects of rapidly occurring events on the markets. This has led to the revisions of the methodologies of all orthodox or heterodox approaches. Because, according to the "ceteris paribus principle", which is another basic assumption of economics theories, all other variables are assumed to remain same. On the other hand, the behaviors of firms, markets, individuals and even the government are in interaction through the speed of advances in technological innovations and social networks in real life and this is obviously visible in economic decisions they take. Even in a controlled laboratory environment, it is thought that the factors, which are not included in the analysis by assuming them as a constant, do not remain same and change in time (Kahneman and Tversky, 1979).

Another factor that influences experimental economics as a preferred method is that experimental economics provides productivity. What make the experiments in the literature unique are the variables themselves, because the subjects involved in the experiments consist of different people. Due to the different demographic and sociological characters that the subjects have because of different age, different gender and even different countries that they are from have an impact on preferences during the experiment (Yavuzaslan, 2018), it is possible to get unique findings in each study.

The focus of scientific studies in economics is on the way of experimental verification and testing of traditional models. Beyond the data obtained from the examination of real economies, the data of the laboratory tests are used, which now include the analyses by induction method instead of deduction. Neuroeconomics and behavioral economics arising from the mentioned studies have become popular and active areas in economics. In this sense, economists adapted the tools and methods of positive sciences to the science of economics. In order to understand the events occurring in real life better, scientists have begun to understand the human behavior that governs economic phenomena especially with the use of modeling strategies that address important aspects of real-life attitudes, and this enables more realistic results.

3 Behavioral Foundations of Individuals' Perception on Income Distribution Inequality

When considered within the framework of traditional economics theory, individuals are defined as human beings who look out for their own interests and do not decide with their emotions. However, the theory has not showed the necessary attention to cognitive and social psychology. In contrast to traditional economists, behavioral economics has also included psychological factors in their economic analysis, and in this process psychology was used in economic analysis. Thus, it has been thought that the power to explain the events by economics theories will increase through behavioral economics (Hey, 1994). Behavioral economics studies have shown that economic decision-making units, in their decisions that include risks, can take decisions not only as the result of mathematical process but also risk taking, loss aversion, expectation, instinct, emotion, altruism even if it is obvious after the mathematical results (Chowdhury and Jeon, 2014).

There are so many studies about the "loss aversion bias", which is the part of the prospect theory of Kahneman and Tversky (1979). Experimental studies that show "doing worse" than others is more important than "doing better" than others in terms of feeling good, point out that individuals who see income as social property, act with the motive to avoid harm (D'Ambrosio and Clark, 2015). Inequality avoidance is contrary to the assumption that individuals are status seekers, as stated by Bolton and Ockenfels (2000). Relative standing anxiety is the focus of another set of contributions in experimental economics (Alpizar et al., 2005; Johansson-Stenman et al., 2002; Yamada and Sato, 2013). The approach here is to allow individuals to make choices on the hypothetical states of the world to understand how absolute and relative outcomes are important to them. In terms of income, they are assessed regarding to their incomes and average social incomes. As the importance of relative income increases, the individual is expected to be willing to give up his own income to achieve a better relative standing.

Most of the people, when they hear the concept of distribution, think on a mechanism that uses some principles or criteria to share some objects. There may be some errors in the distribution process of the shares. There is no centralized distribution in which all resources are controlled by an individual or a group, and how resources are allocated. In a free society, different people control different resources, and new acquisitions are obtained through volunteer exchange and people's activities. The total result is a product of many individual decisions of different individuals involved in the subject (Nozick, 1974).

When the results obtained from the experiments performed with the standard game theories, particularly the dictator game, are evaluated in which the behavioral economics models are tested, findings were compatible with pure altruism or prediction of Rawls' (1971) aversion of inequality theory and even none of the subjects was able to choose an equal income distribution (Michelbach et al., 2003).

Anderson et al (2008) moved their study to a different dimension by changing the level of payments to the subjects and manipulating the distribution to test inequality perception by using the public property game instead of the dictator game. When the public was informed about the determinants of social capital and whether each individual was standing within the group, the perception of the inequality of the subjects changed and the contributions of the group to the society differed for the members of the group. From this point, Akbaş et al.'s (2014) work is also important. Akbaş et al. (2014) concluded that the risk perception of individuals and the social phenomena of the society they live in and the attitudes of income inequality are the determinants of their inequality and aimed to propose a solution for the income inequality in the USA. They concluded that 53 % of Republicans and only 25 % of Democrats believe that the economic system in the country is fair. These findings prove how individual attitudes and behaviors of people with different political views can be differentiated in the same country.

4 Neurological Foundations of Individual Preferences in Income Distribution and Perception of Income Distribution in Minds

Due to behavioral economics, the concept of bounded rationality is developed and brought into the literature of economics. According to the bounded rationality, rationality of human beings whose behaviors are shaped mostly by emotions is restricted. The human brain's ability to compute is not infinite, so it is not possible for humans to have the complete information (Chamberlin, 2002; Konow, 1996; Plott, 2006; Yavuzaslan, 2018). According to Simon (1955), the economic decision-making process actualizes through a complex and at the same time artificial system in the human brain. Neuroeconomics is related to the biological studies of the brain during the economic decision-making process and combines human behavior with the various lobes in the brain to explain that the reasons for rationality deviations are caused by the movement of nerves in the human brain. The neuroeconomics aims reveal the reasons of decision making by individuals and predict behavior in terms of economic behavior by scanning the inside of human brain, which is expressed as "black box" (Glimcher et al., 2009: 10). In other words, it deals with how the different parts of the brain react to the economic problems and how they solve these problems. Camerer et al. (2004) who studied about how different brain types react to a number of economic problems and how to solve them, pointed out that neuroeconomics was first used by Kevin Mc Cabe in 1998.

Neuro-literature is sometimes referred in the literature as a sub-branch of experimental economics. The reason for this is the use of the experimental setup in the field of neuroeconomics. Although there are different research methods in the neuroeconomics, most of the experiments with "Functional MR-fMR" device come to the fore. This method is based upon brain scanning and the examination of brain images. One of the subjects is divided into two experimental groups and the other as control group in neuroeconomics studies. The purpose is to compare the functions of the brain using different functions of the subjects in the groups (Weber and Eric, 2009).

Fliessbach et al. (2007) provide monitoring the effect of relative payments on reward processing in the movements of brain functions by conducting two experiments simultaneously to the determined experimental and control groups under the control of two individual scanner fMR devises. In each experiment, the task was the same for both experiments, and while the subjects knew this, the amount that one individual earned and the amount that the other player earned was altered to create the opposite conditions for them (Fliessbach et al., 2007). When the brain images of the players were examined with this experiment, similar images to the levels of activation in the "ventral striatum" which is the area where neuron bodies are concentrated in processes such as reward, pleasure, dependence, hope or ambition, and revenge were obtained. Even though they were not actively involved in decision-making process, social comparison was seen to affect the "ventral striatum" activity and in this sense, the absolute knowledge of the mind about another person has a significant effect on the brain processes depending on his or her motivation. Thus, one of the assumptions of classical economics, the assumption that the relative income differences under the same conditions is observed by individuals is physiologically tested and the responses of the brain functions to the income distribution have been proved by an important neuroeconomics model. Furthermore, Wu et al. (2012) found evidence of social comparisons in brain activity and suggested that this was seen in later cognitive evaluations rather than initial evaluation. Fliessbach et al. (2012) repeated their experiments in 2007, but this time they grouped the subjects into two groups as male and female. Hence, they aimed to distinguish between advantageous and disadvantageous inequality. They have found that disadvantageous inequality has a much greater effect on brain activity in ventral striatum than advantageous inequality. Dohmen et al. (2011) used the same experiment and in the regression analysis, they calculated that the income of the subject himself and others were equal in their effects on activation in the ventral striatum, but that the predicted effect of both income variables was greater in men. In another study in which brain activation was examined with fMR device, Tricomi et al. (2010) designed a model with advantageous and disadvantageous inequalities by assigning the subjects that they were divided into two groups with random rich (holders of 50 dollars) or poor (without money). When they measured the effects of monetary incentives on the brain activity of both groups, they found that the transfers to the subjects who were determined as poor more strongly responded than the others, and those who were determined as rich seized up the transfers to others more strongly than the transfers made to them.

These experiments that were carried out within the scope of neuroeconomics show that individuals have different preferences in terms of both advantageous and disadvantageous inequalities, and physiologically they give a standard neurological response that supports these preferences. In addition, these experiments show that each individual's sociological position is the determinant in the perception of income distribution in his mind. The important part is to be able to evaluate the findings of these neuroeconomic experiments in terms of social perspective.

5 The Importance of Individuals' Attitude to Income Inequality

While the literature on individuals' attitudes to income inequality shows the attitudes and behaviors and even physiological evidences of the perception of inequality among individuals, it should be taken into account that inequality may increase anti-social behavior as the way Fehr (2018) obtained in his experiment. People may take predictable wrong decisions. Even if people think the worst possible, the expectation of optimism that everything is going to be good lead people to take wrong decisions or inertia. Since the majority consider themselves above average, they think that problems and difficulties will not happen to them. It is necessary for these situations to be urged (Thaler and Sunstain, 2008). In the analysis of experimental macroeconomic indicators, such as the distribution of income, while trying to measure the attitude of individuals in the face of changes in income distribution, most of the experimental findings are from a stylized environment. To make any declaration of welfare, we need to know who is with whom, how different types of income gaps are important, and how much relative income is crucial compared to absolute income. Experimental methods to measure these indicators with any degree of accuracy in existing data may not yield similar results to macro data (Duffy and Puzzello, 2014; Duffy, 1998).

On the other hand, D'Ambrosio and Clark (2015) emphasize that the comparative reference group exists and represents a central component of the attitudes towards inequality in an economy. D'Ambrosio and Clark (2015) argues that individual attitudes to income distribution will be successful in making policy proposals to improve the income distribution inequality by the efforts of distributing the various motivations behind the actions of individuals, that is, to promote social welfare from a normative point of view of the distribution of income inequality.

Experimental economics studies have recently started to be published in Turkey. In order to reveal the behavioral foundations of sharing in Turkish society, Dilek and Keskin (2018) performed an ultimatum game and obtained consistent findings with the results of similar studies. In the experiment they stated that participants were willing to share 33,06 % of their possessions with others. If the share is not fair, participants have rejected offers even at the risk of returning with zero gain. The findings reveal how justice is important in the distribution of income in terms of ensuring social peace and tranquility. Considering similar studies, the regulation of the perception of inequality between personal income or the groups' income will affect both individual and social welfare (D'Ambrosio and Clark 2015). As in the method that Anderson et al. (2008) performed, subjects can be directed to more equitable attitudes by manipulating the perception of inequality and policy suggestions can be made to prevent income distribution inequality of the countries.

6 Conclusion

In addition to the analysis of the countries in terms of income inequality, the attitude of the society towards the injustice in income distribution must be addressed individually. In recent years, the importance of rapidly developing experimental economics in micro and macroeconomics has become indispensable. Empirical studies reveal how much individuals are conscious about income distribution and personal characteristics that can lead to income distribution inequality. Moreover, in an environment with rapid decision making and unpredictable risks brought by technological innovations, it requires macroeconomic analysis to be done on micro foundations.

Individuals' attitudes to income inequality depend not only on how much they earn, but also on how much they earn compared to others. The perception that the distribution of wealth is unbalanced may deteriorate the social belonging of the individuals and may negatively affect the economic development by causing them to lose their faith in the corporate structure. As a result, experimental findings obtained from individual and group decisions may contribute to the adoption of a more equitable approach to the increase of social welfare and the sharing of this increase. For this reason, when policymakers establish economic processes, taking into consideration the attitude of individuals towards their inequality of income distribution is critical for increasing social benefit.

References

Akbaş, M., Ariely, D., & Yüksel, S. (2014). When is inequality fair? An experiment on the effect of procedural justice and agency. *SSRN Electronic Journal*, 84(919), 487–492.

Alpizar, F., Carlsson, F., & Johansson-Stenman, O. (2005). How much do we care about absolute versus relative income and consumption? *Journal of Economic Behavior & Organization*, 56, 405–421.

Anderson, L. R., Mellor, J. M., & Milyo, J. (2008). Inequality and public good provision: An experimental analysis. *The Journal of Socio-Economics*, 37, 1010–1028.

Bett, Z., Poulsen, A., & Poulsen, O. (2016). The focality of dominated compromises in tacit coordination situations : Experimental evidence. *Journal of Behavioral and Experimental Economics*, 60, 29–34. Retrieved from https://doi.org/10.1016/j.socec.2015.11.004 (11.09.2018).

Bolton, G. E. & Ockenfels, A. (2000). ERC: A theory of equity, reciprocity, and competition. *The American Economic Review*, 90(1), 166–193.

Camerer, C., Loewenstein, G., & Prelec, D. (2004). Neuroeconomics: Why economics needs brains. *Scandinavian Journal of Economics*, 106(3), 555–579.

Chamberlin, E. H. (2002). An experimental imperfect market. *Journal of Political Economy*, 56(2), 95–108. Retrieved from https://doi.org/10.1086/256654 (10.08.2018).

Chowdhury, S. M. & Jeon, J. Y. (2014). Impure altruism or inequality aversion?: An experimental investigation based on income effects. *Journal of Public Economics*, 118, 143–150. Retrieved from https://doi.org/10.1016/j.jpubeco.2014.07.003 (05.10.2018).

D'Ambrosio, C. & Clark, A. E. (2015). Attitudes to income inequality: Experimental and survey evidence. In *Handbook of Income Distribution*, edited by Atkinson, A. B. François Bourguignon, 1148–1201 Vol. 2A. Paris: Elsevier.

Dilek, S. & Kesgingöz, H (2018). Paylaşmak Güzeldir: Bir Ültimatom Oyunu Uygulaması, *BMIJ*, 6(4): 822–834. Retrieved from doi: http://dx.doi.org/10.15295/bmij.v6i4.334 (12.04.2019).

Dohmen, T., Falk, A. Huffman, D., Sunde, U., Schupp, J., & Wagner, G. (2011). Individual risk attitudes: Measurement, determinants and behavioral consequences. *Journal of the European Economic Association*, 9(3), 522–550.

Duffy, J. & Puzzello, D. (2014). Experimental evidence on the essentiality and neutrality of money in a search model. *Research in Experimental Economics*, 17, 259–311. Retrieved from https://doi.org/10.1108/S0193-230620140000017008 (08.03.2019).

Duffy, J. (1998). Monetary theory in the laboratory. *Federal Reserve Bank of St. Louis Economic Review*, 80 (September/October), 9–26.

Fehr, D. (2018). Is increasing inequality harmful? Experimental evidence. *Games and Economic Behavior*, 107, 123–134. Retrieved from https://doi.org/10.1016/j.geb.2017.11.001 (12.03.2019).

Fliessbach, K., Weber, B., Trautner, P., Dohmen, T., Sunde, U., Elger, C. E., & Falk, A. (2007). Social comparison affects reward-related brain activity in the human ventral striatum. *Science*, 318(5854). Retrieved from 1305–1308. https://doi.org/10.1126/science.1145876 (12.04.2019).

Fliessbach, K., Phillipps, C. B., Trautner, P., Schnabel, M., Elger, C. E., Falk, A., & Weber, B. (2012). Neural responses to advantageous and disadvantageous inequity. *Frontiers in Human Neuroscience*. doi: 10.3389/fnhum.2012.00165 Source: PubMed.

Glimcher, P. W., Colin, F., Camerer, E., Fehr, R., & Poldrack, A. (2009). A brief history of neuroeconomics. In *Neuroeconomics: Decision Making and the Brain*, edited by P. W. Glimcher, C. Camerer, E. Fehr, and A. Poldrack, 1–11. Cambridge: Elsevier.

Hey, J. D. (1994). *Introduction and Overview, Experimental Economics: Studies in Empirical Economics*. Heidelberg: Physica-Verlag.

Johansson-Stenman, O., Carlsson, F., & Daruvala, D. (2002). Measuring future grandparents' preferences for equality and relative standing. *Economic Journal*, 112, 362–338.

Kahneman, D. & Tversky, A. (1979). Prospect Theory: An Analysis of Decision under Risk. *Econometrica*, 47(2), 263–292.

Konow, J. (1996). A positive theory of economic fairness. *Journal of Economic Behavior and Organization*, 31(1), 13–35. Retrieved from https://doi.org/10.1016/S0167-2681(96)00862-1 (22.03.2019).

Michelbach, P. A., Scott, J. T., Matland, R. E., & Bomstein, B. H. (2003). Doing rawls justice: An experimental study of income distribution norms. *Faculty Publications, Department of Psychology*, 186. Retrieved from http://digitalcommons.unl.edu/psychfacpub/186 (12.03.2019).

Nozick, R. (1974). *Anarchy, State, and Utopia*. Oxford: Blackwell Publishers.

Piketty, T. (2014). *Yirmi Birinci Yüzyılda Kapital*. İstanbul: Türkiye İş Bankası Kültür Yayınları.

Plott, C. R. (2006). Will economics become an experimental science? *Southern Economic Journal*, 57(4), 901. Retrieved from https://doi.org/10.2307/1060322 (04.03.2019).

Rawls, John. (1971). *A Theory of Justice*. Cambridge, MA: Harvard University Press.

Riedl, Arno. (2010). Behavioral and experimental economics do inform public policy. *Finanzarchiv*, 66(1), 65–95.

Simon, H. A. (1955). A behavioral model of rational choice. *The Quarterly Journal of Economics*, 69(1), 99–118.

Thaler, R. H. & Sunstain, C. R. (2008). *Nudge: Improving Decisions About Health, Wealth, and Happiness*. New Haven & London: Yale University Press.

Tricomi, E., Rangel, A., Camerer, C., & O'Doherty, J. (2010). Neural evidence for inequality averse social preferences. *Nature*, 463, 1089–1091.

Tsang, E. (2007). Computational intelligence determines effective rationality. Centre for Computational Finance and Economic Agents, Working Paper Series, WP 015-07: 1–9.

Vriend, N. J. (1996). Rational behavior and economic theory. *Journal of Economic Behavior and Organization*, 29, 263–285.

Weber, E. U. & Eric, J. J. (2009). Decisions under uncertainty; psychological, economic and Neuroeconomic explanations of risk preference. In *Neuroeconomics: Decision Making and the Brain*, edited by P. W. Glimcher, C. Camerer, E. Fehr, and A. Poldrack, 127–144. Cambridge: Elsevier.

Wu, Y., Zhang, D., Elieson, B., & Zhou, X. (2012). Brain potentials in outcome evaluation: When social comparison takes effect. *International Journal of Psychophysiology*, 85, 145–152.

Yamada, K. & Sato, M. (2013). Another avenue for anatomy of income comparisons: Evidence from hypothetical choice experiments. *Journal of Economic Behavior & Organization*, 89, 35–57.

Yavuzaslan, K. (2018). Experimental Economics as a method in the new paradigm of economics. *Kafkas University Journal of the Faculty of Economics and Administrative Sciences*, 9, 641–657. Retrieved from https://doi.org/10.9775/kauiibfd.2018.030 (12.03.2019).

Mustafa Şit and Erdal Alancıoğlu

16. Macroeconomic Factors Determining Income Distribution: An Analysis on Mist Countries

The forces in a capitalist society, if left unchecked, tend to make the rich richer and the poor poorer.

— *Jawaharlal Nehru*

Abstract In this chapter, macroeconomic factors which determine the income distribution for the period of 1990–2015 were investigated. Panel data analysis was used as the methodology. In the econometric analysis, the Theil index was considered as dependent variable. The macroeconomic factors determining the distribution of income such as growth, inflation, tax burden, exchange rate, total factor productivity and human capital index were analyzed. As a result of the study, economic growth, total factor productivity, human capital and inflation were found as significant determinants of income inequality.

Keywords: Theil index, Income distribution, MIST countries, Panel data analysis

1 Introduction

One of the fundamental problems that economists are interested in is the question of how income inequality is formed and how it develops over time. According to Tinbergen (1956), inequality in income distribution is important because it is the basis of the most important economic problems. The problem of income inequality is on the agenda of both developed and developing countries since the implemented policies and strong economic growth have not significantly reduced global poverty (Adams and Klobodu, 2017: 169).

The factor revenues are obtained as a result of the contribution of production factors to production processes. Realization of this income process within the scope of equality can be expressed as justice in income distribution. The issue of income distribution is fundamentally influenced by society and individuals, as well as economic, social and political processes. Therefore, it is always up to date and along with the analysis on this subject, it serves as a guide. This is because the analysis on income distribution is a feedback of current mechanism. Furthermore, it has a strategic importance in terms of directing the policies to be formed (Bükey, 2017: 104).

According to the World Economic Forum 'Global Risk' report (2014), the increase in total debt to GDP ratio in developed countries results in income distribution inequalities. The report also states that the problem that will threaten global stability in the next 10 years will be the unemployment issue. Employment is considered to be one of the most effective income distribution methods. Therefore, income distribution in the upcoming periods will continue to be one of the most discussed issues in the world (Özdemir, 2011: 4).

Jim O'Neill from Goldman Sachs, who established the BRIC (Brazil, Russia, India and China), has announced the existence of the group of countries "MIST", which consisted of Mexico, Indonesia, South Korea and Turkey in February 2011. Probit as Partners, one of the US investment firms, also included MIST countries in their report (2011) and mentioned that MIST countries could be considered as future rising market leaders. Although there are some uncertainties due to being a very new entity, the existence of MIST countries needs to be taken into account. The factor that brings the MIST countries together is that they are rising economically (Narin and Kutluay, 2013: 37).

In this study, macroeconomic factors that determine the distribution of income in the period of 1990–2015 for the MIST country group are investigated. Panel data analysis was used as the method. The findings were interpreted as economic. In this study, estimated Theil index data for these countries were used in the scope of UTIP (the University of Texas Inequality Project). The study consists of two parts. In the first part of the study, the income distribution was evaluated theoretically. In the second section, an empirical analysis was made to examine the factors affecting income distribution in MIST countries.

When the factors affecting the distribution of income are analyzed, from the theoretical point of view, economic growth is expected to have a negative impact on income inequality. Macroeconomic growth in most cases increases the income of the poor and reduces the number of people below the poverty line (Angelsen and Wunder, 2006: 1). In terms of inflation, low and fixed income groups, who cannot protect their income against price increases, will lose their income due to inflation. As a result, the effect of inflation on income distribution is expected to be negative (Dişbudak ve Süslü, 2007: 5).

A clear way in which the tax burden can affect income inequality is due to changes in the redistribution amount of the state. For example, a higher labor tax reduces the net fee. Too much tax burden reduces workers' savings. This may be another factor that potentially increases inequality (Ciminelli et al., 2018: 2). The effect of exchange rate on income distribution can be explained as follows: Alexander (1952) implies that income will be shifted if workers' wages do not comply with the inflationary effects of the devaluation. Because, in this case, there will be a monetary shift from workers to producers, or from poor to rich. If devaluation shifts income from poor to rich, it worsens the distribution of income (Bahmani-Oskooee and Motavallizadeh-Ardakani, 2018: 266). Higher total factor productivity means more production and higher income with the same amount of capital. In this case, an increase in productivity leads to an improvement in the income of individuals. Therefore, income inequality is expected to decrease. Likewise, with the increase of human capital, productivity and income increase are expected. Hence, it is thought to affect the distribution of income positively.

2 Empirical Analysis of Factors Affecting Income Distribution in MIST Countries

In this part of the study, the factors affecting income distribution in MIST countries were analyzed empirically. First, a literature review was given, then the method of the study was explained and finally the findings obtained from the analysis were discussed.

2.1 Literature

There are many empirical studies examining the relationship between income distribution and economic growth. Some studies (Alesina and Rodrick, 1994; Perotti, 1996; Barro, 2000; Knowles, 2005; Ostry et al. 2014) find a negative relationship between the two variables. However, there are other studies that found a positive correlation between these two variables (Benabou, 2000; Forbes, 2000; Deininger and Olinto, 2000; Dişbudak and Süslü, 2007; Yang and Greaney, 2017; Acaravcı et al. 2018), which confirmed the theory. As it is seen, the studies demonstrate inconsistent results about the relationship between variables. This lack of consensus can be attributed to differences in the data sets used, the countries studied, time and the methodology selected (Babu et al., 2016: 99).

The relation between income distribution and inflation was examined by Gottschalk and Moffitt (2009), Jensen and Shore (2015), Shin and Solon, (2011), Dişbudak and Süslü, (2007) and Latner, (2019). These studies found that inflation has a negative effect on income distribution by applying panel data analysis. Glomm and Ravikumar (1992), Mahmood and Noor (2014), Destek (2018) examined the impact of human capital and total factor productivity on income distribution. These researchers found the corrective effect of both variables on income distribution. Papanek and Kyn (1986), Anand and Kanbur. (1993), Li

Tab. 16.1: Explanation of Variables. Source: Authors' own construction

Variables	Sign	Description	Source
Theil Index	TH	Represents the Inequality of Income Distribution	UTIP
Economic Growth	GROW	Growth Rate (%)	World Bank
Inflation	INF	Inflation Rate (%)	World Bank
Total Factor Productivity	TFP	Total Factor Productivity Index (USA=1)	World Bank
Human Capital	HC	Human Capital Index	World Bank
Exchange Rate	EXC	Dollars/National Money	World Bank
Tax Burden	TAX	Total Taxes/GDP	World Bank

et al. (1998), Barro (2000), Dollar and Kraay (2002), Bahmani-Oskooee et al. (2008) and Bahmani-Oskooee and Motavallizadeh-Ardakani (2018) researched the effect of the exchange rate on income distribution by carrying out a panel data analysis. The findings indicate that an increase in exchange rate worsens the income inequality.

The impact of tax policies was researched by Piketty and Saez (2007), Poterba (2007), Adam et al. (2015), Ciminelli et al. (2018). As a result of the studies, it was determined that there was a relationship between the tax structure and income inequality. It was concluded that the tax burden on labor and capital disrupts income distribution.

2.2 Explanation of variables

The symbols, descriptions and sources of the variables used in the analysis are shown in Tab. 16.1. In the econometric analysis, Theil index will be considered as dependent variable as a proxy of income distribution. The macroeconomic factors that determine the income distribution will be analyzed as growth, inflation, total factor productivity index, human capital index, exchange rate and tax burden. The analysis period was determined as 1990–2015 because of the availability of data.

2.3 Findings

In the analysis part of this study, the homogeneity of the series and the cross-sectional dependence of the series were first tested. Then, according to the findings obtained, appropriate unit root test, cointegration analysis and coefficient estimation tests were applied.

Tab. 16.2: Swamy S homogeneity test results. Source: Authors' own calculations

TH	Coef.	Std. Err.	z	P>|z|	95 % Conf.	Interval
grow	.0007118	.0010919	0.65	0.514	-.0014283	.0028518
Inf	.0005491	.0004489	1.22	0.221	-.0003308	.001429
tfp	.0567556	.0582224	0.97	0.330	-.0573581	.1708693
hc	-.0317328	.0780641	−0.41	0.684	-.1847356	.12127
exc	.0091727	.0094092	0.97	0.330	-.0092689	.0276143
tax	.0019717	.0015274	1.29	0.197	-.001022	.0049654
_cons	.0104419	.1487861	0.07	0.944	-.2811735	.3020573

Test of parameter constancy: $\chi^2(24) = 1097.11$ Prob$>\chi^2 = 0.0000$

2.3.1 Swamy S homogeneity test findings

Cointegration tests and estimation methods are selected depending on whether the constant and slope parameters are homogeneous or heterogeneous with respect to the units. Therefore, it is important to conduct homogeneity tests before selecting the methods to be used. The hypothesis to be tested is established in the form H_0: $\beta_i = \beta$ and indicates that the parameters are homogeneous. In this test, which is derived from Swamy (1971) and is a Hausman type test, the statistic is defined as in Equation 1.

$$\hat{S} = \chi^2_{k(N-1)} = \sum_{i=1}^{N} \left(\hat{\beta}_i - \bar{\beta}^* \right)' \hat{V}_i^{-1} \left(\hat{\beta}_i - \bar{\beta}^* \right) \tag{1}$$

In the Equation 1, $\hat{\beta}_i$ indicates OLS estimators derived from the regressions according to the units, $\bar{\beta}^*$ refers to the weighted estimator (WE), and $\hat{V}_i$ is the difference between the variances of the two estimators. The test statistic has a K (N-1) degree of freedom and a χ^2 distribution. If the test statistic is greater than the critical value, it is concluded that the parameters are heterogeneous (Yerdelen, 2017: 247).

According to the findings in Tab. 16.2, H_0 hypothesis was rejected because the probability value was greater than 'chi2'. It was determined that the parameters were not homogeneous and changed from unit to unit (heterogeneous). In this case, it would be appropriate to rely on the results of those heterogeneous from the cointegration tests, and to use the proposed estimation methods for heterogeneous panels.

Tab. 16.3: Horizontal Cross-Sectional Dependence Test Results. Source: Authors' own calculations

Test	Statistic	p-value
LM	10.24	0.1150
LM adj*	1.978	0.0479
LM CD*	1.998	0.0458

Notes: *Two-sided test

2.3.2 Findings of horizontal cross-sectional dependency test

Bias-Adjusted CD test yields meaningful results in small samples, while it provides consistent and strong results in asymptotic distributions. Bias-Adjusted CD test statistic is as follows:

$$CDLM_{adj} = \sqrt{\frac{2}{N(N-1)}} \sum_{i=1}^{N-1} \sum_{j=i+1}^{N} \rho_{ij}^2 \frac{(T-K-1)\left(\hat{\rho}_{ij} - \hat{\mu}_{Tij}\right)}{V_{Tij}} \quad (2)$$

The statistics in Equation 2 show the standard normal distribution as asymptotically with $CDLM_{adj} \sim N(0,1)$ (Pesaran et al. 2008). Zero and alternative hypothesis;

H_0: No cross-sectional dependency.
H_1: Horizontal cross-sectional dependence.

Breusch-Pagan LM test of the cross-sectional dependence tests, gave consistent results in "T> N" and Pesaran CD test in "N> T" conditions. Breusch-Pagan LM test was taken into consideration in this study because of T> N. According to the results of the cross-sectional dependency test in Tab. 16.3, there is a cross-sectional dependence for MIST economies data since the null hypothesis was rejected at 5 % significance level. For this reason, the second-generation Covariate Augmented Dickey-Fuller (CADF) unit root test was used for stability analysis.

2.3.3 CADF unit root test findings

Pesaran (2007) used $\bar{y}_t$ and $\Delta \bar{y}_t$ delayed values in order to eliminate the cross-sectional dependency problem ($\Delta \bar{y}_{t-1}$, $\Delta \bar{y}_{t-2}$, ...). In this case, the CADF process can be reduced to the OLS estimate of the following equation:

$$\Delta \bar{y}_{it} = a_i + b_i y_{i,t-1} + \sum_{j=1}^{p_i} c_{ij} \Delta y_{i,t-j} + d_i t + h_i \bar{y}_{t-1} + \sum_{j=0}^{p_i} \eta_{ij} \Delta \bar{y}_{t,t-j} + \varepsilon_{i,t} \quad (3)$$

For $H_0^i: b_i = 0$ $H_A^i: b_i < 0$ $i=(1,2,....N)$,

In the CADF test, t values for the b_i coefficients are obtained. With Pesaran Monte Carlo simulations, it was found that the CADF test was valid for both N> T and T> N.

CIPS statistics are based on the average of CADF statistics (Pesaran et al. 2008).

$$CIPS = \frac{\sum_{i=1}^{N} CADF_i}{N} \qquad (4)$$

According to the findings in Tab. 16.4, TH, GROW and INF series are stable in fixed model at level values. The TFP and HC variables are still in the level values, but they are stable in fixed and trend models. The EXC and TAX series became stationary with the first differences.

2.3.4 Westerlund cointegration analysis findings

Westerlund (2007) proposed four panel cointegrated tests with error correction model to determine the cointegration of panel data analysis. Two of them are group average statistics (Gt, Ga), the other two are panel (Pt, Pa) statistics. If the panel is heterogeneous, "group average statistics" are more reliable, however, in the case of homogeneity, "panel statistics" are more reliable (Yerdelen, 2017: 200–203).

The slope parameters of the variables used in the study are heterogeneous, and the variables have horizontal cross-sectional dependence. For this reason, in Tab. 16.5, which shows the results of Westerlund (2007) panel cointegration test, the resistance values of the group mean statistics (Gt, Ga) should be taken into account. When we examine the resistance values of these tests, the H_0 hypothesis ("No Co-integration") were rejected for GROW, TFP and HC series, and it was concluded that the series were cointegrated. Since the H_0 hypothesis for the INF, EXC and TAX series is not rejected, there is no cointegration relationship.

2.3.5 The panel dynamic OLS (PDOLS) estimation findings

The PDOLS estimator is one of the several appropriate estimators that can be applied in the case of a cointegration relationship. Other estimators include bias-corrected least-squares and least-modified least-squares. The PDOLS test has better properties compared to these estimators.

The PDOLS estimator is shown in equation 5:

Tab. 16.4: CADF Unit Root Test Results. Source: Authors' own calculations

Variables	Countries	LEVEL		1.DIFFERENCE
		CADF (Constant/ Fixed)	CADF (Constant/ Fixed+Trend)	
TH	Mexico	−1,80 (2)		
	Indonesia	−4,16**(2)		
	S. Korea	−2,35 (2)		
	Turkey	−1,77 (2)		
	CIPS	**−2,52****		
GROW	Mexico	−2,66(2)		
	Indonesia	−2,45(2)		
	S. Korea	−2,26(2)		
	Turkey	−2,85(2)		
	CIPS	**−2,55****		
INF	Mexico	−5,96* (5)		
	Indonessia	−1,63(2)		
	S. Korea	−2,43(3)		
	Turkey	−2,08(2)		
	CIPS	**−3,03***		
TFP	Mexico	−2,02 (2)	−2,89 (3)	
	Indonesia	−1,58 (5)	−2,53 (3)	
	S. Korea	−2,58 (2)	−3,33 (2)	
	Turkey	−1,72 (3)	−3,54 (5)	
	CIPS	−1,97	**−3,07****	
HC	Mexico	−3,72** (2)	−3,63*** (2)	
	Indonesia	−1,60(3)	−5,75* (4)	
	S. Korea	−3,01*** (2)	−2,92 (2)	
	Turkey	−0,36(2)	−1,06(2)	
	CIPS	−2,17	**−3,34***	
EXC	Mexico	−0,92 (2)	−1,38 (4)	−3,57** (2)
	Indonesia	−2,73(2)	−3,31*** (2)	−3,53**(2)
	S. Korea	−1,99 (4)	−1,66 (4)	−2,59 (4)
	Turkey	−2,37 (2)	−2,79 (2)	−2,31 (2)
	CIPS	−2,01	−2,29	**−3,00***
TAX	Mexico	−0,70 (3)	0,341 (3)	−3,25*** (5)
	Indonesia	−3,27***(2)	−3,44(2)	−3,09*** (2)
	S. Korea	−1,81 (2)	−2,44 (2)	−3,67**(4)

Tab. 16.4: (continued)

Variables	Countries	LEVEL		1.DIFFERENCE
		CADF (Constant/ Fixed)	CADF (Constant/ Fixed+Trend)	
	Turkey	−1,31 (3)	−1,27 (3)	−2,76 (2)
	CIPS	−1,42	−1,70	**−3,19***

Note: In the fixed model for the CADF test, the table values are −4.35, −3.43, and 3.00 for the significance levels of 1, 5, and 10 %, respectively. In the fixed and trendy model, the table values are −4.97, −3.99, −3.55 for the significance levels 1, 5, 10 %, respectively. For the CIPS values, the table values in the fixed model are −2,60, −2,34, −2,21. In the fixed and trendy model, it is −3,15, −2,88, −2,74, respectively. Values in parentheses indicate the delay length. *, **, *** notations are significant at levels of 1, 5, 10 % respectively.

$$\hat{\beta}_{PDOLS} = \left[N^{-1} \sum_{i=1}^{N} \left(\sum_{t=1}^{T} z_{it} z_{it} \right) \left(\sum_{t=1}^{T} z_{it} \tilde{y}_{it} \right) \right]_{1} \quad (5)$$

In the formula, z_{it} 2x(q+1)x1 regressor vector, $z_{it} = [x_{it} - \overline{x}_i, \Delta x_{i,t-q}, \ldots, \Delta x_{i,t+q}]$, $\tilde{y}_{it} = y_{it} - \overline{y}_i$ and the vector 1 which is outside the parentheses indicate that we only consider the first elements of the vector. Details of the Wald test for constraint distribution and parameter constraints can be found in Kao and Chiang (2000) (Allen and Liu 2007: 233).

The second generation PDOLS estimator, which considers the cross-sectional dependence and panel heterogeneity, was used to estimate the long-term relationship between the series. The PDOLS estimation findings are shown in Tab. 16.6.

According to the PDOLS estimation findings in Tab. 16.6, GROW, INF, TFP and HC variables were statistically significant at 95 % confidence level.

Accordingly, economic growth, total factor productivity and human capital series positively affect income distribution. Inflation series, on the other hand, have a negative effect on income distribution. According to the findings in the analysis, the results of exchange rate and tax burden variables are insignificant. This result is consistent with the findings of cointegration analysis.

Tab. 16.5: Westerlund Cointegration Analysis Results. Source: Authors' own calculations

Variables: TH and GROW				
Test Stat.	Value	Z Value	Probability	Resis. Probability
Gt	−3.012	−2.742	0.003 *	0.080 ***
Ga	−10.827	−1.338	0.091 ***	0.300
Pt	−9.645	−6.609	0.000	0.120
Pa	17.919	−5.912	0.000	0.190
Variables: TH and INF				
Test Stat.	Value	Z Value	Probability	Resis. Probability
Gt	−2.510	−1.614	0.053***	0.270
Ga	−10.073	−1.060	0.145	0.370
Pt	−8.833	−5.808	0.000	0.100
Pa	−17.539	−5.746	0.000	0.160
Variables: TH and TFP				
Test Stat.	Value	Z Value	Probability	Resis. Probability
Gt	−3.361	−3.527	0.000 *	0.050**
Ga	−11.985	−1.766	0.039**	0.160
Pt	−9.876	−6.837	0.000	0.140
Pa	−15.765	−4.973	0.000	0.210
Variables: TH and HC				
Test Stat.	Value	Z Value	Probability	Resis. Probability
Gt	3.643	−4.161	0.000 *	0.040**
Ga	−5.692	0.557	0.711	0.840
Pt	−9.241	−6.211	0.000	0.330
Pa	−10.863	−2.836	0.002	0.600
Variables: TH and EXC				
Test Stat.	Value	Z Value	Probability	Resis. Probability
Gt	2.325	1.197	0.116	0.290
Ga	8.249	0.387	0.350	0.450
Pt	9.049	6.022	0.000	0.240
Pa	15.517	4.865	0.000	0.290
Variables: TH and TAX				
Test Stat.	Value	Z Value	Probability	Resis. Probability
Gt	2.899	2.488	0.006*	0.160
Ga	9.761	0.945	0.172	0.400
Pt	8.611	5.589	0.000	0.150
Pa	13.930	4.173	0.000	0.220

Tab. 16.6: PDOLS Estimation Results. Source: Authors' own calculations

Variables	Coefficient (Beta)	T-Stat
GROW	−0,0021	**−2,724**
INF	0,0069	**5,068**
TFP	−0,0916	**−3,370**
HC	−0,1034	**−2,980**
EXC	2,5607	−0,380
TAX	−0,0007	−0,341

Note T statistic value is 1.96 for $\alpha = 0.05$.

3 Conclusion

In this study, macroeconomic factors affecting income distribution were analyzed for the MIST countries group for the years between 1990 and 2015. Since the recent analysis is conducted for the first time for MIST group, this chapter contributes to the literature.

According to horizontal cross-sectional dependence and homogeneity tests, the model is cross-sectional and the coefficients to be calculated are heterogeneous. Westerlund (2007) panel cointegration test was performed due to the heterogeneity of the slope parameters of the variables and the fact that the variables had horizontal cross-sectional dependence. According to the findings, the long-term relationship of TH series with GROW, TFP and HC series was determined. However, there was no long-term relationship between the TH series and the INF, EXC and TAX series.

A 1 % increase in economic growth improves income distribution by 0.002 %. A 1 % increase in inflation leads to a deterioration of approximately 0.007 % in income distribution. A 1 % increase in total factor productivity improves income distribution by 0.09 %. Likewise, the increase in human capital also improves income distribution by 0.10 %. The results are consistent with economic theory and similar to the literature (Forbes, 2000; Deininger and Olinto, 2000; Gottschalk and Moffitt, 2009; Mahmood and Noor, 2014; Bahmani-Oskooee et al. 2008; Piketty and Saez, 2007).

Income inequality issue is a concern both in high-income countries (South Korea) and middle-income countries (Mexico, Turkey, Indonesia). Especially in middle-income countries such as Mexico and Turkey, when the increase in revenue generated by economic growth is transferred to the low-income segment (through social transfers), it makes a positive impact on income distribution.

The distorting effect of inflation is more noticeable on income distribution in economies with high inflation rates, such as Indonesia and Turkey. In South Korea, where total factor productivity and human capital increase are high, an increase in these variables leads to an improvement in the income of individuals. Therefore, income inequality decreases.

As a result, although the economic structures of the analyzed countries were different, economic growth, total factor productivity, human capital and inflation were determined as important determinants of income inequality. This result will guide the MIST group in determining the policies to be applied in combating income distribution inequality.

References

Acaravcı, A., Erdoğan, M. S., & Artan, S. (2018). Gelir Dağılımı, Demokrasi, Reel Gelir ve Dışa Açıklık İlişkisi: Balkan Ülkeleri İçin Ampirik Bir Uygulama. *Uluslararası İktisadi ve İdari İncelemeler Dergisi*, 167, 73–81.

Adam, A., Kammas, P., & Lapatinas, A. (2015). Income inequality and the tax structure: Evidence from developed and developing countries. *Journal of Comparative Economics*, 43(1), 138–154.

Adams, S. & Klobodu, E. K. M. (2017). Capital flows and the distribution of income in sub-Saharan Africa. *Economic Analysis and Policy*, 55, 169–178.

Alexander, S. S. (1952). Effects of a devaluation on a trade balance. *IMF Staff Papers*, 2, 263–278.

Allen, J. & Liu, Y. (2007). Efficiency and economies of scale of large Canadian banks. *Canadian Journal of Economics/Revue CanadienneD'économique*, 40(1), 225–244.

Alesina, A. & Rodrick, D. (1994). Redistributive politics and economic growth. *Quarterly Journal of Economics*, 109, 465–490.

Anand, S. & Ravi Kanbur, S. M. R. (1993). The kuznets process and the inequality-development relationship. *Journal of Development Economics*, 40(1), 25–52.

Angelsen, A. & Wunder, S. (2006). Poverty and inequality: Economic growth is better than its reputation. In Dan Banik (ed.), Poverty, Politics and Development: Interdisciplinary Perspectives. Bergen: Fagbokforlaget. Retrieved from https://pdfs.semanticscholar.org/9120/7a92fccea3ef5394e4dcc87b45a3b2ae11ef.pdf (13.08.2019).

Babu, M. S., Bhaskaran, V., & Venkatesh, M. (2016). Does inequality hamper long run growth? Evidence from Emerging Economies. *Economic Analysis and Policy*, 52, 99–113.

Bahmani-Oskooee, M., Hegerty, S. W., & Wilmeth, H. (2008). Short-run and long-run determinants of income inequality: Evidence from 16 countries. *Journal of Post Keynesian Economics*, 30, 463–484.

Bahmani-Oskooee, M. & Motavallizadeh-Ardakani, A. (2018). Exchange rate changes and income distribution in 41 countries: Asymmetry analysis. *The Quarterly Review of Economics and Finance*, 68, 266–282.

Barro, R. J. (2000). Inequality and growth in a panel of countries. *Journal of Economic Growth*, 5, 5–32.

Benabou, R. (2000). Unequal societies: Income distribution and the social contract. *American Economic Review*, 90, 96–129.

Bükey, A. M. & Çetin, B. I. (2017). Türkiye'de Gelir Dağılımına Etki Eden Faktörlerin En Küçük Kareler Yöntemi ile Analizi. *Maliye Araştırmaları Dergisi*, 3(1). Retrieved from file:///Users/halekirer/Downloads/66-126-2-PB.pdf (12.08.2019).

Ciminelli, G., Ernst, E., Merola, R., & Giuliodori, M. (2018). The composition effects of tax-based consolidation on income inequality. *European Journal of Political Economy*, 57, 107–124.

Deininger, K. & Olinto, P. (2000). Asset distribution, inequality, and growth. The World Bank Development Research Group Working Paper No. 2375. Washington, DC: World Bank.

Destek, M. A. (2018). Neoliberal Politikalar Işığında Gelir Dağılımı Adaleti Ve Finansal Krizler: Seçilmiş Ülkeler Üzerine Bir İnceleme. Yayımlanmamış Doktora Tezi. *Gaziantep Üniversitesi SBE, İktisat ABD*, Gaziantep.

Dişbudak, C. & Süslü, B. (2007). Türkiye'de Kişisel Gelir Dağılımını Belirleyen Makroekonomik Faktörler. *Ekonomik Yaklaşım*, 18(65), 1–23.

Dollar, D. & Kraay, A. (2002). Growth is good for the poor. *Journal of Economic Growth*, 7(3), 26–195.

Forbes, K. J. (2000). A reassessment of the relationship between inequality and growth. *American Economic Review*, 90, 869–887.

Glomm, G. & Ravikumar, B. (1992). Public versus private investment in human capital: endogenous growth and income inequality. *Journal of Political Economy*, 100, 818–834.

Gottschalk, P. & Moffitt, R.A. (2009). The rising instability of U.S. earnings. *Journal of Economic Perspective*, 23(4), 3–24.

Jensen, S. T. & Shore, S. H. (2015). Changes in the distribution of earnings volatility. *Journal of Human Resources*, 50 (3), 811–836.

Kao, Chihwa & Min-HsienChiang. (2000). On the estimation and inference of a cointegrated regression in panel data. *Advances in Econometrics*, 15, 179–222.

Knowles, S. (2005). Inequality and economic growth: The empirical relationship reconsidered in the light of comparable data. *Journal of Development Studies*, 41, 135-159.

Latner, J. P. (2019). Economic insecurity and the distribution of income volatility in the United States. *Social Science Research*, 77, 193-213.

Li, H., Squire, L., & Zou, H. (1998). Explaining international and intertemporal variation in income inequality. *Economic Journal*, 108, 26-43.

Mahmood, S. & Noor, Z. M. (2014). Human Capital and Income Inequality in Developing Countries: New Evidence using the Gini Coefficient. *Journal of Entrepreneurship and Business*, 2(1), 40-48.

Narin, M. & Kutluay, D. (2013). Değişen küresel ekonomik düzen: BRIC, 3G ve N-11 ülkeleri. *Ankara Sanayi Odası Yayını*, 30-50. Retrieved from http://www.aso.org.tr/b2b/asobilgi/sayilar/dosyaocaksubat2013.pdf (13.08.2019).

Ostry, J. D., Berg, A., & Tsangarides, C. G. (2014). Redistribution, inequality, and growth. *IMF Staff Discussion Note14/2*, International Monetary Fund Washinton, DC.

Özdemir, M. Ç. (2011). Gelir ve Servet Dağılımına İlişkin Temel Kavramlar. *Sakarya Üniversitesi Yayınları*, Retrieved from http://content.lms.sabis.sakarya.edu.tr/Uploads/50975/45321/gelir_da%C4%9F%C4%B1l%C4%B1m%C4%B1_b%C3%B6l%C3%BCm2.docx (10.02.2019).

Papanek, G. & Kyn, O. (1986). The effect on income distribution of development, the growth rate and economic strategy. *Journal of Development Economics*, 23(1), 56-65.

Perotti, R. (1996). Growth, income distribution, and democracy: What the data say. *Journal of Economic Growth*, 1, 149-187.

Pesaran, M. H. (2007). A simple panel unit root test in the presence of cross-section dependence. *Journal of Applied Econometrics*, 22(2), 265-312.

Pesaran, M. H., Ullah, A., & Yamagata, T. (2008). A bias-adjusted LM test of error cross-section independence. *The Econometrics Journal*, 11(1), 105-127.

Piketty, T. & Saez, E. (2007). How progressive is the U.S. federal tax system? A historical and international perspective. *Journal of Economic Perspective*, 21, 3-24.

Poterba, J. (2007). Income inequality and income taxation. *Journal of Policy Modeling*, 29, 623-633.

Probitas Partners. (2011). Private Equity Deskbook 2011. Retrieved from http://probitaspartners.com/pdfs/probitas_partners_private_equity_deskbook_2011.pdf (22.03.2019).

Shin, D. & Solon, G. (2011). Trends in men's earnings volatility: what does the panel study of income dynamics show? *Journal of Public Economics*, 95, 973–982.

Swamy, P. (1971). *Statistical Inference in Random Coefficient Regression Models*. New York: Springer.

Tinbergen, J. (1956). *On the Theory of Income Distribution*. Weltwirtschaftlichesarchiv, 155–175.

UTIP. (2018). University of Texas Inequality Project. *Theil Index Datas*. Retrieved from http://utip.lbj.utexas.edu/data/utipunidov2017.xlsx (12.03.2019).

Westerlund, J. (2007). Testing for error correction in panel data. *Oxford Bulletin of Economics and Statistics*, 69(6), 709–748.

Worldbank. (2018). PWT 9.0 Datas. Retrieved from https://data.worldbank.org/indicator (12.01.2019).

World Economic Forum Global Risk Report. (2014). Retrieved from http://reports.weforum.org/global-risks-2014/ (15.12.2018).

Yang, Y. & Greaney, T. M. (2017). Economic growth and income inequality in the Asia-Pacific region: A comparative study of China, Japan, South Korea, and the United States. *Journal of Asian Economics*, 48, 6–22.

Yerdelen Tatoğlu, F. (2017). *Panel Zaman Serileri Analizi*. İstanbul: Beta Yayınları.

Selçuk Çağrı Esener

17. The Forgotten Unit of Income, Expenditure and Wealth Chain: Remembering the Taxation of Wealth and Our Fight Against Inequality

Abstract The need of discussing the good standards of a wealth taxation and finding the channels that it affects inequality in various ways are the main reasons behind writing this chapter. By doing so, one has a chance to remember what she or he learned from the past experiences on this subject. For that reason, first presenting its theoretical base and then mentioning relevant literature is important. Therefore, it is possible to argue its fiscal, economic and social effects, and maybe more than that, its realization in the real world. This study stands on not losing hypothetical doctrines on one side but at the same time, adding some quantitative proofs into the game to understand the problem in real terms. According to the general inference of our study, when income and wealth distribution are taken into account, it is easily seen that latter is much more distorted than the former, today. If this postulate is correct, meaning the one that distorted is wealth, then it would be wise to focus directly to the wealth taxes for its solution, again.

Keywords: Wealth tax, Inequality, Fiscal policy, Public finance, Taxation

1 Introduction

The lexical meaning of the wealth can be expressed as "asset, richness, property". In the same way, the definition of the "wealthy person" is "a person with a vast amount of properties; prosperous, rich".[1] Wealth has a different meaning than "estate", a concept frequently mixed with itself. Estate is the whole of the goods, rights and debts that can be measured by the money that a person may have or may be liable to constitute a legal integrity. Since the debts are included, a person can have a negative estate. However, wealth cannot be negative (Öner, 1986: 111).

It is not true to define wealth by capital alone; it is a wider concept than capital. As it is known, capital is the part of the income generated by the participation of production factors in the production activity *(referred to as 'wage, interest, rent, profit')* allocated to the investment and the savings, instead of consumption.

---

1 See, *Merriam-Webster Dictionary*, https://www.merriam-webster.com/dictionary/wealth; Türk Dil Kurumu, http://www.tdk.gov.tr/.

Wealth refers to all the assets (both movable properties and real estates) that a person has at a given moment. As in capital, 'conditions for participation in production' and 'income generation' are not sought. In summary, wealth belongs to the *individual*, while capital is a concept related to *business* (Öner, 1986: 110, 111).

Although taxation of wealth goes a long way back in the history of humanity, it is seen that it is not discoursed sufficiently, it is outdated, it is perceived as "hate against wealth" and this situation, particularly *in developing countries*, plays a role in increasing the gap between direct and indirect taxes. However, it is inevitable that an economic or financial system, in which income, expenditure and wealth chain cannot be captured, will have disruptive effects on long-term income distribution. This can ultimately lead to the loss of social peace and reconciliation.

In this study, we tried to discuss the good standards of wealth taxation and the channels that it affects inequality. In this manner, the organization of the remainder of the paper is as follows. In the second, third and fourth sections, we present the theoretical base and relevant literature on wealth. The fifth section provides fiscal, economic and social effects of taxation of wealth. The sixth section offers some interpretations on wealth taxation with the numbers. Lastly, the seventh section shows a brief conclusion on our subject.

2 The Concept of Wealth Tax and Its Historical Development Process

Wealth refers to all economic assets of real or legal persons in a certain period. Therefore, wealth taxes[2] excise *all* the economic assets a person has (Erdem, Şenyüz and Tatlıoğlu, 1998: 169). In the same way, these taxes are the indirect ones received from the value of all, or part of all kinds of movable properties and real estates within the scope of wealth (Turhan, 1982: 265). The common character of all wealth taxes, which vary significantly from country to country and over time, is that the existence of various economic assets (belonging to a person) is taken into account, rather than their economic activities (Edizdoğan, 1998: 204). If it is stated as economic asset or value, it is commercial, real estate

---

[2] The term is called as "wealth tax" in Turkish, while it is called as "property tax" in English, "Impôt sur la fortune" in French, and "Vermögensteuer" in German. Essentially, the terms "property" in English, "richess" in French and "reichtum" in German are the expression of wealth. However, in all three languages, it is seen that the words "real estate, goods and chattels" can be preferred as a result of historical development. In Turkish, there is no dichotomy that can lead to such inconvenience (Tuncer, 2003: 324, 325).

and individual wealth. *Commercial wealth* includes shareholders' equity, *real estate wealth* involves buildings, parcels and lands; *individual wealth* includes goods such as automobiles, household goods, jewelry, and other intangible goods such as stocks, bonds, cash and other rights (Batırel, 1990: 169).

While examining the historical development of wealth taxes, the first forms can be found in the First and Medieval Age Europe. The wealth tax that is the oldest form of emergency and war taxes was presented in ancient Greece and even in Rome, probably in the time of Monarchy. According to the simple structure of wealth at that time, the center of gravity of the tax was real estate. However, herd animals, slaves that are seen as goods at that time, household goods, jewelry and coins were also in the subject of tax (Tuncer, 2003: 332).

The Franks who dominated Western and Central Europe in the 8th century, widened their lands during the reign of the famous king Charlemagne and established the most powerful state in Europe. Charlemagne created an empire from the Teutonic Principality and declared himself the Holy Roman-German emperor. In this period, when Europe was dominated by the Roman-Germanic State, wealth taxes lost their characteristic of being 'taxes of public authorities', and became taxes that the masters of slaves, in other words the landowners, would pay. These taxes, which were used to finance the Crusades towards the end of the 12th century, have been implemented as a measure for the prevention of distress and depressions (and for the first time in cities) again since the 13th century (Tuncer, 2003: 333).

A wealth tax, which is charged annually has been imposed on many European cities in order to cover the increasing military expenses as a result of the continuous armies starting from the 15th century. However, from the 17th century onwards, these taxes became insufficient in the face of demands for a fair distribution of the increasing local needs (Tuncer, 2003: 333, 334). After all these developments, tax subjects such as land, building and working capital were arisen by dividing the taxes. The development continued, not only by taxing the wealth itself, but also by proceeding with the taxation of revenue.

Parallel to the development of capitalism in the 19th century, the idea of regulating taxation in a manner to minimize the obstruction of development of wealth and capital accumulation has come to the fore. In this understanding, it is considered that a state budget would be formed, which would be kept as small as possible, and which would be fed with taxes on income and expenses instead of wealth taxes. Because, the idea that wealth taxes have risk of disrupting private wealth and drying the resources of social wealth was becoming widespread (Uluatam, 2003: 351). In last two centuries, taxes on expenditures and then taxation of income elements have gained importance, and as a result, wealth taxes have lost their importance (Öner, 1986: 110) and they were attributed to a rather 'complementary' role. Nowadays, although the revenue generated by wealth taxes is low, it is seen that countries continue to include them in taxation systems for social purposes.

3 The Main Reasons Behind the Taxation of Wealth

Wealth taxes are principally considered as an important tool in the elimination of inequalities in wealth distribution among the three main objectives of taxation (financial, economic and social) (Edizdoğan, 1998: 203, 204). These taxes, beyond the functions of providing income to the state, serve to control and complete the income tax; and they carry out the duty of 'taxing political power and reputation which does not generate income'. In this framework, it would not be wrong to say that the wealth taxes, which have a significant share in the old financial systems, have lost their importance *especially in terms of "quantity"*[3] in favor of taxes on income and expenditures today.

As a result of the acceptance that wealth is an important indicator of the taxpaying ability, direct tax practices is applied because of the wealth or the transfer of wealth (Akdoğan, 2003: 249). In this sense, wealth has an appeal in terms of measurement of additional payment power (Goode, 1984: 133). If one of the two people who has the same amount of annual income does not have any wealth, while the other one has significant cash in the bank, a large house in which he or she lives, valuable furniture and paintings and automobiles, it will not be straight to argue that these two people have the same economic power. Therefore, it can be claimed that wealth provides significant advantages to its owner. Examples of it can be given as having a wider spectrum for decision-making or choice, having different alternatives, a sense of security and independence (Sandford, 1992: 218). On the other hand, having a great wealth –naturally –brings respect and political power to the person owning it (Goode, 1984: 133).

In addition to the benefits of implementing the 'ability to pay principle', wealth taxes can also help to implement the 'principle of utilization'. As it is known, the principle *"taxation or claiming compensation by taking into account the level of benefiting of people from the public services"*, is explained by the principle of utilization (see, Akdoğan, 2003: 202–203). It can be thought that at least some of

---

3 Taxes can be defined as basic or subsidiary taxes according to their place in the tax system. If a tax has a significant share in total tax revenues, it is considered in terms of quantity; if it is an independent tax with a specific purpose in the tax system, it is considered as fundamental tax in terms of quality (Turhan, 1982: 274).

the services provided by the state are more oriented towards wealth owners. It is clear that services such as roads, parking, and street lighting services are mainly addressed to building, land or automobile owners. It is even possible to think that public services such as national defense or internal security have benefited mainly the wealth owners. In this sense, wealth taxes provide the opportunity to implement the principle of utilization by helping to finance the public service for wealth (Uluatam, 2003: 352).

Wealth taxes serve the "financial purpose" of taxation as an additional resource for the state along with economic and social goals, such as reducing interpersonal income-wealth disparities, achieving progress towards equal opportunity, mobilizing the economic values that remain vacant. As a result, it is possible to summarize the reasons for the taxation of wealth as follows:

* Wealth taxes help to correct the problems that may arise from income tax application (Rosen, 1995: 494, 495).
* In case other sources of income are the same, more wealth ownership means having more payment power.[4]
* Taxation of wealth is also important in terms of decreasing concentration on wealth (Rosen, 1995: 495).
* The main functions of the state (exact public goods) are national defense, justice and internal security services. These services are made to protect the wealth of individuals. In fact, the state has emerged to protect individual property (Batırel, 1990: 170). Since the state produces a variety of public goods and services to ensure the safety of individuals and property (wealth), it is very normal for the individuals with wealth to participate in the financing of its expenditures to produce such goods and services (Erdem et al., 1998: 169).[5]

---

4   There are several main reasons for the taxation of wealth, according to those who argue that the ability to pay principle will ensure justice in taxation:

   I.   The idea that wealth ownership is superior to income in the form of earnings,
   ii. Increasing economic productivity by regulating wealth taxes to ensure more efficient use of wealth,
   iii. Preventing large wealth ownership by ensuring a balanced redistribution,
   iv. Providing justice in the wealth tax by treating different to the wealth that are obtained through inheritance, savings or donations (Tekir, 1990: 158, 159).

5   This view, which prioritizes benefit and efficiency, emerged towards the end of the 17th century and was influenced by Locke's theory of the state. According to Locke and natural law theorists, one of the basic functions of the state is the protection of

* Another reason for the taxation of wealth is due to the "Mutual Relationship" between the wealth elements that individuals hold in their hands and the public utilities of the local nature (road construction, lighting, etc.).
* Wealth taxes can be used as a tool for directing wealth to more efficient areas. These taxes can encourage individuals to dispose of their assets or to convert them into efficient professional investments.

Today, the reasons for justifying wealth taxation are collected especially around these reasons: The 'taxation principle according to "the ability to pay" and the "supervisory and complementary function of taxes on income and expenditure" '(Turhan, 1982: 268). Within the framework of these functions, it can be said that in many countries, the aim of taxes is the *'realization of a specific economic or social goal'*, and so the wealth taxes are included in the plans as *'taxes for specific purposes by their effects'* (Schmölders, 1976: 71-79).

4 Types of Wealth Taxes

4.1 General and private wealth taxes

General wealth taxes are the taxes levied on all the wealth elements that are considered under commercial wealth, real estate wealth and personal wealth types. Real and legal persons are the taxpayers of the 'general wealth tax', which is a subjective (personal) direct tax that is targeted to the purchasing power. The subject of the general wealth tax is the total wealth of real and legal persons; and the real and net amount of the wealth is taken into consideration. These taxes, today, are not very common in market economies except for extraordinary situations (Edizdoğan, 1998: 207, 208). However, two types of general wealth tax arise according to whether the total wealth is taken as purely or grossly, in other words, "whether the debts are taken into account or not". These are the general property tax and the net wealth tax (Öner, 1986: 112). The tariffs of these taxes may be single-rate and incremental.

Private wealth taxes are objective taxes that impose only part of the wealth on taxation, rather than the entire wealth. In countries which do not apply general wealth tax, they are generally applied as "building and land taxes" that include real estate in taxation. The subject of the tax is the buildings and lands owned by real and legal persons; the base is their gross value. However, in practice,

---

property (Musgrave and Musgrave, 1994: 411) and for the same reason wealth taxes are legitimate and necessary (Tekir, 1990: 158, 157).

sometimes it is seen that the wealth is calculated on the basis of the elements of wealth (Edizdoğan, 1998: 209–213). Another important special wealth tax in these countries is the motor vehicles tax.

4.2 Continuous (ordinary) and temporary (extraordinary) wealth taxes

Continuous wealth taxes are tax items that are periodically collected at regular intervals. Real estate tax and motor vehicles tax in many countries can be shown as examples of these continuous wealth taxes. It is important that a tax that is directed to wealth is payable from the income of the individuals in order to be able to apply a tax continuously (Uluatam, 2003: 358).

Some wealth taxes, on the other hand, are put into practice for certain periods of time for certain reasons. Capital Tax in Turkey, which was implemented in 1942 due to the exceptional circumstances of the Second World War, and "National Solidarity Tax (Impôt de Solidarité Nationale)" implemented in 1945 in France (more or less similar reasons to that of Turkey), are concrete examples of temporary wealth taxes received for once. After applying such taxes, the legal validity of the relevant legislation is terminated (Nadaroğlu, 1996: 348, 349).

Some tax laws, however, can be applied only once for the same event, although they remain in action. Taxes such as Inheritance and Gift Tax, Vehicle Purchase Tax are exemplary of this type of wealth tax that is collected once in the event that gives rise to the tax.

4.3 Apparent (artificial) and real (actual) wealth taxes

Although the subject and base of the taxes are the "amount" of the wealth, when the taxes are personalized by taking into account the age and family status of the taxpayer, type and source of wealth, and the nature and amount of the debts; if the source is generally the *yield* or the *inheritance* of wealth, these are called "Apparent (Artificial) Wealth Taxes" (Sayar, 1975: 194). However, the basis and also the source of some taxes is wealth itself. Such taxes, called "Real (Actual) Wealth Taxes" tax the wealth on real and material terms. 'Inheritance and Gift Tax', which is paid in the event that the yield of the inheritance tax cannot be paid by the revenue of it; 'Real Estate Tax', which is taken for owning a building which the owner lives at (which also does not generate revenue), and Motor Vehicle Tax are among the Real (Actual) wealth taxes (Nadaroğlu, 1996: 349–350).

4.4 Property tax and the taxes on transfer of wealth

Wealth tax can be collected from persons either because the person has the elements of wealth, or because he or she has acquired wealth elements. The first case is the wealth tax that includes property, and the event that gives rise to tax in these taxes is the possession of wealth elements of the person (Erdem et al., 1998: 171). Taxes on wealth transfers are the taxes on where the transfer of ownership (in one way or another) of property rights to another person, and/or the transfer of wealth elements like inheritance, donation or purchase transactions (Edizdoğan, 1998: 213). In other words, the subject of this tax is the process of the transfer of the movable property or real estates that a person has as asset to another person both in a way gratuitous (voluntarily) or paid (onerous) (Aksoy, 1998: 353).

4.5 Wealth tax and the taxes on accumulation of wealth

If basis value is taken as the total value of wealth, "wealth tax", and if the basis value is taken as the value increase in wealth, an "accumulation of wealth tax" should be considered. Taxes on accumulation of wealth are the levies on parts of the "incremental" at the end of certain periods, without touching the main bulk of the wealth, in order to protect the wealth and capital that have just begun to evolve. On the contrary to this, in order to tax the strong wealth and capital, the tax can be taken from the increasing part between certain years after the general sum is taxed (Sayar, 1975: 197–198). In other words, the value increases that occur in the wealth stock—which occurred without any labor of the owner—are taxed. If the value increases occurring in the whole wealth are taxed, then they are called "general wealth increase taxes"; and if they tax the increases in certain elements of all wealth, they are called "private wealth increases taxes" (Aksoy, 1998: 354).

5 Economic, Fiscal and Social Effects of Wealth Taxes

Wealth taxes can be considered as effective and administratively simple taxes for both creating public revenue (the fiscal objective of taxation) and eliminating inequalities in income distribution (social purpose). On the other hand, effective taxation of the increases in the values of wealth elements such as buildings and lands can be an important tool in terms of providing resources to the economy (economic purpose), especially in developing countries due to rapid population growth and urbanization (Ataç, 1999: 284). In this sense, efficient investments can be accelerated with the tax that will be applied effectively at an appropriate rate (Şener, 1998: 170). The results of wealth taxes on stability and growth targets are briefly mentioned below.

5.1 The Effects of Wealth Taxes on Allocation of Resources

It can be assumed that a tax on all elements of wealth will have a negative impact on resource allocation. Due to such taxation, the rate of savings, investments, and, as a result, the growth rate may decrease. However, the flexibility of the supply of savings should not be zero in order for the judiciary to be valid. In other words, there must be a change in the supply of savings due to tax. If flexibility is zero, then it means the person is not able to make a substitution between the funds he or she will save and their personal consumption. In this case, the taxing to be done on all elements of wealth will have no impact on the choices between consumption and savings (Batırel, 1990: 170).

5.2 The effects of wealth taxes on cyclical stability

In terms of cyclical (conjuncture) flexibility, the composition and distribution of wealth in the economy plays an important role. Indeed, the greater the share of securities, the more cyclical stabilizing efficiency of wealth tax increases. The fact that large wealth occupies more space and the tax burden on such wealth also increases the cyclical flexibility. In terms of implementation, we can say that the general, private, transfer of wealth or wealth increase taxes have a "neutral" effect in terms of cyclical fluctuations (Turhan, 1982: 411–412). This, depending on the country, can be explained by the fact that the wealth tax is real estate-intensive due to the composition of wealth taxes, the large amount of wealth occupying less space in the economies of the country or the tax burden is not wanted to be concentrated on wealth for various reasons.

5.3 The effects of wealth taxes on distribution of income

Wealth taxes are generally applied to promote savings. It can be said that if the tax issue is the wealth itself and a relative and high tax is levied on the wealth, the tax will increase the consumption and decrease the savings. Taxation of revenue from wealth can lead to more active use of productive wealth elements in the economy (Brown and Jackson, 1978: 317–327). When we examine in terms of work effort, for people who are not able to save money and do not make a choice between work and rest, more work may result in less rest (income effect) (Tekir, 1990: 159). In contrast, a general wealth tax covering all elements of wealth is neutral in the choice of individuals between work and rest. In contrast to the situation in the income tax, there is no negative impact on the work effort. The final taxpayers of general wealth taxes are the owners of capital. Since all assets owned by individuals are taxed, there is no way to divide savings into other wealth elements to avoid taxes. The exception is when wealth holders find the opportunity to move their investments out of the country. In the event of such a possibility, the owners of wealth and capital have the opportunity to reflect this tax to the consumer (Batırel, 1990: 170).

6 Recent Outlook of the Taxation of Wealth

The most common form of wealth taxes since ancient times is the "property tax" that is being taken from the real estate (Bulutoğlu, 1997: 474). The 'real estate tax' (buildings, parcels, and lands), which was taken first in China in 2000 B.C. and which was taken also in the Roman State, is one of the oldest taxes in history (Turhan, 1982: 280, 281). Another reason why the history of this type of tax stretches so far in the past has been the ease of comprehension of the buildings, parcels and lands that constitute the subject of the tax (Aronson and Hilley, 1986: 128; Nadaroğlu, 1996: 351). This easy-to-reach tax also has controllable features (Goode, 1984: 135). In ancient times when economic activity was defined primarily by agriculture, land ownership was the most important indicator of wealth. Therefore, for centuries, the "land tax" has maintained its importance in ancient Roman, Greek and Arabic civilizations, in Western and Eastern countries. In the last two centuries, with the development of industrial and commercial activities,[6] the progression of civilization and the accumulation of the population in cities, the "building tax" has gained more importance (Sayar, 1975: 202–205).

Today, in addition to building, parcel and land taxes (Property Tax), "Inheritance and Gift Tax" on acquisitions or donations arising from inheritance law and "Motor Vehicle Tax", which occurs parallel to the development of motor vehicles used in transportation, constitute the most important wealth tax items. Although almost every country has its own sub-policy implementations

---

6 For instance, in the French finance literature, two of the so-called "four old ladies (les quatre vieilles)" building and land taxes, together with the real estate tax, were enacted in 1870. These taxes, which were included in the French Income Tax in sedimentary character as a separate sediment in 1917, took their final forms with the reform that started in 1959 and completed 17 years later in 1976 (see Nadaroğlu, 2001: 159–161; Piketty, 2018).

in relation to wealth taxes, as is the case in most of its fiscal policies, the most basic financial instruments used in defining wealth can be stated as such. Today, the total tax revenues of these taxes and their outlook in GDP are presented in the Tab. 17.1.

As can be seen in Tab. 17.1, taxes on property occupy a very small proportion of total tax revenues. In a significant part of the developed and developing countries in the table, these taxes are almost non-existent, and a significant proportion of countries are still below the OECD average of 5.6 %. When we compare it with the GDP values, it is seen that the situation is not different from the first column and that the wealth taxes have a share of 1.9 % in the total economy. It can be thought that this situation arises from the intertwined structure of wealth and capital, or at least that it is perceived in this way by the countries. Considering that indirect taxes constitute roughly two-third of the system, it can be argued that in the developing countries it will be difficult to understand the income-expenditure-wealth chain in the future. This tax structure, which includes negativities in terms of justice, may cause some structural problems in the GINI coefficient and may obstruct the advancement of policies in terms of income distribution in the future.[7]

Tab. 17.2 shows the percentage of wealth increases in the study of Piketty (2018). Accordingly, the average for annual wealth of the top 1 % was around 6.5–7 %, while the average for the average world citizen was 1.5–2 %. In such a scheme where the economic growth trend is 3.3 %, it is clear that the wealth tax/GDP size (1.9 %) in Tab. 17.1 cannot be a solution to wealth inequality in particular.

According to Tab. 17.3, wealth inequality is twice the level of income inequality which means the wealthiest 10 % of households hold 52 % of the total net wealth compared with the 24 % of the total income held by the top 10 %. From this perspective, it is easy to defend that wealth is more unequally distributed than income, at least, in OECD countries. On the other hand, as it seems from the Tab. 17.3 that most of the countries find a way (maybe a fiscal one) to trim the differences as a percentage but the fluctuations in wealth across those same countries are still more visible and volatile.

---

7 According to the S80/S20 quintile share ratio used by OECD -as GINI coefficients-, there is a huge difference between top and lowest classes for some countries (please see Annex I).

Tab. 17.1: Taxes on Property as a Common Indicator of Taxation on Wealth. Source: OECD (2018), Revenue Statistics 1965–2017.

Countries	Taxes on property, total taxation, %, 2017	Taxes on property, GDP, %, 2017
Australia	10.8 *(2016)*	3 *(2016)*
Austria	1.3	0.5
Belgium	7.9	3.5
Canada	11.9	3.8
Chile	5.4	1.1
Czech Republic	1.4	0.5
Denmark	4	1.8
Estonia	0.7	0.2
Finland	3.6	1.5
France	9.5	4.4
Germany	2.7	1
Greece	8.1	3.2
Hungary	2.8	1.1
Iceland	5.4	2
Ireland	5.6	1.3
Israel	10	3.3
Italy	6	2.6
Japan	8.3 *(2016)*	2.5
Korea	11.7	3.1
Latvia	3.3	1
Lithuania	1.3	0.4
Luxembourg	9.6	3.7
Mexico	1.9 *(2016)*	0.3 *(2016)*
Netherlands	4	1.6
New Zealand	6	1.9
Norway	3.3	1.3
Poland	3.6	1.2
Portugal	4	1.4
Slovak Republic	1.3	0.4
Slovenia	1.8	0.6
Spain	7.5	2.5
Sweden	2.2	1
Switzerland	7.1	2
Turkey	4.5	1.1
United Kingdom	12.6	4.2
United States	15.4	4.2
OECD	**5.6**	**1.9**

Tab. 17.2: The Growth Rate of Top Global Wealth, 1987–2013. Source: Please see Piketty, T. (2018) for more information, piketty.pse.ens.fr/capital21c.

Average real growth rate per year *(after deduction of inflation)*	1987–2013
The top 1/(100 million) highest wealth holders *(about 30 adults out of 3 billion in 1980s, and 45 adults out of 4,5 billion in 2010s)*	6.8 %
The top 1/(20 million) highest wealth holders *(about 150 adults out of 3 billion in 1980s, and 225 adults out of 4,5 billion in 2010s)*	6.4 %
Average world wealth per adult	2.1 %
Average world income per adult	1.4 %
World adult population	1.9 %
World GDP	3.3 %
Between 1987 and 2013, the highest global wealth fractiles have grown at 6 %-7 % per year, vs. 2.1 % for average world wealth and 1.4 % for average world income. All growth rates are net of inflation (2.3 % per year between 1987 and 2013).	

7 Concluding Remarks

The fight against the inequality of income and wealth in the world and countries is not only a recent phenomenon. From the past to the present, academics and economic thinkers have put forward many ideas about how this can be achieved. While some of them search the solution of the problem in reforming intrasystem instruments, others have seen the solution in architecture as a harsh alternative.

Beyond the theoretical debates on the economic literature about the 'part' or 'whole', 'equality' or 'efficiency', it will be recalled that one of the most important instruments is 'wealth taxes' when approaching the issue from the 'public finance' field. Although today wealth taxes are seen at a point far from its importance in the past, it continues to be the most important instruments for the inclusion of income, expenditure and wealth as a whole.

According to the general judgments of our study, when the income and wealth distribution from various sources are analyzed, it is seen that wealth is much more distorted than income today. If this proposition is correct, then the place to seek for solution should be close to the source. In other words, if the one which is distorted is the wealth, then it would be significant to reconsider the wealth taxes for its solution. At least, this is the most fundamental result of this study.

Countries	Wealth (Top 10%)	Income (Top 10%)
United States	79%	29%
Netherlands	68%	25%
Denmark	64%	21%
Latvia	63%	26%
Germany	60%	23%
Chile	58%	36%
Estonia	56%	25%
Austria	56%	22%
Ireland	54%	23%
New Zealand	53%	27%
United Kingdom	52%	28%
Portugal	52%	26%
OECD 27	**52%**	**24%**
Norway	51%	21%
Canada	51%	24%
France	51%	24%
Luxembourg	49%	23%
Slovenia	49%	20%
Hungary	48%	22%
Australia	46%	26%
Spain	46%	25%
Finland	45%	21%
Italy	43%	24%
Belgium	42%	21%
Greece	42%	25%
Poland	42%	23%
Japan	41%	24%
Slovak Republic	34%	20%

Tab. 17.3: Distribution of Wealth and Income. Source: For more information, please see Balestra and Tonkin, 2018. The OECD Data covers the years between 2012 and 2015, which is subject to the space and time availability. Data refers to the share held by the richest 10 % of households in the case of wealth and by the richest 10 % of individuals in the case of income.

References

Akdoğan, A. (2003). *Kamu Maliyesi*. Gazi Kitabevi, 9. Ankara: Baskı.
Aksoy, Ş. (1998). *Kamu Maliyesi*. Filiz Kitabevi, 3. İstanbul: Baskı.

Aronson, J. R., & Hilley, J. L. (1986). *Financing State and Local Governments*. Studies of Government Finance—The Brookings Institution, fourth edition. Washington, DC.

Ataç, B. (1999). *Maliye Politikası*. Anadolu Üniversitesi Eğitim, Sağlık ve Bilimsel Araştırma Çalışmaları Vakfı Yayınları; No: 118—Etam A.Ş. Matbaa Tesisleri, 5. Eskişehir: Baskı.

Balestra, C. & Tonkin, R. (2018). Inequalities in household wealth across OECD countries: Evidence from the OECD Wealth Distribution Database. OECD Statistics Working Papers, No. 2018/01. Paris: OECD Publishing.

Batırel, Ö. F. (1990). *Kamu Maliyesi Teorisine Giriş*. Marmara Üniversitesi Yayın No: 492 – İktisadi İdari Bilimler Fakültesi Yayın No: 388, 3. İstanbul: Baskı.

Brown, C. V. & Jackson, P. M. (1978). *Public Sector Economics*. Suffolk: Basil Blackwell Ltd.

Bulutoğlu, K. (1997). *Kamu Ekonomisine Giriş*. Filiz Kitabevi, 5. İstanbul: Baskı.

Edizdoğan, N. (1998). *Kamu Maliyesi II (Kamu Gelirleri ve Vergi Teorisi)*. Ekin Kitabevi, 4. Bursa: Baskı.

Erdem, M., Şenyüz, D., & Tatlıoğlu, İ. (1998). *Kamu Maliyesi*. Ekin Kitabevi Yayınları, 2. Bursa: Baskı.

Goode, R. (1984). *Government Finance in Developing Countries*. Studies of Government Finance—The Brookings Institution. Washington, DC.

Musgrave, R. A. & Musgrave, P. B. (1994). *Public Finance in Theory and Practice*, fifth edition. McGraw-Hill International Editions Finance Series, Literatür Yayıncılık-Dağıtım-Pazarlama Sanayi ve Tic. Ltd. Şti..

Nadaroğlu, H. (1996). *Kamu Maliyesi Teorisi*. Beta Basım Yayım Dağıtım A.Ş., 9. İstanbul: Baskı.

Nadaroğlu, H. (2001). *Mahalli İdareler*. Beta Basım Yayım Dağıtım A.Ş., 7. İstanbul: Baskı.

OECD. (2018). Social and Welfare Statistics: Income Distribution and Inequality Data. Retrieved from https://data.oecd.org/inequality/income-inequality.htm (13.01.2019).

Öner, E. (1986). *Kamu Maliyesi I-Kamu Harcamaları ve Kamu Gelirleri*. Maliye ve Gümrük Bakanlığı Araştırma, Planlama ve Koordinasyon Kurulu Yayın No: 1986/282.

Piketty, T. (2018). *Top Incomes in France in the Twentieth Century: Inequality and Redistribution (1901-1998)*, translated by S. Ackerman. Cambridge, MA; London, England: The Belknap Press of Harvard University Press..

Rosen, H. (1995). *Public Finance*, fourth edition. Chicago: Irwin Publishing.

Sandford, C. T. (1992). *Economics of Public Finance*, fourth edition. Exeter: Pergamon Press Ltd., B.P.C.C. Wheatons Ltd.

Sayar, N. S. (1975). *Kamu Maliyesi-Kamu Gider ve Gelirleri Prensipleri (Cilt I)*. Sermet Matbaası, 5. İstanbul: Baskı.

Schmölders, G. (1976). *Genel Vergi Teorisi*. Translated by Salih TURHAN. İstanbul Üniversitesi Yayınları No: 2149-İktisat Fakültesi No: 374-Maliye Enstitüsü No: 55, 4. İstanbul: Baskı.

Şener, O. (1998). *Kamu Ekonomisi*. Alkım Yayınları, 6. İstanbul: Baskı.

Tekir, S. (1990). *Vergi Teorisi*. İzmir: Aklıselim Ofset.

Tuncer, S. (2003). *Vergi Hukuku ve Uygulaması*. Ankara: Yaklaşım Yayınları.

Turhan, S. (1982). *"Vergi Teorisi"*, İstanbul Üniversitesi Yayınları No: 2913, İktisat Fakültesi No: 480, İstanbul.

Uluatam, Ö. (2003). *"Kamu Maliyesi"*, İmaj Yayınevi, 8. Ankara: Baskı.

Appendix I.

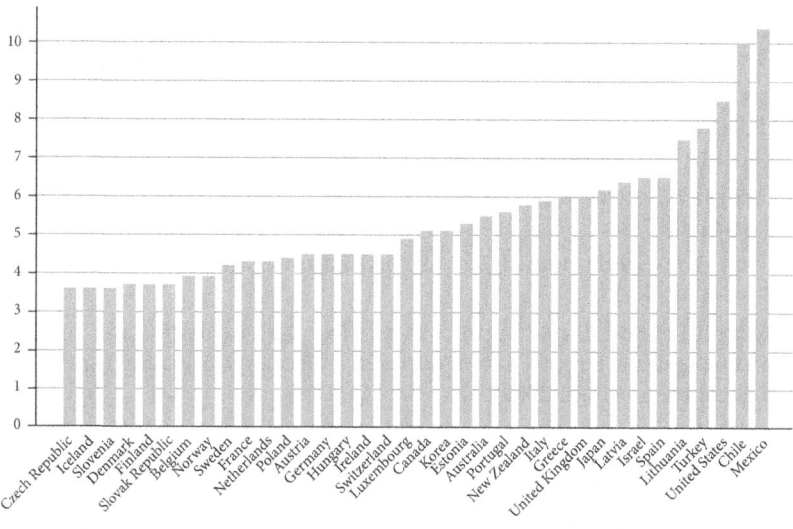

Graph 17.1: How Many Times the Rich Is Richer Than the Poor? *(2014–2017, S80/S20 Quintile Share Ratio)*. Source: OECD (2018), Social and Welfare Statistics: Income distribution Data, https://data.oecd.org/inequality/income-inequality.htm.

Nilüfer Yörük Karakılıç

18. Social Perspectives of Gender-Based Discrimination Perception: An Empirical Study on University Students

Abstract Discrimination is a phenomenon encountered in every aspects of life. Being precluded because of many factors such as age, gender, race, lineage, income, seniority or the exposure to non-deserved behavior is defined as discrimination. When considered socially, the most prominent type of discrimination is gender-based discrimination. The aim of this study is to reveal the demographic factors affecting the perception level of discrimination based on gender and gender-based discrimination. For this purpose, a questionnaire was applied to the students studying at Afyon Kocatepe University. The questionnaires used in the study were derived from the questionnaires obtained from the literature review. A total of 600 questionnaires were distributed, and the questionnaires with 491 reliable data were analyzed for validity and reliability by using Structural Equality Model through running the SPSS 16 and AMOS 19 softwares, and the hypotheses only by running SPSS 16.

Keywords: Discrimination, Gender-based discrimination, Gender

1 Introduction

Nowadays, there is a continuous change in the economic, social, cultural and political aspects. The most affected elements by the change process are undoubtedly those related to human, social structure and cultural values. The process of change regarding to people and society has accelerated with globalization and increased the interactions of societies. In this process of rapid change, the place of women in society and especially their position in the business life has also changed. In the society, a woman is accepted as a housewife, the person who babysits, cooks, cleans, tidies up the house; nowadays this situation has changed as the women appear in the business life and receive titles.

In fact, male and female represent two types that differ by a biological basis. This situation is an unchangeable and indisputable fact of their creation. This is called as gender in the literature. The human being is divided into two by its nature and a gender-based classification takes place as men and women. No matter how education, revolutions and evolution happen, there is a fact that there is a distinction between men and women. When women and men have very different structures due to their nature, social roles loaded on them in their growth processes, social stereotypes take shape, the differences become completely clear. In this study, it is aimed to reveal whether these differences turn into gender discrimination in the social sense or not. A gender discrimination survey was conducted for university students. The main reason why university students are selected as samples is because as the level of education increases, we expect them to have a high cultural change and development.

2 Literature Research

Snizek and Neil (1992) examined the underlying processes of gender discrimination in their study. By the surveys on 625 women and 512 men participants in Australia, they concluded that 37 % of women and 41 % of men experienced daily discrimination. Sipe et al. (2009) tried to determine students' perceptions of expected gender discrimination. In their surveys on 1,373 university students, they argued that girls' perception of discrimination would be higher. Zeher (2011) emphasized the importance of gender inequality and gender discrimination. She stated that stereotypes about organizational factors and gender can constitute discrimination. She also emphasized the importance of the effects of cultural, structural and interaction processes on gender discrimination after 219 case studies on gender discrimination. Jung and Choi (2017) examined gender discrimination in employee selection. Harnois and Bastos (2018) examined the effect of discrimination and harassment on gender inequalities. In the surveys they conducted on 3,724 people in 2006, 2010 and 2014, they showed the relationship between sex discrimination and harassment with mental and physical health. Heilman and Caleo (2018) examined gender inequality and gender discrimination. They focused on two basic strategies. The first one is to reduce the perception that women are not suitable for male type positions, while the second is to avoid negative performance expectations due to the perception of inappropriateness. Khan et al. (2018) measured university students' perceptions of gender discrimination. According to the questionnaires conducted on 150 students from different universities, it was concluded that the gender factor had a significant difference in the perception of gender discrimination, and that the education and age factors did not make a significant difference in the perception of gender discrimination. Skewes et al. (2018) investigated the relationship between gender equality perception and gender role in their surveys in two different countries.

In domestic literature, Gözütok et al. (2017) aimed to develop a gender scale in their work. İçli (2017) studied equality policies in his study. Pesen et al. (2016) tried to determine a gender perception and violence, conflict-related awareness levels in university students and found a significant difference in terms of gender variable. Korkmaz (2014) investigated the effect of gender discrimination on women's participation in senior positions in business life. Türeli and Dolmacı (2013) examined the relationship between gender discrimination and mobbing against women in business life. In their study on academicians, they found a significant relationship between being a woman and mobbing. Altinova and Duyan (2013) developed the "Social Gender Perception Scale" for adults to measure how gender roles are perceived. Urhan and Etiler (2011) analyzed university students in terms of gender discrimination. The "Social Gender Role Attitude Scale" developed by Zeyneloğlu and Terzioğlu (2011) was prepared for university students and focused on gender roles in marriage. Demir (2011) investigated the effect of demographic characteristics on discrimination in work life. Onay (2009) examined the factors affecting gender discrimination. Dedeoğlu (2009) evaluated how gender equality policies affect women's employment in Turkey from a critical perspective. She evaluated the effect of the regulation of laws on women workers on issues such as equal payment, equal treatment, maternity leave, parental leave, night work, flexible work and child care. The Ambivalent Sexism, developed by Glick and Fiske and adapted into Turkish by Sakallı-Uğurlu (2002), measures sexism for women. The adaptation of this scale to Turkish was done with university students.

3 Conceptual Framework

3.1 Gender and social gender concept

Gender is defined by the physiological characteristics of the individual, which is shaped by the biological and genetic structure. Gender, which defines a biological structure, is a demographic variable that classifies individuals as men and women (Başçı and Giray, 2016: 119). The gender of the individual is considered as biological gender and social gender. While the gender of the individual genetically symbolizes the physiological structure of congenital birth, social gender is a concept that is formed by the cultural values that the society attribute to men and women.

Social gender contains everything classified as feminine and masculine at culture, behavior, attitudes, interests, goals and values level. Human roles in society are defined on the basis of gender (Esen, 2013: 760). The roles that society attribute to individuals as women and men are called gender roles. Social gender means being a woman or a man in terms of psychological, social and cultural aspects. Gender role for women is called "feminine", while for men "masculine" (Başçı and Giray, 2016: 120).

There are many differences between men and women. Some of these differences arise from gender, however, the main root is in social gender. Biological differences between women and men are called gender differences. For instance, men have beard, while women do not or women can nurse a baby, while men cannot. These differences are the ones which do not change over time and are the same in all societies. Differences due to social gender are about what boys and girls learn from the process of socialization and the differences between emotions, roles, behaviors and attitudes that society approves according to gender (Gözütok et al., 2017: 1038).

3.2 Discrimination concept and social aspects of discrimination

Lexical meaning of discrimination is the act in which a person is victimized because of any personal characteristic. The fact that a person is treated differently from other people because of an individual characteristic means that he/she is victimized due to this different characteristic (Manav, 2013: 732).

Discrimination is defined as any action taken to prevent a person to have equally a right, freedom, and benefit.

Gender-based discrimination is defined as the behavior of a person, who is more negative or less favorable (an direct discrimination) to a woman than this person would treat a man or a person who has formally egalitarian behavior or practices in appearance, which have a discriminatory effect on women afterwards (indirect discrimination) (Kırel, Kocabaş and Özdemir, 2010: 5).

The stereotypes are important in the formation of discrimination, inequality and attitudes towards women and men. While men in many societies around the world are defined with stereotypes such as strong personalities, powerful, self-confident, fearless, independent, realistic, strong, self-reflecting, women are characterized by weak personalities such as dependent, passive, emotionally unstable. In culture, men and women are not only seen as different from birth; men are seen to be congenitally superior to women about subjects such as rationality, independence, leadership, and these characteristics associated with men are generally the characteristics that culture values most. The characteristics that are associated with women such as emotionality, sensitivity, and co-operation are the features that culture values the least. In these expectations, there is a woman's role who does not cause problems, compatible, unquestioning, obedient wife, serving to her husband, both a good wife and a good business woman, a good mother and the idea that this is the reasonable image is dominant (Altınova and Duyan, 2013: 11). In fact, it can be seen that discrimination in terms of society starts from the process of raising children according to their gender and shaping

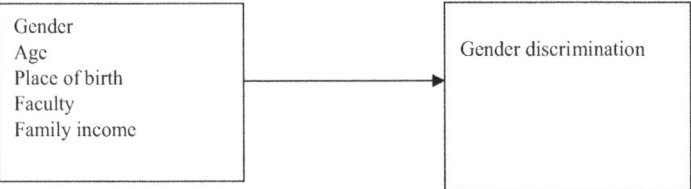

Fig. 18.1: Model of the Study. Source Author's construction

them according to gender. This social structure, which imposes gender-based preferences on children even in colors and constantly takes the differences into consideration, also contains many cultural values that serve the concept of discrimination. This is not only observed in Turkey but also in all countries in the world. And even in recent years there have been a number of attempts by countries to combat gender-based discrimination.

4 Methodology

4.1 Research model and hypotheses

Hypothesis 1. There is a positive correlation between gender and gender-based discrimination.

Hypothesis 2. The gender of students has a significant effect on gender-based discrimination perception.

Hypothesis 3. The students' perception of discrimination based on gender varies significantly according to their gender.

Hypothesis 4. Students' perception of discrimination based on gender varies significantly according to their age.

Hypothesis 5. Students' perception of discrimination based on gender shows a significant difference according to their place of birth.

Hypothesis 6. Students' perception of discrimination based on gender varies significantly according to their faculties.

Hypothesis 7. Students' perception of discrimination based on gender varies significantly according to their income.

4.2 Variables of research

The literature was reviewed in order to measure the perception of discrimination in the study. The statements developed from "Social Gender Perception Scales" that are generated by Altınova and Duyan (2013), Kahraman (2010), Özden (2018)

Tab. 18.1: Cronbach's Alpha Value of the Scale. Source: Author's construction

Cronbach's Alpha	Cronbach's Alpha Based on Standarized Items	N
,816	,828	16

are used. The scale consists of 16 statements. On the scale formed as a five-point Likert scale, the participants are asked to submit a five-point opinion: Strongly Disagree (1), Disagree (2), Partially Agree (3), Agree (4), Totally Agree (5). Six of the expressions are negative, ten are positive. Accordingly, the scores that can be obtained from the scale are in the range of 16–80 and high scores indicate that gender perception is positive.

4.3 Data collection method

The questionnaires of the study were conducted to the students studying in Afyon Kocatepe University. Six hundred questionnaires were distributed and 491 reliable data were obtained. Demographic data from the participants were requested but not the personal data.

4.4 Data analysis method

Validity and reliability tests were performed to ensure the construct validity of the scale. After the data were entered to SPSS 16.0 package program, confidence and exploratory factor analysis were performed in SPSS and structural equation modeling was carried out in AMOS 19 program. The hypotheses of the study were tested by running T test and Anova, Variance analysis in SPSS 16.

4.5 Confidence levels of scales used in research

Cronbach's Alpha and KMO values were calculated and interpreted in the validity and reliability analysis of the scale used in the study by entering the data into SPSS 16.

In Tab. 18.1, the Cronbach's Alpha value of the Gender Perception Scale was 0,816. The scale is highly reliable.

If $0.00 \leq \alpha < 0.40$, the scale is not reliable,
If $0.40 \leq \alpha < 0.60$, the scale has low reliability,
If $0.60 \leq \alpha < 0.80$, the scale is highly reliable,
If $0.80 \leq \alpha < 1$, the scale is a highly reliable scale.

Tab. 18.2: KMO and Bartlett's Test. Source: Author's own calculations

Kaiser-Meyer-Olkin Measure of Sampling Adequacy.		
Bartlett's Test of Sphericity	Approx. Chi-Square	2,258E3
	Df	120
	Sig.	,000

KMO: 1,00≥KMO≥0,90 =>Perfect
KMO: 0,90>KMO≥0,80 =>Good
KMO: 0,80>KMO≥0,70 => Intermediate level
KMO: 0,70>KMO≥0,60 =>Poor
KMO: KMO<0,60 => Bad

When the criteria were examined, the result of the study for KMO value is good value.

4.6 Evaluation of the reliability of scales by structural equality model

When the structural equity model results examined in Tab. 18.3, it is seen that good adaptive values increase after eliminating the statements that disrupt the scale.

4.7 Demographic characteristics of participants

40.5 % of the participants who answered the study questionnaires were male and 59.5 % were female. 6.1 % were between 17 and 18 years old, 25.5 % were between 19 and 20 years old, 38.9 % were between 21 and 22 years old, 22.2 % were between 23 and 24 years old and 7.3 % were 25 years old and over. In terms of the place of birth, %10.6 % were born in a village, 4.5 % in a town/borough, 25.1 % in a county, 36 % in a city and 23.8 % in a metropolitan. 4.9 % of the students are from tourism department, 63.1 % economics and administrative sciences, 18.1 % vocational school, 3.1 % science and letter, 1 % law, 4.7 % education and technology, 0.8 % veterinary, 1.8 % engineering, 0.6 % Islamic science, 0.4 % fine arts and 0.4 % graduate students. 8.2 % of them have a family income level between 2000 and 3000 TRY, 16.1 % 3001–4000 TRY, 11.4 % 4001–5000 TRY, 7.3 % 5001–6000 TRY and 6.9 % 60001 TRY and over.

Tab. 18.3: Structural Equality Model Results. Source: Author's own calculations

	x2	Df	X2/dj	GFI	CFI	RMSEA
Compliance Values without excluding the expressions	343,116	100	3,43	0,92	0,88	0,07
When the first expression (C3) is excluded	260,575	86	3,03	0,93	0,91	0,06
When the second expression (C9) is excluded	202,601	73	2,77	0,94	0,93	0,06
When the third expression (C5) is excluded	165,096	63	2,62	0,95	0,94	0,05
Good Compliance Values			#3	$ 90	$ 0,9	#0,0
Acceptable Compliance Values			#5	0,89–0,85	$ 0,9	0,06–0,08

Notes: p>.05, x2 =Chi- Square; df= Degree of Freedom; GFI= Goodness of Fit Index; CFI= Comparative Fit Index; RMSEA= Root Mean Square Error of Approximation (Meydan and Sesen, 2015: 37).

5 Testing the Hypotheses

Hypothesis 1. There is a positive correlation between gender and gender-based discrimination.

Pearson Correlation Analysis was used to examine the hypothesis. The results are given in Tab. 18.5.

As expressed in Tab. 18.5, P is the value of Pearson Correlation test, which is performed to answer the questions if there is a correlation between gender and gender discrimination perception, and if there is, what's the direction. It is found as 0.000.

Since P value is less than 0.05, it is interpreted that there is a relationship between them. As the correlation coefficient is 0.464, we can say that there is a moderate positive relationship between them. H1: There is a significant positive relationship between gender and gender-based perception of discrimination (ACCEPT). This hypothesis was accepted because there was a significant relationship between the variables.

Hypothesis 2. The gender of students has a significant effect on gender-based discrimination perception.

Linear Regression Analysis was used to analyze the hypothesis. The results are given in Tabs. 18.6, 18.7 and 18.8.

After the relationships between the dimensions were determined, linear regression analysis was performed which showed that the students were affected

Tab. 18.4: Demographic Characteristics of Participants. Source: Author's own calculations

		Frequency	Percentage%
Gender	Male	199	40,5
	Female	292	59,5
Age	17–18	30	6,1
	19–20	125	25,5
	21–22	191	38,9
	23–24	109	22,2
	25 and over	36	7,3
Place of birth	Village	52	10,6
	Town/Borough	22	4,5
	County	123	25,1
	City	177	36,0
	Metropolitan	117	23,8
Faculty	Tourism	24	4,9
	Economics and administrative sciences	310	63,1
	Vocational	89	18,1
	Science and letters	15	3,1
	Laws	5	1,0
	Education and Technology	23	4,7
	Veterinary	4	0,8
	Engineering	9	1,8
	Islamic sciences	3	0,6
	Medical School	6	1,2
	Fine arts	1	0,2
	Master programs	2	0,4
İncome	Between 2000 and 3000 TRY	286	58,2
	Between 3001 and 4000 TRY	79	16,1
	Between 4001 and 5000 TRY	56	11,4
	Between 5001 and 6000 TRY	36	7,3
	Between 6001 and over	34	6,9
Total		491	100

by gender and gender discrimination perception. The effect of gender on gender discrimination perception is modeled by Linear Regression Analysis (Tab. 18. 6).

The R value found in the model is the correlation value between dependent and independent variable, which indicates a moderate relationship between the

Tab. 18.5: Analysis of Relationships Between Individuals' Perception of Gender and Gender. Source: Author's own calculations

Pearson Correlation Analysis	Job Assurance	Performance
Correlation coefficient	1	0,464**
Gender P value	.	0,000
Pearson's N coefficient	491	491
Gender Discrimination P value	0,464**	1
N	0,000	.
	491	491

Tab. 18.6: Results of Linear Regression Analysis. Source: Author's own calculations

Model	R	R2	Adjusted R2	Estimated by Std. Error
1	a 0,464	0,216	0,214	0,63159

Tab. 18.7: Linear Regression Anova Table. Source: Author's own calculations

Model	Sum of Squares	sd	Square of averages	F	P
Regression	53,662	1	53,662	134,522	,000[b]
Residual	195,068	489	0,399		
Total	248,730	490			

Tab. 18.8: Linear Regression Analysis Coefficients Table. Source: Author's own calculations

Coefficients[a]						
Model	Non standarized coefficients		Standarized coefficients	T	Significance	
	B	Std. Sapma	Beta			
1 (constant)	1,345	0,086		15,568	0,000	
Gender	0,673	0,058	0,464	11,598	0,000	

two variables. According to the result of R2, 21 % change in the dependent variable is explained by the independent variable included in the model.

a. Dependent variable: Gender Discrimination Perception
b. Determinant: Gender

Tab. 18.9: T-Test Results of Gender Discrimination Perceptions of Individuals by Gender. Source: Author's own calculations

		t	Df	Significance	Mean Difference	Difference Std. error
The Perception of Gender Discrimination	The Assumption of the Mean Variances Equality	-11,598	489	0,000	-0,67338	0,05806
	Assumption of Inequality of Variances	-11,101	358,409	0,000	-0,67338	0,06066

The model of the effect of gender independent variable on gender discrimination perception dependent variable was statistically significant (p = 0.000 <0.01) (Tab. 18.7). Hypothesis 2: The gender of students has a significant effect on gender-based discrimination perception. (ACCEPT). This hypothesis was accepted because there was a significant effect between the variables.

a. Dependent Variable: gender discrimination perception

Significance value (p = 0.000 <0.01) indicates that the calculated beta coefficient is significant. If we interpret the beta coefficient, a standard deviation change in the independent variable causes a change in the standard deviation ratio of 0.464 in the dependent variable. In Tab. 18. 8, the significance value of both constant and the independent variable is less than 0.001 is interpreted as having a significant effect on gender discrimination perception.

Hypothesis 3. The students' perception of discrimination based on gender varies significantly according to their gender.

The T-Test was used to analyze the hypothesis. The results are given in Tabs. 18.9 and 18.10.

As stated in Tab. 18.9, according to the T-Test, P value is found 0.000 and since the value is less than 0.05, there is a significant difference between the gender discrimination perception levels of the participants and their gender. When these findings are taken into account together with the results of variance analysis, H3: The students' perception of discrimination based on gender varies significantly according to their gender (ACCEPT). The research hypothesis is accepted and it is concluded that there is a significant difference between the gender discrimination perception levels of the participants and their gender.

Tab. 18.10: Mean and Standard Deviation Values According to Gender and Gender Discrimination Perception. Source: Author's own calculations

	Gender	N	Mean	Standard Deviations	Standard Error
Gender Discrimination Perception	Male	199	2,6922	0,71653	0,05079
	Female	292	2,0188	0,56657	0,03316

Tab. 18.11: Results of the Analysis of Variance of Individuals According to Their Age Status. Source: Author's own calculations

	Squares	DF	Mean Square	F	Significance
Between groups	3,696	4	0,924	1,833	0,121
In groups	245,034	486	0,504		
Total	248,730	490			

The gender-discrimination perception of male participants in the research institution was found 2.6922, which is higher than the average value of female participants.

Hypothesis 4. Students' perception of discrimination based on gender varies significantly according to their age.

One-way analysis of variance was used to analyze the hypothesis. The findings are given in Tabs. 18.11 and 18.12.

As the result of analysis of variance in Tab. 18.11, the significance value of 0.121 is greater than 0.05. It is understood that the difference between the variables is not statistically significant. H4: Students' perception of discrimination based on gender varies significantly according to their age (REJECT). This hypothesis was rejected because there was no significant difference between the variables.

Tab. 18.12 shows that there is no significant difference between the average scores of the employees at different age levels.

Hypothesis 5. Students' perception of discrimination based on gender shows a significant difference according to their place of birth.

One-way analysis of variance was used to analyze the hypothesis. The results are given in Tabs. 18.13 and 18.14.

As the result of the variance analysis performed in Tab. 18.13, significance value is less than 0.05, which means that the difference between the variables is statistically significant. H5: Students' perception of discrimination based on gender shows a significant difference according to their place of birth (ACCEPT). This hypothesis was accepted as a significant difference between the variables.

Tab. 18.12: Mean and Standard Deviation Values by Age Status and Gender Discrimination Perception. Source: Author's own calculations

	N	Mean	Standard Deviation	Standard error
17–18	30	2,2938	0,67717	0,12363
19–20	125	2,3655	0,74866	0,06696
21–22	191	2,3416	0,69054	0,04997
23–24	109	2,1468	0,70707	0,06773
25 and over	36	2,2083	0,70994	0,11832
Total	491	2,2918	0,71247	0,03215

Tab. 18.13: Results of Variance Analysis of Gender Discrimination Perception According to Their Status. Source: Author's own calculations

	Squares	DF	Mean Square	F	Significance
Between groups	13,174	4	3,294	6,795	0,000
In groups	235,556	486	0,485		
Total	248,730	490			

Tab. 18.14: Average and Standard Deviation Values According to Discrimination Perception and Gender. Source: Author's own calculations

	N	Mean	Standard Deviation	Standard Error
Village	52	2,3906	0,60859	0,08440
Town/Borough	22	2,9290	0,39636	0,08450
County	123	2,3653	0,68526	0,06179
City	177	2,1674	0,68132	0,05121
Metropolitan	117	2,2388	0,80075	0,07403
Total	491	2,2918	0,71247	0,03215

Tab. 18.14 shows that there is a significant difference between the average scores of the employees from different birth places. When the mean data is taken into consideration, the mean value of the birth places in towns/boroughs was found as 2.9290, and it can be interpreted that the perception of gender discrimination is high for the ones who were born in towns/boroughs.

Hypothesis 6. Students' perception of discrimination based on gender varies significantly according to their faculties.

One-way analysis of variance was used to analyze the hypothesis. The results are given in Tabs. 18.15 and 18.16.

Tab. 18.15: Results of Variance Analysis Regarding Gender Discrimination Perceptions of Individuals According to Their Situation. Source: Author's own calculations

	Squares	DF	Mean Square	F	Significance
Between groups	29,211	11	2,656	5,794	0,000
In groups	219,519	479	,458		
Total	248,730	490			

Tab. 18.16: Mean and Standard Deviation Values According to Faculty Locations and Gender Discrimination Perception. Source: Author's own calculations

	N	Mean	Standard Deviation	Standard error
Tourism	24	2,1328	0,70285	0,14347
Economics and administrative sciences	310	2,4290	0,69144	0,03927
Vocational	89	1,8687	0,56886	0,06030
Science and letters	15	2,3042	0,83915	0,21667
Laws	5	2,8875	0,79844	0,35707
Education and Technology	23	2,2418	0,70265	0,14651
Veterinary	4	2,4219	0,53370	0,26685
Engineering	9	1,7083	0,53765	0,17922
İslamic sciences	3	2,9167	0,25259	0,14583
Medical School	6	2,1146	0,89041	0,36351
Fine arts	1	3,0000	.	.
Master programs	2	2,3438	1,19324	0,84375
Total	491	2,2918	0,71247	0,03215

As a result of the variance analysis performed in Tab. 18.15, significance value 0.000 is less than 0.05, which means that the difference between the variables is statistically significant. H6: students' perception of discrimination based on gender varies significantly according to faculties (ACCEPT). This hypothesis was accepted as a significant difference between the variables.

Tab. 18.16 shows that there is a significant difference between the mean scores of the employees in different faculties. When the mean data is taken into account in terms of faculty locations, Islamic sciences and fine arts can be interpreted as having a higher perception of gender discrimination.

Tab. 18.17: Results of the Analysis of the Variance of Gender Discrimination According to Income Status Source: Author's own calculations

	Squares	DF	Mean Square	F	Significance
Between groups	2,689	4	0,672	1,328	0,258
In groups	246,041	486	0,506		
Total	248,730	490			

Tab. 18.18: Mean and Standard Deviation Values by Income Status and Gender Discrimination Perception. Source: Author's own calculations

	N	Mean	Standard Deviation	Standard error
Between 2000 and 3000 TRY	286	2,2970	0,72261	0,04273
Between 3001 and 4000 TRY	79	2,1503	0,69773	0,07850
Between 4001 and 5000 TRY	56	2,3962	0,71473	0,09551
Between 5001 and 6000 TRY	36	2,4080	0,65661	0,10943
Between 6001 and over	34	2,2813	0,69771	0,11966
Total	491	2,2918	0,71247	0,03215

Hypothesis 7. Students' perception of discrimination based on gender varies significantly according to their income.

One-way analysis of variance was used to analyze the hypothesis. The results are given in Tabs. 18.17 and 18.18.

As a result of the analysis of variance in Tab. 18.17, the difference between the variables is not statistically significant since the value in the significance column 0.258 is greater than 0.05. H7: Students' perception of discrimination based on gender varies significantly according to their income (REJECT). This hypothesis was rejected because there was no significant difference between the variables.

On the other hand, it is observed that there is no significant difference between the mean scores of the income factors and the mean scores of the factors displayed in Tab. 18.18.

6 Conclusion

Being "a man or a woman", which is based on the biological origin of man, generates gender. This basic phenomenon actually contains many features that shape all the values in society. The characteristics of being a woman and the characteristics of being a man have been defined from the past to the present in line with the cultural structure of the society, and have been affected from the changes. Since they include sociological and social characteristics, they brought some rooted factors later. It is a fact that the judgments created in the gender-based society cannot be easily demolished, no matter what kind of changes and developments are experienced but an effort can be done to make people have a favorable perception.

The aim of this study is to evaluate the perception of gender discrimination from a social point of view. For this purpose, a survey is conducted on students studying at Afyon Kocatepe University. The hypothesis of the study is evaluated according to the results of the survey conducted to reveal the perception of gender discrimination. According to this:

Hypothesis 1. There is a positive correlation between gender and gender-based discrimination. This hypothesis was accepted. Gender perception of the participants is related to gender discrimination.

Hypothesis 2. The gender of students has a significant effect on gender-based discrimination perception. This hypothesis is accepted.

Hypothesis 3. The students' perception of discrimination based on gender varies significantly according to their gender. This hypothesis is accepted. The gender of the participants is a difference in the perception of gender-based discrimination.

Hypothesis 4. Students' perception of discrimination based on gender varies significantly according to their age. This hypothesis is rejected. Age has no effect on discrimination based on gender.

Hypothesis 5. Students' perception of discrimination based on gender shows a significant difference according to their place of birth. This hypothesis is accepted. The birth places of the participants change the perception of discrimination based on gender. The perception of gender discrimination is found to be higher in those who were born in towns/boroughs.

Hypothesis 6. Students' perception of discrimination based on gender varies significantly according to their faculties. This Hypothesis is accepted. Participants from the Islamic sciences and fine arts faculty have a high perception of gender discrimination.

Hypothesis 7. Students' perception of discrimination based on gender varies significantly according to their income. The hypothesis was rejected. The income situation does not make a significant difference on the perception of gender discrimination.

When the hypothesis results are examined, there is a significant relationship, influence and difference between gender and gender perception. Depending on the place of birth and the faculty students are studying in, there is a significant difference with the perception of discrimination on the basis of gender. No significant difference was found between the age and the income level of the student's family and the perception of discrimination based on gender.

References

Altınova, H. H. & Duyan, V. (2013). Toplumsal Cinsiyet Algısı Ölçeğinin Geçerlik Güvenirlik Çalışması. *Toplum ve Sosyal Hizmet*, 24(2), 9-22.

Başçı, B. & Giray, S. (2016). Üniversite Öğrencilerinin Toplumsal Cinsiyet Rollerine İlişkin Tutumlarının Çok Değişkenli İstatistiksel Tekniklerle Analizi. *Journal of Life Economics*, 3(4), 117-142.

Dedeoğlu, S. (2009). Eşitlik mi Ayrımcılık mı? Türkiye'de Sosyal Devlet, Cinsiyet Eşitliği Politikaları ve Kadın İstihdamı. *Çalışma ve Toplum Dergisi*, 2, 41-54.

Demir, M. (2011). İş Yaşamında Ayrımcılık: Turizm Sektörü Örneği. *Uluslararası İnsan Bilimleri Dergisi*, 8(1), 760-784.

Esen, Y. (2013). Eğitim Süreçlerinde Cinsiyet Ayrımcılığı: Öğrencilik Deneyimleri Üzerinde Yapılmış Bir Çözümleme. *International Online Journal of Educational Sciences*, 5(3), 757-782.

Gözütok, D., Toraman, Ç. & Erdol, T. (2017). Toplumsal Cinsiyet Eşitliği Ölçeğinin Geliştirilmesi. *Elementary Education Online*, 16(3), 1036-1048.

Harnois, C. E. & Bastos, J. L. (2018). Discrimination, Harassment, and Gendered Health Inequalities: Do Perceptions of Workplace Mistreatment Contribute to the Gender Gap in Self-reported Health? *Journal of Health and Social Behavior*, 59(2), 283-299. doi: 10.1177/0022146518767407.

Heilman, M. E. & Caleo, S. (2018). Combating Gender Discrimination: A Lack of Fit Framework. *Group Processes & Intergroup Relations*, 21(5), 725-744. doi: 10.1177/1368430218761587.

İçli, G. (2017). Toplumsal Cinsiyet Eşitliği Politikatlari ve Küreselleşme. *Pamukkale University Journal of Social Sciences Institute*, Retrieved from https://www.journalagent.com/pausbed/pdfs/PAUSBED_2018_30_133_143.pdf (13.08.2019).

Jung, J. H. & Choi, K. S. (2017). Gender Discrimination and Worker Selectivity: A Comparison of the Self-Employed and Wage Earners. *Global Economic Review*, 46(3), 251-270. Retrieved from https://doi.org/10.1080/12265 08X.2017.1305284 (13.02.2019).

Kahraman, S. D. (2010). Kadınların Toplumsal Cinsiyet Eğitliğine Yönelik Görüşlerinin Belirlenmesi. *DEUHYO ED*, 3(1), 30-35.

Khan, M. J., Kehkashan, A., Hafsah, A., Noreen, N., & Mehwash, N. (2018). Attitude of Male and Female University Students Towards Gender Discrimination. *Pakistan Journal of Psychological Research*, 33(2), 429-436.

Kırel, Ç., Kocabaş, F. & Özdemir, A. A. (2010). İşletmelerde Algılanan Cinsiyet Temelli Ayrımcılık: Eskişehir'de Özel Sektörde Bir Alan Araştırması. *Çimento İşveren Dergisi*, 24(3), Retrieved from http://www.ceis.org.tr/dergiDocs/makale143.pdf (13.08.2019).

Korkmaz, H. (2014). Yönetim Kademelerinde Kadına Yönelik Cinsiyet Ayrımcılığı ve Cam Tavan Sendromu. *Akademik Sosyal Araştırmalar Dergisi*, 2(5), 1-14.

Manav, A. E. (2013). H2000/43, 2000/78, 2006/54 Sayılı Ab Direktifleri Çerçevesinde İş Hukukunda Ayrımcılıkla Mücadele ve Türkiye'deki Uygulamalar. AB Direktifleri Çerçevesinde İş Hukukunda Ayrımcılıkla Mücadele, *Dokuz Eylül Üniversitesi Hukuk Fakültesi Dergisi*, 15(Özel S), 731-779 (Print Year: 2014).

Meydan, C. H. & Şeşen, H. (2015). *Yapısal Eşitlik Modellemesi Amos Uygulamaları*. Detay yayıncılık, Ankara.

Onay, M. (2009). Algılanan Cinsiyet Ayrımcılığının Sonuçları ve Konuyla İlgili Ampirik Bir Araştırma *Ege Akademik Bakış*, 9(4), ss. 1101-1125.

Özden, S. (2018). Sağlık Çalışanlarının Toplumsal Cinsiyet Rollerine İlişkin Tutumlarının Belirlenmesi. Cumhuriyet Üniversitesi, Sağlık Bilimleri Enstitüsü. Yayınlanmamış Yüksek Lisans Tezi.

Pesen, A., Kara, İ., Kale, M., & Abbak, B. S. (2016). Üniversite Öğrencilerinin Toplumsal Cinsiyet Algısı İle Çatışma ve Şiddete İlişkin Farkındalık Düzeylerinin İncelenmesi, *Uluslar arası Toplum Araştırmaları Dergisi*, 6(11), 325-340.

Sakallı-Uğurlu, N. (2002). Çelişik Duygulu Cinsiyetçilik Ölçeği: Geçerlik ve Güvenirlik Çalışması. *Türk psikoloji dergisi*, 11(21), 1-11. Retrieved from https://toad.halileksi.net/sites/default/files/pdf/erkeklere-iliskin-celisik-duygular-olcegi-toad.pdf (13.08.2019).

Sipe, S., Johnson, C. D., & Fisher, D. (2009). University Students' Perceptions of Gender Discrimination in the Workplace: Reality Versus Fiction. *Journal of Education for Business*, 339-349.

Skewes, L., Fine, C., & Haslam, N. (2018). Beyond Mars and Venus: The Role of Gender Essentialism in Support for Gender Inequality and Backlash. PLOS ONE. Retrieved from https://doi.org/10.1371/journal.pone.0200921 (18.02.2019).

Snizek, W. E. & Neil, C. C. (1992). Job Characteristics, Gender Stereotypes and Perceived Gender Discrimination in the Workplace. *Organization Studies*, 13(3), 403-427.

Türeli, N. Ş. & Dolmacı, N. (2013). İş Yaşamında Kadın Çalışana Yönelik Ayrımcı Bakış Açısı ve Mobing Üzerine Ampirik Bir Çalışma. *Ekonomi ve Yönetim Araştırmaları Dergisi*, 2(2), 83-104.

Urhan, B. & Etiler, N. (2011), Sağlık Sektöründe Kadın Emeğinin Toplumsal Cinsiyet Açıdan Analizi. *Çalışma ve Toplum*, 2, 191-216.

Zeher, D. B. (2011). Gender Discrimination at Work: Connecting Gender Stereotypes, Institutional Policies, and Gender Composition of Workplace. *Gender & Society*, 25(6), 764-786. doi: 10.1177/0891243211424741.

Zeyneloğlu, S. & Terzioğlu, F. (2011). Toplumsal Cinsiyet Rolleri Tutum Ölçeğinin Geliştirilmesi ve Psikometrik Özellikleri. *Hacettepe Üniversitesi Eğitim Fakültesi Dergisi*, 40, 409-420.

Hale Kirer Silva L. and Rüya Eser

19. Energy Consumption, Carbon Emissions and Income Inequality in Turkey

Abstract Income inequality is one of the most studied topics since the existence of Economics as a science. Furthermore, energy consumption and environmental pollution issues remain on the agenda nowadays. The aim of this chapter is to investigate the relationship of income inequality with energy consumption and carbon emissions in Turkey for the years between 1987 and 2016. We employ energy use (kg of oil equivalent per capita), CO_2 emissions, and Gini coefficient as income distribution indicator. We run times series analysis to examine the causality relationships. According to the results, our findings do not support a causality relationship between the variables.

Keywords: Energy Consumption, Emissions, Income Inequality

1 Introduction

Income inequality issue is a hot topic since the society exists in all over the world no matter if the country is developed or not. The researches about the subject trace back to Pareto's 80/20 rule and still continue this way. Turkey is ranked among the high-income inequality countries with the 0.40 Gini index (coefficient)[1] in 2015 (OECD, 2019). In 2016, while top 1 % received 23.4 % of total income, bottom 50 % received 14.6 % of income in Turkey (World Inequality Database). The distortion of income distribution causes income distribution to be considered as a crucial economic and social issue. One of the main objects of economy policies is to improve the income inequality and to take measures to enhance the economic welfare. A crucial determinant to meet this object is access to the energy. The International Energy Agency points out that energy access is a "golden threat" together with economic growth, human development and environmental sustainability (IEA, 2017). The economic thoughts that the spread of energy consumption causes an increase in income inequality lead to new studies about the relationship between energy consumption and income inequality. The population of the world is increasing day by day and this situation brings concomitantly a lot of problems and environmental pollution is one of the

---

1   The Gini index measures the distribution of income and a Gini index of zero indicates perfect equality, while 1 represents perfect inequality.

most severe issues. Unfortunately, we all know that income inequality has been worsening with the growing population and we should investigate the casual relationship of income inequality both with energy consumption and environmental pollution.

The analyses mainly focus on the impact of economic growth on energy consumption. Kraft and Kraft (1978) is one of the pioneer studies analyzing the relationship between energy consumption and economic growth. And there are numerous subsequent studies that investigate this subject in the literature (Lee, 2005; Soytas, Sari and Ewing 2007; Huang, Hwang, and Yang, 2008; Odhiambo 2009; Belke, Dobnik, and Dreger, 2011; Akpan and Akpan, 2012; Chen, Chen, and Chen, 2012; Ogundipe and Apata, 2013; Herrerias, Joyeux, and Girardin, 2013).

As the number of the world population goes up, energy consumption also increases and in consequence of this environmental pollution increase as well. Kuznets (1955) developed a hypothesis that economic inequality first increases and then after a certain level of development it decreases and propose an inverted-U shape relationship between income and growth. In this context, Grossman and Krueger (1991) propose a similar relationship between growth of income and environmental pollution. This inverted-U shape relationship is called as Environmental Kuznets Curve (EKC). They claim that during the initial years of the growth, environmental pollution increases; however, environmental pollution starts to decrease during the following years through a sustainable growth. Grossman and Krueger (1995, 1996), Shafik and Bandyopadhyay (1992) and Selden and Song (1994) are the subsequent studies that investigated EKC in the 1990s. Bruvoll and Medin (2003) prove the correlation between income and pollution, while Perman and Stern (2003) do not support the EKC theory.

However, the relationship between income inequality and energy consumption is not highly examined. Ravallion, Heil and Jalan (2000) find that income distribution is among the determinant factors that affect carbon emissions. Padilla and Serrano (2006) indicate that an increase in income inequality causes inequality in CO_2 emission distribution across the countries. Coondoo and Dinda (2008) determine that income inequality among the countries has a considerable effect on the emission levels for most of the countries examined. Zhang and Zhao (2014) examine the impact of income inequality on carbon dioxide emissions and find that higher incomes deteriorate environmental quality. Robert (2018) points to a strong negative causality between energy use and income inequality by employing panel data analysis. He demonstrates the significant impact of energy use on income inequality. Khan and Heinecker (2018) investigate the relationship between disparity of income distribution and energy use from a complex systems perspective and they use a scaling indicator as a measure of the income distribution. Following this literature, our aim is to examine the relationship between income inequality, energy consumption and CO_2 emissions in Turkey by employing time series analysis techniques.

Within this context in the following section of this chapter, we introduce the data we use in the analysis and give the results. We finalize our chapter with findings and recommendations in the conclusion section.

2 Analysis

In this section, we analyze empirically the relationship between energy consumption, CO_2 emissions and income inequality for the years between 1987 and 2016.

2.1 Data and methodology

Economy is a complex system itself and requires simultaneous equations to explain complex relations between the variables. This makes difficult to determine the endogenous and exogenous variables in an economic model. This issue can be solved with Vector Autoregressive (VAR) model. The VAR model is introduced by Sims (1980) and suggests that each variable used in an econometric model can affect another variable, and that these variables can also be influenced by other variables. VAR models give dynamic relationships without any restrictions on the structural model and can be used frequently for time series.

When we consider the economic system, it is actually a sub-system of larger human environment system (Stern, 2012) and this social system is a sub-system of ecological system. And energy is one of the main inputs of the natural environment. Our aim in this study is to find out the relationship among income distribution, energy use and carbon emissions. When we evaluate the data to determine the dependent variable, we see that all variables can affect each other. For this reason, we employ the VAR model to see the dynamics between the variables. We also run Granger Causality Test. In this manner, we use energy use (kg of oil equivalent per capita), CO_2 emissions, and Gini coefficient as a proxy of income inequality. We gather the energy use from World Bank, CO_2 emission data from OECD, and Gini Index data from the Standardized World Income Inequality Database (Solt, 2019). We determine the analysis period according to the recent availability of the data.

2.2 Empirical results

We start the empirical part by investigating the stationary of variables. In this context, we run Augmented Dickey-Fuller (ADF), Phillips-Perron (PP), and

Tab. 19.1: Unit Root Tests. Source: Authors' own calculations by using EViews 10 statistical package program.

Variable	ADF	PP	KPSS	Decision
GINI	Level −0.995(1) 2nd Difference −4.656(0)	Level −1.127544[b] 2nd Difference −4.640003[b]	2nd Difference 0.3280[b]	I(2)
CO_2	Level −0.052(0) 1st Difference −5.196(0)	Level 9.523 1st Difference −5.199	2nd Difference 0.284[b]	I(1)
LENERGY	Level −3.827(7) 1st Difference −4.518(0)	Level 3.455 1st Difference −4.633	1st Difference 0.052[b]	I(1)

Note: The numbers in parenthesis (.) are the lag lengths, which are determined by AIC in ADF test. Bartlett Kernell estimation method is used in PP, bandwidth is determined as Newey-West. [a] indicates regression contains both constant term and trend; [b] denotes regression contains constant term.

Kwiatkowski-Phillips-Schmidt-Shin (KPSS) Unit Root tests to determine whether the variables have unit root or not. The results of the Unit Root Tests are shown in Tab. 19.1.

According to unit root test results that are shown in Tab. 19.1, Gini variable is second order stationary, while CO_2 and LENERGY variables are first order stationary. To select the lag length, we apply the Information Criteria. The findings indicate that the optimum lag length is 2 (Tab. 19.2).

After the optimal lag determination, we estimate the Standard VAR model. Following this, we perform LM Autocorrelation and White Heteroskedasticity tests. The results are displayed in Tabs. 19.3 and 19.4 respectively. Accordingly, there is no autocorrelation and no heteroskedasticity at 5 % significance level.

Following this, we apply the impulse-response analysis. Fig. 19.1 shows the findings.

Fig. 19.1 illustrates the response of Gini index to both carbon emissions and energy use, the response of carbon emissions to both Gini and energy use and the response of energy use to Gini and carbon emissions. The findings indicate that one standard deviation shock in carbon emissions affect Gini positively until eighth period, then it changes, while one standard deviation shock in energy use affect Gini index increasingly negative. One standard deviation shock in Gini

Tab. 19.2: Lag Selection. Source: Authors' own calculations by using EViews 10 statistical package program.

Lag	LogL	LR	FPE	AIC	SC	HQ
0	-1.268241	NA	0.000279	0.328326	0.473491	0.370128
1	79.97982	137.4967	1.08e-06	-5.229217	-4.648557	-5.062008
2	95.02465	21.98859*	7.04e-07*	-5.694204*	-4.678049*	-5.401588*
3	99.06586	4.973797	1.12e-06	-5.312758	-3.861108	-4.894736
4	105.4913	6.425440	1.62e-06	-5.114715	-3.227570	-4.571286

Note 1* indicates lag order selected by the criterion. LR: Sequential Modified LR Test Statistic, FPE: Final Prediction Error, AIC: Akaike Information Criteria, SC: Schwarz Information Criteria, HQ: Hannan-Quinn Information Criteria.

Tab. 19.3: LM Autocorrelation Test. Source: Authors' own calculations by using EViews 10 statistical package program.

Lag	LRE* stat	df	Prob.	Rao F-stat	df	Prob.
1	5.363001	9	0.8016	0.582846	(9, 39.1)	0.8029
2	3.169165	9	0.9572	0.335520	(9, 39.1)	0.9575
3	2.129787	9	0.9892	0.222718	(9, 39.1)	0.9893
4	10.08107	9	0.3440	1.159937	(9, 39.1)	0.3467
5	12.50170	9	0.1865	1.481777	(9, 39.1)	0.1889
6	11.24585	9	0.2592	1.312520	(9, 39.1)	0.2619
7	5.215576	9	0.8151	0.565824	(9, 39.1)	0.8164
8	22.14890	9	0.0084	2.962628	(9, 39.1)	0.0088

Tab. 19.4: White Heteroskedasticity Test. Source: Authors' own calculations by using EViews 10 statistical package program.

Chi-sq	Df	Prob.
79.20447	72	0.2622

index affect carbon emissions negatively until eighth period and then it shifts, while one standard deviation shock in energy use affect in a positive trend. One standard deviation shock both in carbon emissions and in Gini index affects energy use positively.

We also examine the Variance Decomposition Analysis. The results are displayed in Tab. 19.5.

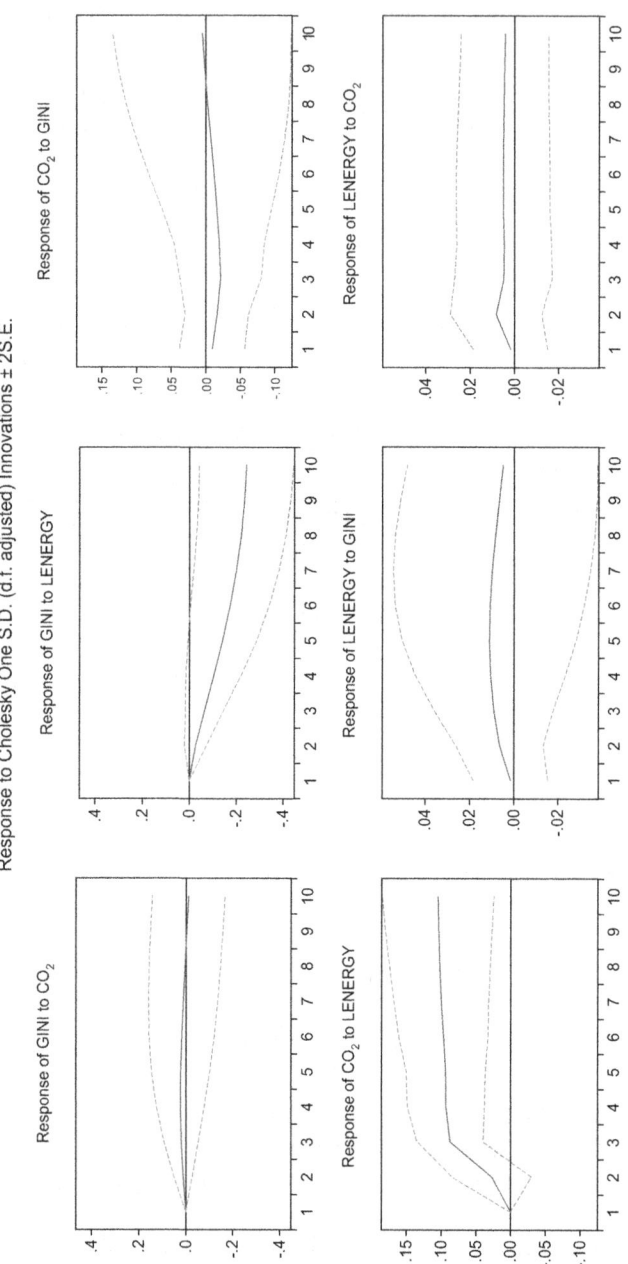

Fig. 19.1: Impulse-Response Functions. Source: Authors' own calculations by using EViews 10 statistical package program

Tab. 19.5: Variance Decomposition Analysis. Source: Authors' own calculations by using EViews 10 statistical package program.

	Gini				Carbon emissions				Energy Use			
Period	S.E.	GINI	CO2	LENERGY	S.E.	GINI	CO2	LENERGY	S.E.	GINI	CO2	LENERGY
1	0.102432	100.0000	0.000000	0.000000	0.125336	0.590031	99.40997	0.000000	0.044555	0.108601	0.146726	99.74467
2	0.204969	97.98953	0.295525	1.714943	0.135044	1.984451	94.06210	3.953451	0.056274	1.375170	2.241940	96.38289
3	0.306686	94.24724	0.556061	5.196697	0.162153	3.149140	65.25119	31.59967	0.066857	2.790415	2.122810	95.08677
4	0.399726	89.56359	0.664177	9.772238	0.188002	3.456004	48.58541	47.95858	0.076294	4.050833	1.995666	93.95350
5	0.480850	84.02648	0.670829	15.30269	0.210570	3.348286	38.84710	57.80462	0.084247	5.065089	2.000880	92.93403
6	0.549947	77.72135	0.619112	21.65954	0.232179	3.021756	32.05654	64.92171	0.091069	5.741551	2.027459	92.23099
7	0.608712	70.88807	0.541237	28.57069	0.252986	2.630765	27.10660	70.26264	0.097029	6.078840	2.047237	91.87392
8	0.659649	63.90424	0.464368	35.63139	0.272920	2.271088	23.40618	74.32274	0.102319	6.134634	2.056219	91.80915
9	0.705275	57.20996	0.409636	42.38040	0.292035	1.985775	20.56274	77.45149	0.107119	5.987444	2.051908	91.96065
10	0.747564	51.19767	0.389571	48.41276	0.310377	1.783381	18.32746	79.88916	0.111590	5.715931	2.033996	92.25007

Tab. 19.6: Pairwise Granger Causality Test. Source: Authors' own calculations by using EViews 10 statistical package program.

Null Hypothesis:	Obs	F-Statistic	Prob.
CO_2 does not Granger Cause GINI	28	0.61381	0.5499
GINI does not Granger Cause CO_2		1.80891	0.1864
LENERGY does not Granger Cause GINI	28	2.13137	0.1415
GINI does not Granger Cause LENERGY		0.28948	0.7513
LENERGY does not Granger Cause CO_2	28	3.62837	0.0427
CO_2 does not Granger Cause LENERGY		0.36789	0.6962

In the first period, 100 % of the prediction error variance of the Gini index variable is explained by itself, in the tenth period this percentage decline to 51 %. Particularly in the first three periods, Gini index is almost only described by itself. 99 % of the prediction error variance of the carbon emissions are explained by itself in the first period, while 32 % of this rate is described by energy consumption in the third period and this value goes up to approximately 80 % in the tenth period. When we look at the prediction error variance of energy use, it is described by itself almost for all the periods.

Finally, we perform the Pairwise Granger Causality test. According to findings, there is bilateral causality between neither income distribution and energy use, nor income distribution and carbon emissions. The results point out only a unidirectional relationship from energy use to carbon emissions, which mean energy use cause environmental pollution. The results are shown in Tab. 19.6. These findings are consistent with variance decomposition analysis.

3 Conclusion

Inequality in income distribution and poverty is one of the main issues in today's world. Nowadays, it is not only an economic problem but also a politic and social issue. There are numerous studies in the literature that are investigating the relationship between income inequality and other macroeconomic and social variables. In the literature as the economic growth is succeeded, income inequality first worsens, and then gets better, which means become more equal. This hypothesis is called Kuznets curve. According to this, a sustainable growth is a kind of remedy for income inequality.

In today's world, as the economies grow, energy consumption is increasing day by day together with the population explosions. And all these circumstances bring the environmental pollution issues on the agenda. In this respect, environmental Kuznets Curve hypothesis is proposed and accordingly there is a similar pattern between economic growth and environmental pollution. So that growth first increases the pollution; however, if it sustains it decreases the pollution. In the literature, studies mostly investigate either the relationship between economic growth and energy consumption or economic growth and environmental pollution. However, the number of studies investigating the relationship of income inequality with energy consumptions and environmental pollution are not high. These limited number of studies points to causality relationship between income inequality and energy use and also income inequality and CO_2 emissions. Nevertheless, in our study, the findings do not support a causality relationship both between "income distribution and energy use" and "income distribution and CO_2 emissions" in Turkey. We only find a unidirectional causality from energy use to carbon emissions, as it is obvious. Although energy consumption and carbon emissions have casual relationship with economic growth in Turkey, the finding in this study is not surprising in terms of inequality. This study employs Gini coefficient as income distribution indicator. We believe that Theil index, which is one of the generalized entropy indices, may give more insights about income distribution. For this reason, our findings may not support a causality relationship. In this context, our aim is to include Theil index to our model in our further study.

References

Akpan, G. E. & Apkan, U. F. (2012). Electricity consumption, carbon emissions and economic growth in Nigeria. *International Journal of Energy Economics and Policy*, 2(4), 293–308.

Belke, A., Dobnik, F., & Dreger, C. (2011). Energy consumption and economic growth:New insights into the cointegration relationship. *Energy Economics*, 33, 782–789.

Bruvoll, A. & Medin, H. (2003). Factors behind the environmental kuznets curve. Adecomposition of the changes in air pollution. *Environmental and Resource Economics*, 24(1), 27–48. https://doi.org/10.1023/A:1022881928158 (12.05.2019).

Chen, P.-Y., Chen, S.-T., & Chen, C.-C. (2012). Energy consumption and economic growth– new evidence from meta analysis. *Energy Policy*, 44, 245–255.

Coondoo, D. & Dinda, S. (2008). Carbon dioxide emission and income:A temporal analysis of cross-country distributional patterns. *Ecological Economics*, 65(2), 375–385.

Belke, A., Dobnik, F. & Dreger, C. (2011). Energy consumption and economic growth:New insights into the cointegration relationship. *Energy Economics*, 33(5), 782–789.

Grossman, G. M. & Krueger, A. B. (1991). Environmental impacts of a North American free trade agreement. *National Bureau of Economic Research* (No. w3914). Retrieved from https://www.nber.org/papers/w3914.pdf (13.04.2019).

Grossman, G. M. & Krueger, A. B. (1995). Economic growth and the environment. *The Quarterly Journal of Economics*, 110(2), 353–377.

Grossman, G. M. & Krueger, A. B. (1996). The inverted-U:What does it mean? *Environment and Development Economics*, 1(01), 119–122.

Herrerias, M., Joyeux, R. & Girardin, E. (2013). Short- and long-run causality between energy consumption and economic growth:Evidence across regions in China. *Applied Energy*, 112, 1483–1492.

Huang, B.-N., Hwang, M., & Yang, C. (2008). Causal relationship between energy consumption and GDP growth revisited:A dynamic panel data approach. *Ecological Economics*, 67,41–54.

International Energy Agency—IEA. (2017). Energy Access Outlook 2017, Executive Summary. Retrieved from https://webstore.iea.org/weo-2017-special-report-energy-access-outlook (12.04.2019).

Khan, F. & Heinecker, P. (2018). Inequality and Energy:Revisiting the Relationship Between Disparity of Income Distribution and Energy Use from a Complex Systems Perspective. *Energy Research & Social Science*, 42, s.184–192.

Kraft, J. & Kraft, A. (1978). On the relationship between energy and GNP. *Journal of Energy and Development*, 3, 401–403.

Kuznets, S. (1955). Economic growth and income inequality. *The American Economic Review*, 45(1), 1–28. Retrieved from http://www.jstor.org/stable/1811581 (12.05.2019).

Lee, C.-C. (2005). Energy consumption and GD in developing countries:A cointegrated panel analysis. *Energy Economics*, 27, 415–427.

Odhiambo, N. M. (2009). Electricity consumption and economic growth in South Africa:A Trivariate causality test. *Energy Economics*, 31, 635–640.

OECD. (2019). Income inequality (indicator). doi:10.1787/459aa7f1-en (18.06.2019).

Ogundipe, A. A. & Apata, A. (2013). Electricity consumption and economic growth in Nigeria. *Journal of Business Management and Applied Economics*, II(4). Retrieved from http://eprints.covenantuniversity.edu.ng/1777/1/Ogundipe%20Adeyemi%20A.%203.pdf (12.05.2019).

Padilla, E. & Serrano, A.(2006). Inequality in co2 emissions across countries and its relationship with income inequality:A distributive approach. *Energy Policy*, 34, 1762–1772.

Perman, R. & Stern, D. I. (2003). Evidence from panel unit root and cointegration tests that the environmental Kuznets curve does not exist. *The Australian Journal of Agricultural and Resource Economics*, 47, 325–347.

Ravallion, M., Heil, M., & Jalan, J. (2000). Carbon emissions and income inequality. *Oxford Economic Papers*, 52(4), 651–669.

Robert, S. (2018). A Panel Analysis of Income Inequality and Energy Use, MPRA Paper No. 92171, posted 21February2019 14:34 UTC.https://mpra.ub.uni-muenchen.de/92171/1/MPRA_paper_92171.pdf (12.05.2019).

Selden, T. M. & Song, D. (1994). Environmental quality and development:Is there a Kuznets Curve for air pollution emissions? *Journal of Environmental Economics and Management*, 27(2), 147–162. ISSN 0095-0696, https://doi.org/10.1006/jeem.1994.1031 (19.05.2019).

Shafik, N. & Bandyopadhyay, S. (1992). Economic growth and environmental quality:Time series and cross-country evidence. Policy Research Working Paper Series, 904, the WorldBank.

Sims, C. A. (1980). Macroeconomics and Reality. *Econometrica*, 48(1), 1–48. Retrieved from https://www.pauldeng.com/pdf/Sims%20macroeconomics%20and%20reality.pdf (12.05.2019).

Solt, F. (2019). Measuring income inequality across countries and over time:The standardized world income inequality database. SWIID Version 8.1. Retrieved from https://fsolt.org/swiid/ (21.05.2019).

Soytas, U., Sari, R., & Ewing, B. T. (2007). Energy consumption, income, and carbon emissions in the United States. *Ecological Economics*, 62, 482–489.

Stern, D. I. (2012). Ecological Economics. Crawford School Research Paper No.17, Crawford School of Public Policy. Canberra: Australian National University, Retrieved from https://crawford.anu.edu.au/pdf/crwf_ssrn/crwfrp_1203.pdf (10.06.2019).

World Inequality Database. Income Inequality Data, Retrieved from https://wid.world/country/turkey/ (19.06.2019).

Zhang, C. & Zhao, W. (2014). Panel estimation for income inequality and CO2 emissions:A regional analysis in China. Applied Energy, 136, 382–392, ISSN 0306-2619, https://doi.org/10.1016/j.apenergy.2014.09.048 (11.04.2019).

List of Figures

Begüm Erdil Şahin and Deniz Dilara Dereli
An Evaluation on Middle-Income Trap in Turkey
Fig. 3.1: Middle-Income Trap Process. Source: Ünlü and Yıldız, 2018a: 4. 47
Fig. 3.2: Stages of Industrialization and Middle-Income Trap.
Source: Ohno, 2009: 28; Çaşkurlu and Arslan, 2014: 74. 48

Habibe Günsel Doğrul, Mediha Mine Çelikkol, and İlhan Korkmaz
**The Analysis of the Relationship Between Creative Class,
Financial Development and Regional Innovativeness in Turkey**
Fig. 6.1: Moran Scatter Plots Regarding to Innovation, Creative Class
and Financial Development Level. Source: Authors' own
construction. 104
Fig. 6.2: LISA Map for Innovation Level. Source: Authors' own
construction. 105
Fig. 6.3: LISA Map for Creative Class. Source: Authors' own
construction. 105
Fig. 6.4: LISA Map for Financial Development. Source: Authors' own
construction. 105

Umut Akduğan and Seyhun Doğan
Youth Unemployment: An Empirical Analysis
Fig. 7.1: Impulse-Response Functions. Source: Authors' own construction. 127

Ufuk Bingöl
**Employment Policy Goals in Turkish Government Programs in
Terms of Social Policies: Justice and Development Party Governments**
Fig. 12.1: In the Government Programs, the expression analysis of
the most frequently used words related to 'employment'.
Source: This figure is written by the author with the support
of CAQDAS, resulting from Text Search Query filtered by
Treemap. 205

Nilüfer Yörük Karakılıç
**Social Perspectives of Gender-Based Discrimination
Perception: An Empirical Study on University Students**
Fig. 18.1: Model of the Study. Source Author's construction. 299

Hale Kirer Silva L. and Rüya Eser
Energy Consumption, Carbon Emissions and Income Inequality in Turkey
Fig. 19.1: Impulse-Response Functions. Source: Authors' own
    calculations by using EViews 10 statistical package program. 320

List of Graphs

Semanur Soyyiğit
China's Neo-Mercantilist Policies and Growth Process
Graph 2.1: China's Growth Rate Between 2000 and 2017 (Annual %). Source: The World Bank, https://data.worldbank.org/indicator/NY.GDP.MKTP.KD.ZG?locations=CN (18.02.2019). 37

Begüm Erdil Şahin and Deniz Dilara Dereli
An Evaluation on Middle-Income Trap in Turkey
Graph 3.1: Turkey GDP Growth Rates (2003-2018). Source: General Directorate of Budget and Financial Control, * Data for 2018 were taken from TurkStat. 51
Graph 3.2: The Ratio of Domestic Savings to GDP in Turkey (19902017). Source: World Bank. 54

Ufuk Bingöl
Employment Policy Goals in Turkish Government Programs in Terms of Social Policies: Justice and Development Party Governments
Graph 12.1: The distribution of social policy core components commitment in Government Programs. Source: This graphic is produced by the author with the support of CAQDAS, according to the Reference Coding value. 204
Graph 12.2: Employment commitments by Categorical Pattern coding study. Source: This graphic is produced by the author with the support of CAQDAS, according to the Reference Coding value. 204
Graph 12.3: Training-Employment commitments according to Categorical Pattern coding study. Source: This graphic is produced by the author with the support of CAQDAS, according to the Reference Coding value. 210

Selçuk Çağrı Esener
The Forgotten Unit of Income, Expenditure and Wealth Chain: Remembering the Taxation of Wealth and Our Fight Against Inequality
Graph 17.1: How Many Times the Rich is Richer than the Poor? *(2014 2017, S80/S20 Quintile Share Ratio)*. Source: OECD (2018), Social and Welfare Statistics: Income distribution Data, https://data.oecd.org/inequality/income-inequality.htm. 294

List of Tables

Semanur Soyyiğit
China's Neo-Mercantilist Policies and Growth Process
Tab. 2.1: Foreign Direct Stock in China (20142017). Source: United Nations Conference on Trade and Development, https://unctad.org/sections/dite_dir/docs/wir2018/wir18_fs_cn_en.pdf (3.03.2019). 35

Begüm Erdil Şahin and Deniz Dilara Dereli
An Evaluation on Middle-Income Trap in Turkey
Tab. 3.1: Comparison of Turkey and the US Per Capita GDP (19902017). Source: World Bank, GDP per capita (current US $). 53
Tab. 3.2: The Comparison of Turkey and Developed Countries in terms of Selected Indicators. Source: World Bank 2016, *United Nations Human Development Reports, ** TURKSTAT 2017. 55

Sertaç Hopoğlu
Are Youth Labor, Trade Openness and Foreign Direct Investment Effective in Economic Growth of Civets Countries? A Panel Causality Analysis
Tab. 4.1: Literature Summary. Source: Author's own construction. 65
Tab. 4.2: Causality Relationships Tested in the Analysis. Source: Author's own construction. 68
Tab. 4.3: Results of Cross-Sectional Dependence Test. Source: Author's own calculations. 70
Tab. 4.4: Results of PY(2008) Homogeneity Test. Source: Author's own calculations. 71
Tab. 4.5: Pesaran (2007) CIPS Test Results. Source: Author's own calculations. 73
Tab. 4.6: Results of EK (2011) Test. Source: Author's own calculations. 75

Tuncer Gövdeli
Tourism, Oil Prices and Economic Growth in the Mediterranean Countries: Bootstrap Panel Granger Causality Analysis
Tab. 5.1: Top 10 Countries by International Number of Tourists. Source: UNTWO, 2017. 83
Tab. 5.2: Top 10 Countries by International Tourism Revenue. Source: UNTWO, 2017. 83

Tab. 5.3: Horizontal Cross-Section Dependence and Homogeneity Tests. Source: Author's own calculations. 91
Tab. 5.4: Economic Growth—Tourism Income Causality Results. Source: Author's own calculations. 91
Tab. 5.5: Economic Growth—Crude Oil Prices Causality Results. Source: Author's own calculations. 92
Tab. 5.6: Tourism Revenues and Crude Oil Prices—Economic Growth Causality Results. Source: Author's own calculations. 93

Habibe Günsel Doğrul, Mediha Mine Çelikkol, and İlhan Korkmaz
The Analysis of the Relationship Between Creative Class, Financial Development and Regional Innovativeness in Turkey
Tab. 6.1: Selected Variables. Source: Authors' own construction. 100
Tab 6.2: Descriptive Statistics. Source: Authors' own calculations. 100
Tab. 6.3: LISA Clusters by Variables. Source: Authors' own construction. 107
Tab. 6.4: OLS Regression Results. Source: Authors' own calculations. 108
Tab. 6.5: Results of Spatial Delay Model. Source: Authors' own calculations. 109

Umut Akduğan and Seyhun Doğan
Youth Unemployment: An Empirical Analysis
Tab. 7.1: Results of Panel Unit Root Tests. Source: Authors' own calculations. 133
Tab. 7.2: Estimation of Panel VAR Model. Source: Authors' own calculations. 134
Tab. 7.3: Results of Panel Granger Causality Test. Source: Authors' own calculations. 134

Mehmet Güçlü
Industrial Revolution and Labor Market
Tab. 8.1: The Characteristics of Occupational Tasks. Source: Autor and Dorn (2009) and Acemoglu and Autor (2010). 141
Tab. 8.2: Expected Impacts of Technological Change on Employment and Earnings. Source: World Bank (2016). 144
Tab. 8.3: Comparing Skills Demand, 2018 vs. 2022 (top 10). Source: WEF (2018). 145
Tab. 8.4: Examples of Stable, New and Redundant roles (All Industries). Source: WEF (2018). 147

Kemal Eker, Görkem Bahtiyar, and Hasan Bakır
Systemic Policies Towards Attracting Skilled Labor: An Investigation on Turkey
Tab. 9.1: Foreign Academics Working in Turkey During the 20162017 Academic Year-First 10 Nationalities. Source: Republic of Turkey Minsitry of Interior Directorate General of Migration Management (2016). 162
Tab. 9.2: Foreign Academics Working in Turkey as of 2019 Academic Year-First 10 Nationalities. Source: Council of Higher Education-YÖK (2019). 162

H. Işıl Alkan
The Gender Impact of Last Global Crisis on Labour Markets
Tab. 10.1: Unemployment Rates Between 2008 and 2011 by Regions (%). Source: ILOSTAT, 2019 169
Tab. 10.2: Unemployment Rates by Sex after the Global Crisis in Developed Regions (%). Source: ILOSTAT, 2019. 170
Tab. 10.3: Employment Rates in Developed Regions by Sex (%). Source: ILOSTAT, 2019. 172
Tab. 10.4: Unemployment Rates in Developing Regions after Global Crisis (%). Source: ILOSTAT, 2019. 173
Tab. 10.5: Employment Rates in Developing Regions (%). Source: ILOSTAT, 2019. 176

Mehmet Kenan Terzioğlu
Poverty, Income Inequality, Unemployment and Human Capital in the Democratization Process
Tab. 11.1: Democracy Index 2018 by Regime Type. Source: The Economist Intelligence Unit. 189
Tab. 11.2: Democracy Among Regions. Source: The Economist Intelligence Unit. 190
Tab. 11.3: Western-Eastern Europe Regime Structure, 2018. Source: The Economist Intelligence Unit. 191
Tab. 11.4: Unit Root Test Results. Source: Author's own calculations. 192
Tab. 11.5: Random Effects Model Estimation Results. Source: Author's own calculations. 193

Ufuk Bingöl
Employment Policy Goals in Turkish Government Programs in Terms of Social Policies: Justice and Development Party Governments
Tab. 12.1: Cluster Analysis of Government Programs. Source: The author created this table with CAQDAS and Pearson Correlation Coefficient similarity criterion. 211

Ayşe Aylin Bayar, Bengi Yanık-İlhan and Nebile Korucu-Gümüşoğlu
The Effects of Elders' Earnings on Turkish Income Inequality
Tab. 13.1: General Characteristics of the Samples. Source: Authors' calculations based on SILC 2006, 2011 and 2016. 224
Tab. 13.2: Inequality Measures. Source: Authors' calculations based on SILC 2006, 2011 and 2016. 227
Tab. 13.3: Within Inequality Measures of Different Earnings. Source: Authors' calculations based on SILC 2006, 2011 and 2016. 227
Tab. 13.4: Impacts of Different Earnings to Overall Inequality. Source: Authors' calculations based on SILC 2006, 2011 and 2016. 229

Mehmet Akif Destek
Liberalization, Globalization and Income Inequality in Emerging Economies
Tab. 14.1: Correlation Between the Variables for Liberalization-Inequality Nexus. Source: Author's own calculations. 241
Tab. 14.2: GMM Estimation Results on Liberalization-Inequality Nexus (2000-2014). Source: Author's own calculations. 242
Tab. 14.3: GMM Estimation Results on Liberalization-Inequality Nexus (2000-2008). Source: Author's own calculations. 243
Tab. 14.4: GMM Estimation Results on Liberalization-Inequality Nexus (20092014). Source: Author's own calculations. 244
Tab. 14.5: Correlation Between the Variables for Globalization-Inequality Nexus. Source: Author's own calculations. 245
Tab. 14.6: GMM Estimation Results on Globalization-Inequality Nexus (20002014). Source: Author's own calculations. 245
Tab. 14.7: GMM Estimation Results on Globalization-Inequality Nexus (20002008). Source: Author's own calculations. 246
Tab. 14.8: GMM Estimation Results on Globalization-Inequality Nexus (20092014). Source: Author's own calculations. 247

Mustafa Şit and Erdal Alancıoğlu
Macroeconomic Factors Determining Income Distribution: An Analysis on Mist Countries
Tab. 16.1: Explanation of Variables. Source: Authors' own construction. 266
Tab. 16.2: Swamy S Homogeneity Test Results. Source: Authors' own calculations. 267
Tab. 16.3: Horizontal Cross-Sectional Dependence Test Results. Source: Authors' own calculations. 268
Tab. 16.4: CADF Unit Root Test Results. Source: Authors' own calculations. 270
Tab. 16.5: Westerlund Cointegration Analysis Results. Source: Authors' own calculations. 272
Tab. 16.6: PDOLS Estimation Results. Source: Authors' own calculations. 273

Selçuk Çağrı Esener
The Forgotten Unit of Income, Expenditure and Wealth Chain: Remembering the Taxation of Wealth and Our Fight Against Inequality
Tab. 17.1: Taxes on Property as a Common Indicator of Taxation on Wealth. Source: OECD (2018), Revenue Statistics 19652017. 290
Tab. 17.2: The Growth Rate of Top Global Wealth, 19872013. Source: Please see Piketty, T. (2018) for more information, piketty.pse.ens.fr/capital21c. 291
Tab. 17.3: Distribution of Wealth and Income. Source: For more information, please see Balestra and Tonkin, 2018. The OECD Data covers the years between 2012 and 2015, which is subject to the space and time availability. Data refers to the share held by the richest 10 % of households in the case of wealth and by the richest 10 % of individuals in the case of income. 292

Nilüfer Yörük Karakılıç
Social Perspectives of Gender-Based Discrimination Perception: An Empirical Study on University Students
Tab. 18.1: Cronbach's Alpha Value of the Scale. Source: Author's construction. 300
Tab. 18.2: KMO and Bartlett's Test. Source: Author's own calculations. 301
Tab. 18.3: Structural Equality Model Results. Source: Author's own calculations. 302

Tab. 18.4: Demographic Characteristics of Participants. Source: Author's own calculations. 303
Tab. 18.5: Analysis of Relationships between Individuals' Perception of Gender and Gender. Source: Author's own calculations. 304
Tab. 18.6: Results of Linear Regression Analysis. Source: Author's own calculations. 304
Tab. 18.7: Linear Regression Anova Table. Source: Author's own calculations. 304
Tab. 18.8: Linear Regression Analysis Coefficients Table. Source: Author's own calculations. 304
Tab. 18.9: T-Test Results of Gender Discrimination Perceptions of Individuals by Gender. Source: Author's own calculations. 305
Tab. 18.10: Mean and Standard Deviation Values According to Gender and Gender Discrimination Perception. Source: Author's own calculations. 306
Tab. 18.11: Results of the Analysis of Variance of Individuals According to Their Age Status. Source: Author's own calculations. 306
Tab. 18.12: Mean and Standard Deviation Values by Age Status and Gender Discrimination Perception. Source: Author's own calculations. 307
Tab. 18.13: Results of Variance Analysis of Gender Discrimination Perception According to Their Status. Source: Author's own calculations. 307
Tab. 18.14: Average and Standard Deviation Values According to Discrimination Perception and Gender. Source: Author's own calculations. 307
Tab. 18.15: Results of Variance Analysis Regarding Gender Discrimination Perceptions of Individuals According to Their Situation. Source: Author's own calculations. 308
Tab. 18.16: Mean and Standard Deviation Values According to Faculty Locations and Gender Discrimination Perception. Source: Author's own calculations. 308
Tab. 18.17: Results of the Analysis of the Variance of Gender Discrimination According to Income Status. Source: Author's own calculations. 309
Tab. 18.18: Mean and Standard Deviation Values by Income Status and Gender Discrimination Perception. Source: Author's own calculations. 309

Hale Kirer Silva L. and Rüya Eser
Energy Consumption, Carbon Emissions and Income Inequality in Turkey
Tab. 19.1: Unit Root Tests. Source: Authors' own calculations by using EViews 10 statistical package program. 318
Tab. 19.2: Lag Selection. Source: Authors' own calculations by using eviews 10 statistical package program. 319
Tab. 19.3: LM Autocorrelation Test. Source: Authors' own calculations by using EViews 10 statistical package program. 319
Tab. 19.4: White Heteroskedasticity Test. Source: Authors' own calculations by using EViews 10 statistical package program. 319
Tab. 19.5: Variance Decomposition Analysis. Source: Authors' own calculations by using EViews 10 statistical package program. 321
Tab. 19.6: Pairwise Granger Causality Test. Source: Authors' own calculations by using eviews 10 statistical package program. 322

Made in the USA
Monee, IL
04 May 2026

49438492R10187